# Data Analysis
# in Astronomy III

**ETTORE MAJORANA
INTERNATIONAL SCIENCE SERIES**
Series Editor:
**Antonino Zichichi**
European Physical Society
Geneva, Switzerland

---

**(PHYSICAL SCIENCES)**

*Recent volumes in the series:*

---

A Continuation Order Plan is available for this series. A continuation order will bring delivery of
each new volume immediately upon publication. Volumes are billed only upon actual shipment.
For further information please contact the publisher.

# Data Analysis in Astronomy III

Edited by

## V. Di Gesù
University of Palermo and
Institute of Cosmic Physics and Informatics/CNR
Palermo, Italy

## L. Scarsi
University of Palermo and
Institute of Cosmic Physics and Informatics/CNR
Palermo, Italy

## P. Crane
European Southern Observatory
Garching/Munich, Federal Republic of Germany

## J. H. Friedman
Stanford University
Stanford, California

## S. Levialdi
University of Rome "La Sapienza"
Rome, Italy

and

## M. C. Maccarone
University of Palermo and
Institute of Cosmic Physics and Informatics/CNR
Palermo, Italy

Plenum Press ● New York and London

Library of Congress Cataloging in Publication Data

International Workshop on Data Analysis in Astronomy (3rd: 1988: Erice, Sicily)
   Data analysis in astronomy III / edited by V. Di Gesù . . . [et al.].
      p.   cm.
   "Proceedings of the Third International Workshop on Data Analysis in Astronomy,
held June 20–27, 1988 in Erice, Sicily, Italy."—T.p. verso.
   Bibliography: p.
   Includes index.
   ISBN 0-306-43158-0
   1. Astronomy—Data processing—Congresses. 2. Astrophysics—Data processing—
Congresses. I. Di Gesù, V. II. Title. III. Title: Data analysis in astronomy three.
QB51.3.E43I58  1988                                 89-3702
523′.0028′5—dc19                                    CIP

Proceedings of the Third International Workshop on Data Analysis in Astronomy,
held June 20–27, 1988, in Erice, Sicily, Italy

© 1989 Plenum Press, New York
A Division of Plenum Publishing Corporation
233 Spring Street, New York, N.Y. 10013

Printed in the United States of America

# Preface

In the book are reported the main results presented at the Third International Workshop on Data Analysis in Astronomy, held at the Ettore Majorana Center for Scientific Culture, Erice, Sicily, Italy, on June 20-27,1988.

The Workshop was the natural evolution of the two previous ones. The main goal of the first edition (Erice 1984) was to start a scientific interaction between Astronomers and Computer Scientists. Aim of the second (Erice 1986) was to look at the progress in data analysis methods and dedicated hardware technology.

Data analysis problems become harder whenever the data are poor in statistics or the signal is weak and embedded in structured background. Experiments collecting data of such a nature require new and non-standard methodologies. Possibilistic approaches could be merged with the statistical ones, in order to formalize all the knowledge used by the scientists to reach conclusions. Moreover, the last decade has been characterized by very fast developments of Intelligent Systems for data analysis (knowledge based systems, ...) that would be useful to support astronomers in complex decision making.

For these reasons, the last edition of the workshop was intended to provide an overview on the state of the art in the data analysis methodologies and tools in the new frontieres of the astrophysics ($\gamma$-astronomy, neutrino astronomy, gravitational waves, background radiation and extreme cosmic ray energy spectrum).

The book is organized in two sections:

- Data analysis methods and tools,
- New frontieres in astronomy.

The success of the Workshop has been the result of the coordinated efforts of several people from the organizers to those who presented a contribution and/or took part in the discussions. We wish to thank the entire staff of the Ettore Majorana Center for Scientific Culture for its support and invaluable help in arranging the Workshop.

V. Di Gesù
L. Scarsi
P. Crane
J.H. Friedman
S. Levialdi
M.C. Maccarone

# Contents

## Data Analysis Methods and Tools

## New Frontieres in Astrophysics

# Data Analysis Methods and Tools

# ATTRACTOR RECONSTRUCTION AND DIMENSIONAL

# ANALYSIS OF CHAOTIC SIGNALS

H. Atmanspacher, H. Scheingraber, and W. Voges

*Max-Planck-Institut für extraterrestrische Physik*

*D-8046 Garching, FRG*

## 1. Introduction

Traditional techniques of signal analysis are restricted to investigations in the time and frequency domain. They are based on the statistical procedure of deriving the covariance matrix of a time signal as well as the corresponding Fourier transform, the power spectrum. These techniques have been extensively and successfully used in many different scientific branches. In particular, they are very helpful in distinguishing stationary, periodic, and quasiperiodic processes from nonperiodic ones. In the latter case, the temporal correlations generally vanish for $t \to \infty$, and the power spectrum is continuous:

$$P(\omega) > 0 \qquad \forall \omega \tag{1}$$

The correlation function for periodic processes does not vanish, and the power is distributed over a (more or less) discrete spectrum:

$$P(\omega) = \sum_k c_k \delta(\omega - \omega_k) \tag{2}$$

where the $c_k$ are the coefficients of the Fourier series representing the correlation function. The discrete spectrum provides the number $k$ of frequencies equivalent to the number of independent variables (degrees of freedom) of the system.

Using the traditional methods of signal analysis, not much detailed information can be extracted from a continuous spectrum. With respect to this case considerable progress has been achieved during the last decade. It has become common knowledge that a continuous spectrum is often produced by processes with a low number of independent variables if those are coupled in

3

a nonlinear manner. A phase space representation of such processes shows a rather complicated behavior for which the term deterministic chaos has been coined.

The present contribution deals with techniques enabling a characterization of this complicated structure. Of course, these techniques are also suitable to reveal the comparatively simple structures produced by periodic processes. In addition to temporal and spectral properties of a system, the geometrical properties of its phase space representation are the central subject of this article. (In fact, both views are not independent. Ergodic arguments imply a close relationship among them.)

We start with a brief introduction concerning attracting sets for dynamical systems and their geometrical properties. Several methods of extracting these properties from a measured signal will then be described. One of these methods will be used to give a detailed description how the procedure works and which parameters have to be carefully controlled in order to produce reliable results. As an example, the analysis of an attractor obtained from the X-ray emission of the neutron star Her X-1 will be discussed.

## 2. Attractors and their geometrical properties

Dynamical systems as they occur in nature are generally dissipative systems, i.e. they are not closed and there is a non – vanishing net energy flow out of the system. This means that the temporal evolution of a system of the simple form

$$\dot{\mathbf{x}} = \mathbf{F}(\mathbf{x}) \tag{3}$$

provides a shrinking phase space volume,

$$\operatorname{div} \mathbf{F} < 0 \tag{4}$$

whereas the volume would be preserved for Hamiltonian systems. The vector $\mathbf{x}$ in (3) characterizes the variables $(x_1, ..., x_n)$ of the system, and $\mathbf{F}$ is a matrix describing the generally nonlinear coupling among the variables. The number of variables defines the dimension $n$ of the phase space. Note that this definition is not restricted to the Hamiltonian formalism of canonically conjugate variables for each particle within a system. The dimension $n$ may thus be small if a many particle system is described by collective variables (e.g., temperature, pressure, etc.).[1]

Since the phase space volume of a solution of (3) decreases during its temporal evolution, it is quite suggestive that the motion of this solution (the flow $F_t$) in phase space is asymptotically restricted to a finite subspace of the entire phase space. This subspace will be called an attractor for the flow $F_t$.

---

[1]For details, we refer the reader to basic textbooks, e.g. Lichtenberg and Lieberman[1] or Guckenheimer and Holmes[2]. We also recommend the compact review article of Eckmann and Ruelle[3].

This illustrative concept can be obtained as a rigoros result from a linear stability analysis of the flow in phase space. It is possible to define a set of $n$ characteristic exponents $\lambda_i$ for $F_t$ determining the stability properties along the trajectory within the attractor on a temporal average. These exponents (also called the Ljapunov exponents) are derived from the eigenvalues $\Lambda_i(t)$ of a time dependent stability matrix $\mathbf{L}(t)$ whose eigenvectors $\delta\varsigma$ form a comoving coordinate system for $F_t$ (see Fig.1):

$$\mathbf{L}(t)\delta\varsigma = \mathbf{\Lambda}(t)\delta\varsigma \tag{5}$$

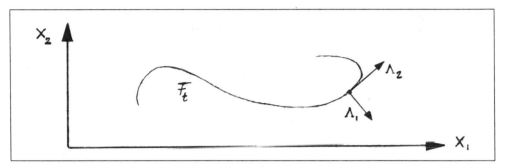

Figure 1. Schematic illustration of the eigenvalues $\Lambda_i$ for a two dimensional flow $F_t$ in phase space. Since $\Lambda_i$ is defined locally, it is a time dependent quantity. In contrast, the Ljapunov exponents $\lambda_i$ are given by the temporal average of $\Lambda_i$.

The Ljapunov exponents are obtained from the temporal average of $\Lambda_i$:

$$\lambda_i = \lim_{T\to\infty} \frac{1}{T} \int_0^T \Lambda_i(t)dt \tag{6}$$

The size of small perturbations $\delta\varsigma$ in the comoving frame thus shows an average evolution according to:

$$\delta\varsigma_i(t) = \delta\varsigma_i(0)e^{\lambda_i t} \tag{7}$$

Using the Ljapunov exponents, the dissipation condition (4) can be reformulated as

$$\sum_i \lambda_i < 0 \tag{8}$$

whereas this sum has to vanish for Hamiltonian systems. Moreover, the $\lambda_i$ are a convenient tool for a classification of different types of attractors. Fixed points, corresponding to stationary solutions of (3) require that $\lambda_i < 0 \ \forall i$. For a limit cycle, corresponding to a periodic system, there is one vanishing exponent into the direction of the flow. This direction is locally accessible by a vanishing vector product $\delta\varsigma \times \dot{x}$. In order to guarantee the stability of the

5

cycle, all the other exponents have to be negative. In case of a $k$ − periodic system the attractor is the surface of a $k$ − torus with $k$ vanishing Ljapunov exponents. For systems with at least three degrees of freedom $n \geq 3$, condition (8) can be satisfied by a combination of positive and negative exponents. This situation defines a chaotic attractor producing chaotic motion due to the sensitive dependence of the evolution on small perturbations. The different types of attractors are summarized in Table 1. A random process is not confined onto an attractor, since (8) is not satisfied.

|  | Ljapunov spectrum | dimension |
|---|---|---|
| fixed point | $\lambda_i < 0 \; \forall i$ | $D = 0$ |
| limit cycle | $\lambda_1 = 0; \; \lambda_{i>1} < 0$ | $D = 1$ |
| k - torus | $\lambda_1 = ... = \lambda_k = 0; \; \lambda_{i>k} < 0$ | $D = k \in N$ |
| chaotic attractor | $\exists j : \lambda_j > 0; \; \sum \lambda_i < 0$ | $D \in R, D < n$ |

Table 1. Specific types of attractors, properties of their Ljapunov spectrum, and the dimension of the attractor as a characterization of its geometrical properties.

There are a lot of examples showing that the structure of chaotic attractors may be highly complicated. In many cases this complexity is of a type known from fractal objects (see Mandelbrot[4]). It should, however, be emphasized that the geometrical property of being fractal is in principle independent from the dynamical property of a chaotic motion. There exist chaotic attractors without a fractal structure and vice versa. [3]

Fractal objects can be characterized by a dimension which is not restricted to integer numbers. The first definition of such a dimension dates back to Hausdorff[5] in 1919. It can be derived from an information theoretical basis. In this framework, the fractal dimension $D$ describes how the information $I_\epsilon$ about the state of the system (its location in phase space) scales with varying spatial resolution $\epsilon$. The dimension $D$ is in principle defined by

$$D = \lim_{\epsilon \to 0} \frac{\log I_\epsilon}{\log 1/\epsilon} \tag{9}$$

thus assuming a power law scaling of $I_\epsilon$ as an inherent property of fractal objects.

In order to specify $I$ formally, consider a partition of the attractor into $m$ boxes of size $\epsilon$. The probability that a point on the trajectory (state of the system) falls into the $i^{th}$ box is then given by

$$p_i = N_i/N \tag{10}$$

6

where $N$ is the total number of points. This definition adopts a discrete distribution of points in phase space. The probability $p_i$ can equivalently be defined using the (continuous) Lebesgue measure $\mu$ of the attractor:

$$p_i = \int_{i^{th}\text{box}} d\mu \tag{11}$$

The probability $p_i$ can now be used to define a generalized information of order $q$ $(q \in R)$[6] :

$$I^{(q)} = \frac{1}{1-q} \log_2 \sum_{i=1}^{m} p_i^q \tag{12}$$

For $q \to 1$, $I^{(q)}$ reproduces the Shannon information. Using (12), a spectrum of generalized dimensions of order $q$ has been introduced by simply inserting (12) into (9)[7]:

$$D^{(q)} = \lim_{\epsilon \to 0} \frac{I^{(q)}}{\log_2 1/\epsilon} \tag{13}$$

The dimensions of low integer order are the Hausdorff (or fractal) dimension $D^{(0)}$, the information dimension $D^{(1)}$, and the correlation dimension $D^{(2)}$. For the total set of generalized dimensions, it can be shown that

$$D^{(q)} \leq D^{(q')} \quad \text{if} \quad q' < q \tag{14}$$

where the equality holds for a completely homogeneous probability distribution $p_i = 1/N$. Hence, the difference between dimensions of different order measures the degree of inhomogeneity of the attractor in the sense of whether its different subsets (boxes) are visited with equal frequency.

In the last column of Table 1, the dimension $D$ is indicated for the typical attractors discussed above. The derivation of $D^{(q)}$ imposes no fundamental problems if the evolution of the relevant variables can be obtained (e.g, theoretically by a complete equation system, or experimentally by measuring all variables as a function of time). However, in many experimental situations neither all variables nor even their total number are known, so that the attractor of the system is not directly accessible. In case of astrophysical systems, problems can be even more serious since observations are not arbitrarily repeatable as in laboratory systems.

## 3. Attractor reconstruction and scaling analysis

The mentioned difficulties can be overcome by methods allowing for a reconstruction of an attractor from an experimental time series (e.g., the photon flux of astrophysical systems). The procedure of reconstruction only requires the temporal evolution of one single variable $x_i(t)$ $(i = 1, ..., n)$, say $x_1(t)$. In the following we shall characterize the measured time evolution by the notation:

$$\underline{x}_1(t) = \{x_1(t_1), x_1(t_1 + \tau), ..., x_1(t_1 + N\tau)\} \tag{15}$$

with

$$\tau = \frac{t_N - t_1}{N} \tag{16}$$

as the temporal resolution (sampling time), and $N$ as the total number of measured data points.

Using this time series, it is possible to construct a high – dimensional data set from the one – dimensional (scalar) time series $\underline{x}_1(t)$. Such a data set in principle contains all the information about the remaining variables contributing to the investigated process. As it has been proposed by Packard et al.[8], the required data set can be constructed by successive time derivatives $\frac{d\underline{x}^{(i)}(t)}{dt^{(i)}}$ up to a sufficient order. However, the differentiation of the signal amplifies noise, and any corresponding procedure suffers strongly from this disadvantage.

Probably the best and easiest way to get a high – dimensional data set is the time – delay method. The construction is simply carried out by successive time shifts $\Delta t$ of the measured time series $\underline{x}(t)$. The resulting data set of dimension $d$ reads:

$$\begin{array}{ccc}
x_1(t_1) & \cdots & x_1(t_N) \\
x_1(t_1 + \Delta t) & \cdots & x_1(t_N + \Delta t) \\
\vdots & & \vdots \\
x_1(t_1 + (d-1)\Delta t) & \cdots & x_1(t_N + (d-1)\Delta t)
\end{array} \tag{17}$$

Alternative to the representation in terms of shifted series $\underline{x}$, this data set can be viewed as consisting of the column vectors:[2]

$$\vec{x}_i = \{x_1(t_i), x_1(t_i + \Delta t), ..., x_1(t_i + (d-1)\Delta t)\} \tag{18}$$

If $d$ is sufficiently large, then the constructed structure (phase portrait) in the time delay space (embedding space) is in a certain mathematical sense

---

[2] We have now three distinct types of vectors: 1. the $n$ – dimensional vector $x$ as the variable vector of the system (3) with a measured component $x_1$; 2. the $N$ – dimensional vector $\underline{x}$ characterizing the measured time series by its $N$ points; and 3. the $d$ – dimensional vector $\vec{x}$ representing a point in the embedding space. This space is constructed by applying the time delay technique to $\underline{x}$.

equivalent to the attractor of the process. A criterion for a sufficient $d$ is intuitively given by the requirement that different points on the original attractor must not be superimposed by the projection into a lower – dimensional embedding space. In other words: the projected trajectory must not intersect itself. Based on a theorem of Whitney (1936)[9], Takens[10] and Mane[11] have shown that a dimension

$$d > 2D + 1 \tag{19}$$

can be considered to be sufficient in the sense discussed above, if $D$ is the dimension of the attractor.

The data set (17) contains the parameters $\tau$ and $\Delta t$ which are still to be determined. The temporal resolution $\tau$ of the measured time series is obviously related to the time scale on which the process of interest appears to take place. If not known *a priori*, a good guess for this time scale can be obtained from the first zero – crossing of the autocorrelation function of $\underline{x}$. There should be at least some tens of points within this time scale, so that the orbit of a process is covered in a sufficiently dense manner. With a too large value of $\tau$, information about the process of interest will be lost, whereas a too small value of $\tau$ often provides a signal dominated by noise.

An optimum choice of $\Delta t$ is much more problematic. In view of the linear independence of the vectors $\underline{x}$ in (17), one would think that the correlation properties of the signal also gives an estimate for $\Delta t$. However, this estimate may be rather poor, since linear independence does not necessarily provide an optimum amount of information about the projected attractor. Concerning this additional condition, Fraser and Swinney[12] have developed an idea to calculate an optimum $\Delta t$ from the mutual information of the measured signal. The first minimum of the corresponding function characterizes that time delay, for which a maximum amount of information about the structure of the attractor is gained, compared with $\Delta t = 0$. This thoughtful approach usually proposes delay times considerably larger than the naive correlation guess would suggest. For this reason, considerably more points $N$ per time series are required, since the vectors $\underline{x}$ in (17) have to be of equal dimension.

Once an attractor is reconstructed in the embedding space, one can proceed to the determination of its scaling behavior, i.e. its dimension. The most popular methods to do so are the "correlation integral method"[13,14] and the "nearest neighbor method"[15,16]. The former one is based on the evaluation of the function:

$$C_\epsilon^{(2)} = \lim_{\epsilon \to 0} \frac{1}{N^2} \sum_{i \neq j} \Theta(\epsilon - |\vec{x_i} - \vec{x_j}|) \tag{20}$$

The Heaviside function $\Theta(\epsilon - |\vec{x_i} - \vec{x_j}|)$ counts how many pairs of points $\vec{x_i}, \vec{x_j}$ are found within the distance $\epsilon$. Equation (20) is an integral pair correlation function

$$C_\epsilon^{(2)} = \int_0^\epsilon d^d\epsilon' c^{(2)}(\epsilon')$$ (21)

which relates the entire formalism to the concept of statistical correlations $c(r)$ of points on the attractor in the embedding space. For integral $q$ – point correlations (corresponding to $q^{th}$ order corrections to a constant number density), the correlation integral $C_\epsilon^{(q)}$ is given by:[17]

$$C_\epsilon^{(q)} = \lim_{\epsilon \to 0} \left[ \frac{1}{N} \sum_{i=1}^{N} \left( \frac{1}{N} \sum_{j=1}^{N} \Theta(\epsilon - |\vec{x_i} - \vec{x_j}|) \right)^{q-1} \right]^{1/q-1}$$ (22)

As a generalization of (20), $C_\epsilon^{(q)}$ counts the number of $q$ – tuples within the distance $\epsilon$. Identifying the logarithm of the correlation integral with the information $I^{(q)}$, the dimension of order $q$ according to (13) turns out to be:

$$D^{(q)} = \lim_{\epsilon \to 0} \lim_{d \to \infty} \frac{\log_2 C_\epsilon^{(q)}}{\log_2 \epsilon}$$ (23)

For the case $q = 2$, the exponents of both summations in (22) are equal to one. This situation considerably facilitates the numerical work and is the reason for the preferred use of the pair correlation integral. For $q \neq 2$ the calculation is more involved since each individual sum has to be raised to the corresponding power. The spectrum of dimensions $D^{(q)}$ from Eq.(23) contains more detailed information about the metric properties of the attractor than each single dimension does. This is understandable as at different order $q$ different subsets on the attractor become dominant in the determination of $C^{(q)}$ and $D^{(q)}$. Illustratively, the variation of $q$ provides a scan through all degrees of point density existing on the attractor.

The nearest neighbor $(nn)$ method is actually very similar to the correlation integral approach. The main difference consists of the fact that the scaling of $C_\epsilon^{(q)}$ is investigated by counting the number of $q$ – tuples as a function of $\epsilon$, whereas the $nn$ procedure analyses, how the distance $\epsilon$ has to be varied in order to contain a varied number $M$ of nearest neighbors. For this situation, the scaling is assumed to follow:

$$M \propto \epsilon(M)^D$$ (24)

equivalent to

$$D = \lim_{\epsilon \to 0} \lim_{d \to \infty} \frac{\log_2 M}{\log_2 \epsilon(M)}$$ (25)

Here the independent variable is $M$, the number of nearest neighbors. This type of counting is particularly advantageous for low point densities, e.g. in case of high embedding dimensions. The reason is that in this case the quasi − continuity of the point distribution is not satisfied. A "continuous" increase in $\epsilon$ thus provides discrete steps of the correlation integral $C$. On the other hand, an increase in the number $M$ of nearest neighbors is naturally adapted to the discreteness of the distribution. In those situations, where this difference is relevant, a plot of $\epsilon(M)$ vs $N$ should in general be smoother than $C(\epsilon)$ vs $\epsilon$, resulting in a smaller error for $D$. For more details on error estimates in the determination of attractor dimensions compare Refs. [18, 19]. Using the $nn$ method, a dimension function $D(\gamma)$ has been defined[16] which is easily related to the spectrum of dimensions $D^{(q)}$ by:

$$\gamma = (1 - q) D^{(q)} \tag{26}$$

The preceding discussion was restricted to dissipative systems with a finite − dimensional attractor. A realistic extension of these considerations has to include noise. Noise can be regarded as a random process, not bounded to an attractor and not satisfying condition (8). For this reason, there will be no embedding dimension as in (19), and the slopes in the expressions (23) and (25) will not converge in the limit $d \to \infty$. However, if the flow on an attractor is noisy to a certain degree, it is still possible to extract information about the attractor. The plots according to (23) and (25) then show a bend at a value $\epsilon_b$ depending on the noise level. We shall give examples in Sec.4, see also Ref. [20] . For $\epsilon < \epsilon_b$, both methods do not provide a limit determining $D$. The relevant scaling range is restricted to $\epsilon > \epsilon_b$. For rather high noise levels, this scaling range may be partially or totally obscured, thus providing results which may be irrelevant for the underlying attractor. As it has recently been shown[21,22], any low pass filtering of chaotic signals (i.e. limited amplitude resolution) can also influence the results of the attractor analysis.

All these problems make it very desirable to have a method which is more stable against the influence of noise. A corresponding procedure has been suggested by Broomhead and King[23]. It has become known as "singular system analysis" or "principal value decomposition". The procedure is based on the diagonalization of the covariance matrix $< \underline{x}_i(t)\underline{x}_j(t) >$ as it is obtained from the data set (17).[3] Its eigenvalues and eigenvectors allow for a decomposition of deterministic flow and random motion in phase space. Consequently, the embedding dimension for the deterministic motion corresponds to the dimension of the deterministic subspace. An investigation of the local properties of the flow in this subspace reveals further information about the structure of the attractor[25]. In particular, one can apply the correlation integral or nearest neighbor methods to the "cleaned" attractor.

---

[3]A modified form of this data set has recently been proposed by Farmer[24]. The successive time series $\underline{x}_i$ are weighted by a factor $\exp(-h(d - 1)\Delta t)$ (where $h$ is the dynamical entropy) in order to achieve an even better suppression of noise.

These properties and prospects look very promising. However, there is not yet much experience with the singular system approach. The nearest neighbor procedure has been theoretically and numerically studied to a considerable extent, but has also not been applied to too many experimental investigations. In spite of the possible advantages of both methods in specific situations, most of the experimental applications have been carried out using the correlation integral method. For some examples, we refer to investigations of hydrodynamic[26] and acoustic[27] turbulence, of chemical reactions[28], of the dynamics of EEG's[29], of chaotic attractor for climatic processes[30], of low – dimensional chaos in laser systems[31,32], and applications in astrophysics[33,34]. In view of this wide – spread experimental use, the next section describes some examples for the correlation integral method, using numerical and experimental data.

## 4. Numerical and experimental examples

Before describing the dimensional analysis for a chaotic attractor, we start with the simpler situation of a periodic attractor (limit cycle). For this case, the expected dimension is $D = 1$ (cf. Table 1). The time series required to construct the data set (17) is simply given by a sinusoidal function. The correlation period of this signal is identical with the inverse frequency $T = \omega^{-1}$. We use a total number $N$ of 1000 points and a temporal resolution $\tau = 0.007T$, which corresponds to 142 points per $T$ (35 points within the first zero – crossing of the autocorrelation function).

A crucial point is the choice of the time delay $\Delta t$. The minimum requirement on $\Delta t$ is the linear independence of the vectors $x_i$ constituting (17), since otherwise the data set would be one – dimensional by definition. For the present demonstrative purposes we choose $\Delta t = 10\tau$ and $\Delta t = 30\tau$, resp. For reasons discussed above, $\Delta t$ cannot be made arbitrarily large, if there are experimental restrictions concerning the length of the time series.[4]

As a further decisive parameter, the length of the measured time series, i.e. the number $N$ of data points, has to be optimized. Since the meaningful resolution $\tau$ is often limited by noise a very large $N$ of more than $10^4$ is not always achievable. It has, however, been demonstrated that even less than 1000 data points can provide useful results[35]. For the sinusoidal signal a number $N = 1000$ turned out to be sufficient. Figure 2 shows (a) the two – dimensional phase portrait of $x_i(t)$ vs $x_i(t + \Delta t)$; (b) a doubly logarithmic plot of the correlation integral $C_\epsilon^{(2)}$ vs $\epsilon$ according to (20); and (c) the slope $\frac{\log_2 C_\epsilon^{(2)}}{\log_2 \epsilon}$ vs $\log_2 \epsilon$. The time delay is $\Delta t = 10\tau$. In Fig.3, the corresponding diagrams are plotted for $\Delta t = 30\tau$.

The most important plots concerning the determination of the attractor dimension are the slope plots (c). They visualize whether there is a linear

---

[4]Many of such restrictions are conceivable: e.g., systems switching among different coexisting attractors, destruction of attracting structures due to self – induced or external changes of control parameters, influence of noise, etc.

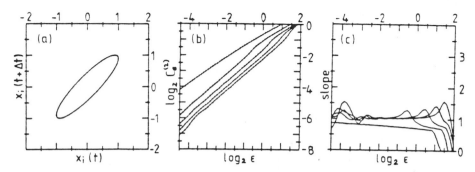

Figure 2. Dimensional analysis for a limit cycle of period $T$ reconstructed with a time delay $\Delta t = 10\tau$. a) Two – dimensional phase portrait of $\underline{x}_i(t)$ vs $\underline{x}_i(t + \Delta t)$; b) correlation integral according to (20); c) plot of the slope $\frac{\log_2 C_\epsilon^{(2)}}{\log_2 \epsilon}$ vs $\log_2 \epsilon$. Embedding dimensions $d = 1$ to $d = 5$ are plotted in (b) and (c); $N = 1000$, $\tau = 0.007\, T$.

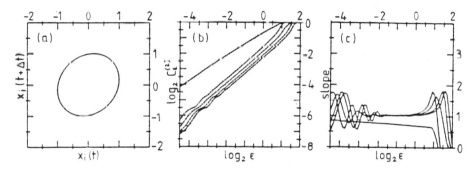

Figure 3. Dimensional analysis for a limit cycle reconstructed with a time delay $\Delta t = 30\tau$. a) - c) as in Fig.2. For embedding dimensions, $N$, and $\tau$ see also Fig.2

scaling range according to (23). Both, Figs. 2c and 3c show such a scaling range with constant slope for $-2 < \log_2 \epsilon < 1$, where the slope clearly indicates a dimension $D = 1$ as expected. For $\log_2 \epsilon < -2$, there are fluctuations of the slope due to the low number of pairs of points at very small distances. (This kind of fluctuations might be smoothed in the nn method.) At the rightmost part of the scaling range a significant hump appears. It can be understood as a boundary effect[36].

In the two – dimensional phase portraits (a) one clearly notices that the projection with $\Delta t = 10\tau$ is still near to the line $\underline{x}_i(t) = \underline{x}_i(t + \Delta t)$ which would represent a linearly dependent situation. For $\Delta t = 30\tau$, a better projection (in the sense of maximum information gain) is obtained, since the projected attractor is covered in a more homogeneous manner. This difference appears again in the slope plots: Fig.2c shows a distinct plateaulike feature with a slope $\approx 0.7$ in the range $0 < \log_2 \epsilon < 1.5$, caused by the artificially rarefied regions on the attractor. For high dimensions this feature disappears since projection effects decrease with increasing dimension. A look at Fig.3c shows that the corresponding feature does not occur if $\Delta t = 30\tau$. Note, however,

13

that for $d = 10$ and $\Delta t = 30\tau$ the vectors $\underline{x}_i$ consist of only $1000 - d\,\Delta t/\tau = 700$ points.

It is worth to be emphasized that for quasiperiodic or chaotic processes projection effects might be even more serious in case of a poor choice of $\Delta t$. The determination of $\Delta t$ as based on the mutual information[12] is thus strongly recommended if the number $N$ of measured data points is sufficiently large. Further below, we shall discuss a sufficiency criterion for the example of a chaotic attractor.

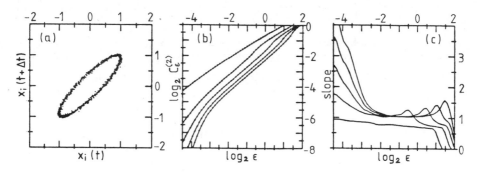

Figure 4. Dimensional analysis for a limit cycle with 6% superimposed Gaussian white noise, reconstructed with a time delay $\Delta t = 10\tau$. a) - c) as in Fig.2. For embedding dimension, $N$, and $\tau$ see also Fig.2.

In Fig.4, a sinusoidal signal with 6% white noise is investigated using a temporal resolution $\tau = 0.007\,T$, $\Delta t = 10\tau$, and $N = 1000$ as in Fig.2. The phase portrait (4a) shows the influence of the noise which appears as a strong increase of the slopes at small $\epsilon$ in (4b) and (4c). The phase portrait (4a) also visualizes, how the projected attractor is inhomogeneously covered by data points. The addition of noise causes a shortening of the linear scaling range. If the noise level is high enough, it can even provide a shift of the slope toward higher values. This effect must be carefully checked in case of experimental signals.

The noise – dominated range of the correlation integral provides no convergence of slopes for successive dimensions $d$, since the underlying random process is not bound to an attractor. The transition from random to deterministic behavior characterizes the noise level. For $d = 3$, this transition occurs at $\log_2 \epsilon_b \approx -2.5$. With respect to the size of the attractor ($\log_2 \epsilon_a \approx 1.7$) one obtains a noise level given by $\frac{\epsilon_b}{\epsilon_a} \approx 0.06$.

In order to illustrate the dimensional analysis in case of an experimental chaotic signal, we consider the X – ray luminosity of the neutron star Her X-1. The detailed analysis has recently been published elsewhere[3,4] The observed signal provides a clearly irregular temporal behavior on time scales in the order of some seconds. The correlation time is approximately 10 sec, and the Fourier spectrum does not provide any periodicities in this range except the rotation period 1.24 sec of the neutron star. Figure 5 shows the temporal evolution of the source signal (a) and the background radiation (b). Apart from the different count rates, it is very hard to state any significant difference between both types of irregular behavior.

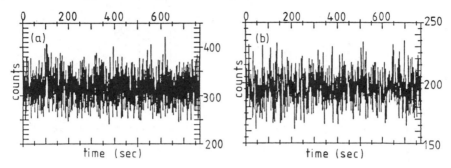

Figure 5. Temporal evolution of the X – ray count rates from Her X-1 (a) and background radiation (b). Both time series show 1000 points with a resolution $\tau = 770$ msec.

For the source signal, the correlation time of 10 sec suggests a temporal resolution of some hundred msec. Since the data were originally sampled in time steps of 9.67 msec, it is easy to produce higher values of $\tau$ simply by integration. As an optimum resolution, $\tau = 770$ msec was obtained. For better resolutions, noise dominates the signal (no convergence of slopes occurs), and for larger values too few data points remain per correlation period. With a resolution of $\tau = 770$ msec, the 1.24 sec rotation period is suppressed in the integrated signal, and it is not necessary to correct the time series for the periodic contribution. (For instance, $\tau = 0.62$ sec provides $D = 1$, the periodicity due to the rotation of Her X-1.)

Figure 6 shows the phase portrait (a), the correlation integral (b), and the slope plot (c) for the source signal in Fig.5a. An increase of $\Delta t$ from $\tau$ up to $10\tau$ did not lead to a significant difference in the phase portraits nor the correlation integrals. Hence, $\Delta t = \tau$ has been choosen. The length of the investigated time series was $N = 1000$. The slope plot for $d = 10$ shows a remarkable amount of $\approx 40$ % noise in the range $\log_2 \epsilon < 7.0$. Linear scaling is obtained for $7.0 < \log_2 \epsilon < 7.5$ where the slope is constantly 2.35 for $d > 10$. The fitting error for this value is surprisingly small, but one should take into account that intrinsic errors as mentioned in Sec.3 might be considerably larger. In view of this point, the analysis has been carried out for different time series and showed the reliability of the results[34].

In addition, a comparison of the source signal (5a) and the background (5b) shows that the background is purely random for the same time scales and parameters. Figures 7a and 7b illustrate that there is no linear scaling range and correspondingly no attractor for the background signal. Although the source signal and the background cannot be distinguished by the time series and the phase portrait, the dimensional analysis provides a clearcut difference.

It is of great practical interest that the spectrum of dimensions $D^{(q)}$ according to (23) gives a criterion for a sufficient length of the analyzed time

15

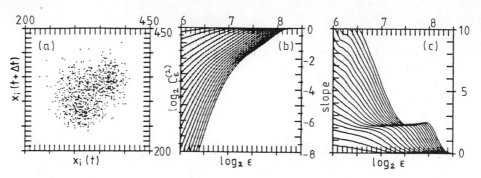

Figure 6: Dimensional analysis for a chaotic attractor, reconstructed with a time delay $\Delta t = \tau$ from the X – ray luminosity of Her X-1. a) - c) as in Fig.2. Embedding dimensions are shown from $d = 1$ to $d = 20$, $N = 1000$, $\tau = 770$ msec.

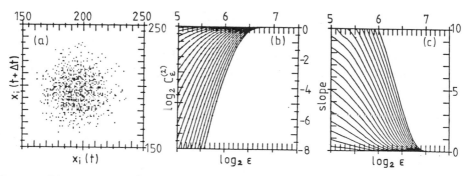

Figure 7. Dimensional analysis of the X – ray background noise, based on a reconstruction with a time delay $\Delta t = \tau$. a) - c) as in Fig.2. Embedding dimensions $d$ from 1 to 20 are shown; $N = 1000$, $\tau = 770$ msec.

series[37]. If a time series does not contain enough data points to reflect the inherent scaling properties of the attractor, it is expected (and has been shown in the analysis of the sinusoidal signal) that the points within the reconstructed attractor are more inhomogeneously distributed than for a sufficient number of data points[17]. The difference in $D^{(q)}$ for different $N$ should vanish in the limit of a sufficiently high $N$. For the Her X-1 attractor, these differences are shown in Fig.8. Apart from slight deviations for negative $q$, the spectra are identical, confirming the proper reconstruction of the attractor already with less than 1000 points.

## 5. Conclusions

The knowledge of the attractor dimension is of undoubtedly high value if one is interested in modeling systems on the basis of the temporal evolution of one observed variable. This is particularly true for chaotic processes which otherwise cannot easily be distinguished from random noise. In various applications[26–34] it has become clear that often a surprisingly low number of variables governs the dynamics of complex systems. The attractor dimension $D$ is a natural lower bound for this number. Therefore, the determination

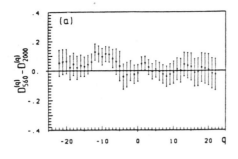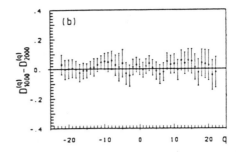

Figure 8. Residua $D_N^{(q)} - D_{N'}^{(q)}$ of the spectrum of dimensions for different numbers $N$ of data points per time series. a) $N = 560$, $N' = 1000$; b) $N = 1000$, $N' = 2000$. Within errors, the dimensions are almost identical for different $N$.

of $D$ allows to judge the degree of complexity of the investigated process. A small dimension $D$ indicates that a simplified mathematical description of the system appears conceivable (although it is still problematic to identify the relevant variables).

However, it must be stressed that an application of the described methods requires a very careful choice of parameters. In many cases an optimum choice is only achievable by (computing time and brain power consuming) trial and error procedures. A thorough study of the system considered is necessary, and traditional methods like correlation and Fourier analysis have to accompany a reasonable investigation. It should by no means be expected neither that reliable results can be obtained by an *en passant* quick look, nor that the methods represent black box algorithms for fool – proof applications.

REFERENCES

1. A. J. Lichtenberg and M. A. Lieberman, *Regular and Stochastic Motion* (Springer, Berlin, 1983)

2. J. Guckenheimer and P. Holmes, *Nonlinear Oscillations, Dynamical Systems, and Bifurcations of Vector Fields* (Springer, Berlin, 1983)

3. J.-P. Eckmann and D. Ruelle, *Rev. Mod. Phys.* **57**, 617 (1985)

4. B. B. Mandelbrot, *The Fractal Geometry of Nature* (Freeman, San Francisco, 1982)

5. F. Hausdorff, *Math. Ann.* **79**, 157 (1919)

6. J. Balatoni and A. Renyi, in *Selected Papers of A. Renyi, Vol.1*, (Akad. Budapest, 1976), p.588

7. H. G. E. Hentschel and I. Procaccia, *Physica* **8 D**, 435 (1983)

8. N. H. Packard, J. P. Crutchfield, J. D. Farmer, and R. S. Shaw, *Phys. Rev. Lett.* **45**, 712 (1980)

9. H. Whitney, *Ann. Math.* **37**, 645 (1936)

10. F. Takens, in *Dynamical Systems and Turbulence, Lecture Notes in*

*Mathematics* **898**, eds. D. A. Rand and L. S. Young (Springer, Berlin, 1981), p.366

11. R. Mane, in *Dynamical Systems and Turbulence, Lecture Notes in Mathematics* **898**, eds. D. A. Rand and L. S. Young (Springer, Berlin, 1981), p.230

12. A. M. Fraser and H. L. Swinney, *Phys. Rev.* **A 33**, 1134 (1986)

13. P. Grassberger and I. Procaccia, *Phys. Rev. Lett.* **50**, 346 (1983)

14. P. Grassberger and I. Procaccia, *Physica* **9 D**, 189 (1983)

15. Y. Termonia and Z. Alexandrovitch, *Phys. Rev. Lett.* **45**, 1265 (1983)

16. R. Badii and A. Politi, *J. Stat. Phys.* **40**, 725 (1985)

17. K. Pawelzik and H. G. Schuster, *Phys. Rev.* **A 35**, 481 (1987)

18. J. Holzfuss and G. Mayer - Kress, in: *Dimensions and entropies in dynamical systems*, ed. G. Mayer - Kress (Springer, Berlin, 1986), p.114

19. W. E. Caswell and J. A. Yorke, in: *Dimensions and entropies in dynamical systems*, ed. G. Mayer - Kress (Springer, Berlin, 1986), p.123

20. A. Ben Mizrachi, I. Procaccia, and P. Grassberger, *Phys. Rev.* **A 29**, 975 (1984)

21. R. Badii and A. Politi, in: *Dimensions and entropies in dynamical systems*, ed. G. Mayer - Kress (Springer, Berlin, 1986), p.67; R. Badii, G. Broggi, B. Derighetti, M. Ravani, S. Ciliberto, A. Politi, and M. A. Rubio, *Phys. Rev. Lett.* **37**, 979 (1988)

22. F. Mitschke, M. Möller, and W. Lange, *Phys. Rev.* **A 37**, 4518 (1988)

23. D.S. Broomhead and G. P. King, *Physica* **20 D**, 217 (1986)

24. J. D. Farmer, "Exploiting chaos to predict the future and reduce noise", preprint 1988

25. D. S. Broomhead, R. Jones, G. P. King, and E. R. Pike, in *Chaos, Noise, and Fractals*, eds. E. R. Pike and L. A. Lugiato (Hilger, Bristol, 1986), p.15

26. A. Brandstater, J. Swift, H. L. Swinney, A. Wolf, J. D. Farmer, E. Jen, and P. J. Crutchfield, *Phys. Rev. Lett.* **51**, 1441 (1983)

27. W. Lauterborn and J. Holzfuss, *Phys. Lett.* **115 A**, 369 (1986)

28. J.-C. Roux, R. H. Simoyi, and H. L. Swinney, *Physica* **8 D**, 257 (1983)

29. A. Babloyantz, J. M. Salazar, and C. Nicolis, *Phys. Lett.* **111 A**, 152 (1985)

30. G. Nicolis and C. Nicolis, *Proc. Ntl. Acad. Sci. USA* **83**, 536 (1986)

31. A. M. Albano, J. Abounadi, T. H. Chyba, C. E. Searle, S. Yong, R. S. Gioggia, and N. B. Abraham, *J. Opt. Soc. Am. B* **2**, 47 (1985)

32. H. Atmanspacher and H. Scheingraber, *Phys. Rev.* **A 34**, 253 (1986)

33. M. Auvergne and A. Baglin, *Astron. Astrophys.* **168**, 118 (1986)

34. W. Voges, H. Atmanspacher, and H. Scheingraber, *Ap. J.* **320**, 794 (1987)

35. N. B. Abraham, A. M. Albano, B. Das, G. de Guzman, S. Yong, R. S. Gioggia, G. P. Puccioni, and J. R. Tredicce, *Phys. Lett.* **114 A**, 217 (1986)

36. A. Brandstater and H. L. Swinney, *Phys. Rev.* **A 35**, 2207 (1987)

37. H. Atmanspacher, H. Scheingraber, and W. Voges, *Phys. Rev.* **A 37**, 1314 (1988)

# THE RAYLEIGH STATISTIC IN THE CASE OF WEAK SIGNALS

# - APPLICATIONS AND PITFALLS

O.C. de Jager[1], B.C. Raubenheimer[1] and J.W.H. Swanepoel[2]

[1] Dept. of Physics, PU for CHE, Potchefstroom 2520, South Africa
[2] Dept. of Statistics, PU for CHE, Potchefstroom 2520, South Africa

## ABSTRACT

The distribution of the Rayleigh statistic for any kind of light curve is derived. It is shown that a two sigma (Gaussian) hole in an ON source region with respect to an adjacent OFF-source region is still acceptable after a claimed periodic sinusoidal signal have been subtracted from the ON-source region. This is mainly due to the large effects of Rayleigh fluctuations. It is also shown that most of the present estimators of the signal strength of sinusoidal light curves are conservative in that they underestimate the signal strength if $p > 1/\sqrt{n}$ (where n is the number of events). If the signal strength is less than this limit, then one should be careful when interpreting results, since the bias increases rapidly as $p \to 0$ which may result in the possibility of identifying a signal where none exists.

## 1. INTRODUCTION

It is well known that $\gamma$-ray astronomy at very high energies (VHE) $E_\gamma$ > 1TeV is confronted with source fluxes which are only a few percent of the isotropic cosmic ray flux. Furthermore the difficulty of calibrating Cerenkov telescopes and the apparent time variability of signal strengths urge us to use sound data analysis techniques. The pioneering work in $\gamma$-ray astronomy below 1GeV was mainly done by the COS-B and SAS II experiments and papers on data analysis techniques followed (see e.g. Gerardi et al., 1982 and Buccheri, 1985). In VHE $\gamma$-ray astronomy it was the Durham group who implemented the well known Rayleigh test (Chadwick et al. 1985a, -b, -c). This test is uniformly most powerful for von Mises densities (Mardia, 1972). De Jager (1987) has shown that the von Mises density is applicable to VHE $\gamma$-ray astronomy only when the concentration parameter of this density is less than one - thus leading to sinusoidal light curves. Consequently the detection of a number of sources with sinusoidal light curves followed when the Rayleigh test was implemented (see e.g. Chadwick et al.1985a, -b, -c, Raubenheimer et al., 1986 and North et al., 1987). The Durham group also made an important contribution to analysis techniques by using this statistic to test for time variability and to estimate signal strengths.

VHE γ-ray astronomers are well aware of the fact that a "clear DC excess" should be visible (if the data permitted such an analysis) before looking for a periodic signal. However, it was never stated how large such a DC excess should be. We shall attempt to answer this question by subtracting the pulsed events from the ON-source data and calculating the probability to get such a remainder in the ON-source region relative to the OFF-source data, given the periodicity is true. It will become clear that Rayleigh fluctuations are much larger than Poissonian fluctuations in the case of weak signals, so that a two sigma hole in the remaining ON-source data need not cause any alarm

Finally, we discuss the estimation of signal strength for sinusoidal light curves. It will be shown that all such estimators are biased (especially when the signal strength $p \to 0$). From the information inequality it follows that the variance of the estimators discussed is already close to the lower limit attainable for the variance of any signal strength estimate. The aim should be to use the estimator with the smallest bias when discussing time variability of weak sources.

## 2. THE DISTRIBUTION OF THE RAYLEIGH POWER FOR A SIGNAL

Let $f(\theta)$ be the probability density function of the continuous random variable $\theta \in [0, 2\pi]$. The means of $\cos(m\theta)$ and $\sin(m\theta)$ (with $m = 1, 2, \ldots$, the harmonics of the periodicity under investigation) are given respectively by

$$\alpha_m = \int_0^{2\pi} f(\theta) \cos m\theta d\theta \quad \text{and} \quad \beta_m = \int_0^{2\pi} f(\theta) \sin m\theta d\theta \tag{1}$$

with a corresponding covariance matrix of

$$\sigma^2(m) = \begin{bmatrix} \sigma_x^2(m) & \sigma_{xy}(m) \\ \sigma_{yx}(m) & \sigma_y^2(m) \end{bmatrix} = \begin{bmatrix} \frac{1}{2}(1 + \alpha_{2m} - 2\alpha_m^2) & \frac{1}{2}(\beta_{2m} - 2\alpha_m\beta_m) \\ \frac{1}{2}(\beta_{2m} - 2\alpha_m\beta_m) & \frac{1}{2}(1 - \alpha_{2m} - 2\beta_m^2) \end{bmatrix}$$

and a correlation coefficient of $\rho(m) = \sigma_{xy}(m)/\sigma_x(m)\sigma_y(m)$ where the subscripts x and y refer to the cosine and sine components respectively. Let $\theta_1, \theta_2, \ldots, \theta_n$ be a random sample drawn from $f(\theta)$. Unbiased estimators of the trigonometric moments in relation (1) will then be given by the following

$$\hat{\alpha}_m = \frac{1}{n} \sum_{i=1}^{n} \cos m\theta_i \quad \text{and} \quad \hat{\beta}_m = \frac{1}{n} \sum_{i=1}^{n} \sin m\theta_i \tag{2}$$

It follows then from the central limit theorem that the distributions of the random variables

$$z_x = \frac{\hat{\alpha}_m - \alpha_m}{\sigma_x(m)/\sqrt{n}} \quad \text{and} \quad z_y = \frac{\hat{\beta}_m - \beta_m}{\sigma_y(m)/\sqrt{n}}$$

approaches that of the standard normal distribution as $n \to \infty$. From the multivariate central limit theorem (Lehmann, 1983) it follows that $\hat{\alpha}_m$ and $\hat{\beta}_m$

22

are asymptotically binormally distributed. The general asymptotic $(n\to\infty)$ density function of the Rayleigh power $U_m = n\bar{R}^2_m = n(\hat{\alpha}_m{}^2 + \hat{\beta}_m{}^2)$ is obtained after a suitable transformation and integration over all phases $\theta$:

$$h_{U_m}(u) \simeq \frac{1}{4\pi\sigma_x(m)\sigma_y(m)\sqrt{1-\rho^2(m)}} \int_0^{2\pi} \exp\left\{ -\frac{n}{2(1-\rho^2(m))}\left[ \left(\frac{\sqrt{u/n}\,\cos m\theta - \alpha_m}{\sigma_x(m)}\right)^2 \right.\right.$$
$$\left.\left. - 2\rho(m)\frac{(\sqrt{u/n}\,\cos m\theta - \alpha_m)(\sqrt{u/n}\,\sin m\theta - \beta_m)}{\sigma_x(m)\sigma_y(m)} + \left(\frac{\sqrt{u/n}\,\sin m\theta - \beta_m}{\sigma_y(m)}\right)^2 \right]\right\} d\theta$$

(3)

In the absence of any signal (i.e. $\alpha_m = \beta_m = 0$) it can be easily shown that the distribution of $2n\bar{R}^2_m$ is approximately equal to that of the $\chi^2_2$-distribution for sufficiently large n (>100). In the case of a sinusoidal signal with signal strength p and position $\mu$ of the peak maximum, the light curve is given by

$$f(\theta) = \frac{(1 + p\cos(\theta - \mu))}{2\pi}$$

(4)

so that

$$\alpha_m = \frac{1}{2}p\cos\mu \quad \text{and} \quad \beta_m = \frac{1}{2}p\sin\mu$$

(5)

if m=1 but these trigonometric moments are equal to zero whenever m>1. We shall refer to the fundamental harmonic (m=1) when dropping the subscript of $U_m$. Since U is rotaion invariant we can take $\mu$=0 so that the correlation coefficient $\rho(1)$ of the fundamental harmonic equals zero and the integral over $\theta$ can be evaluated analytically if we neglect terms of the order of $p^2$ relative to one. Then

$$h_U(u) = \exp\left(-u - \frac{np^2}{4}\right)I_0(p\sqrt{nu})$$

(6)

which is also exactly equal to the expression given by Linsley (1975). The term $I_0$ is the modified Bessel function of the zeroth order. This expression is very useful to predict the range of Rayleigh powers obtainable for a given DC-excess. The latter is quantified by the parameter $x=p\sqrt{n}$ which will be discussed in more detail in the following section.


## 3. WHAT DC EXCESS IS NECESSARY IF A PERIODICITY IS PRESENT?

In VHE $\gamma$-ray astronomy there are two basic modes to observe a source: (1) The drift scan technique whereby a source is allowed to drift into the field of view of the telescope. In this mode a source will be "visible" for only $\approx$10 minutes whereby less than 1000 events are collected ON-source. (2) The tracking technique whereby a source is continuously tracked in Right Ascension. In this case no direct OFF-source comparison is possible unless one has a system which collects data both ON- and OFF-source simultaneoulsy. Alternatively one can switch between ON-source and OFF-source at regular time intervals. In the energy regions $E_\gamma > 10^{14}$ eV and

$E_\gamma$<10 GeV it is possible to obtain skymaps so that it becomes easier to compare ON- and OFF-source regions. In all the cases whereby OFF-source data are available, one normalises the OFF-source counts relative to the ON-source region. The difference between the ON-source counts and the normalised OFF-source counts gives one the "DC excess counts". Dowthwaite et al. (1983) and Li and Ma (1983) gave the correct method to quantify the DC excess in probabilistic terms and the way it is usually done is to quote the Gaussian standard deviations of the ON-source region above (or below) the OFF-source region. However, to compare the DC excess with any periodic effect, we shall use an estimate $\hat{x}$ of the parameter $x=p\sqrt{n}$ where n denotes the number of ON-source counts and p the corresponding signal strength of gamma rays. In this case we can link the DC excess with relation (6) to obtain the distribution of the Rayleigh power. It is interesting to notice that the estimate $\hat{x}=\hat{p}\sqrt{n}$ will also be standard normally distributed in the absence of a gamma-ray source if the time ON-source does not differ from the OFF-source time by more than a factor of two (Li and Ma, 1983).

The necessity of the existence of a clear DC excess above the cosmic ray background as a prerequisite for a periodic analysis has been acknowledged by most γ-ray astronomers during the latest years (see e.g. Lamb and Weekes, 1987 and also Chardin, 1986, 1987). However, it was never said at least how large the DC excess should be to validate a periodic effect found. With the tools available we shall subtract the pulsed events from the ON-source region and compare the remainder with the OFF-source counts. This will give one an idea of how much smaller the remaining ON-source count rate may be relative to the OFF-source count rate.

For sinusoidal light curves one can calculate the moments of the Rayleigh power from relation (6):

$$E(n\overline{R}^2) = \frac{x^2}{4} + 1 \quad \text{with} \quad var(n\overline{R}^2) = \frac{x^2}{2} + 1$$

If one estimates the DC excess x from the observed Rayleigh power, calling s as its estimator, then

$$s = 2\sqrt{n\overline{R}^2 - 1} \quad \text{for} \quad n\overline{R}^2 > 1 \tag{7}$$

and s = 0 when $n\overline{R}^2<1$. Assume that there are no aperiodic γ-rays from the source. Any fluctuations in s will then be due to (I) fluctuations of the γ-ray count rate, (II) fluctuations in the cosmic ray background and (III) Rayleigh fluctuations expressed by equation (6). The DC excess $\hat{x}$ estimated from a single ON-OFF comparison will also fluctuate, but only due to the first two mentioned effects and can then be written as the following random variable.

$$\hat{x} = \hat{p}\sqrt{n} = \frac{n_\gamma}{\sqrt{n_\gamma + n_c}} = \frac{\overline{n}_\gamma + z_\gamma\sqrt{\overline{n}_\gamma}}{\sqrt{\overline{n}_c + \overline{n}_\gamma + z_\gamma\sqrt{\overline{n}_\gamma} + z_c\sqrt{\overline{n}_c}}} \tag{8}$$

where $\overline{n}_\gamma$ is the mean γ-ray counts and $\overline{n}_c$ is the mean cosmic ray counts ON-source. The counts $n_\gamma$ and $n_c$ are normally distributed if $\overline{n}_\gamma$ and $\overline{n}_c \gtrsim$

50 and $z_\gamma$ and $z_c$ will then be standard normal variables. Otherwise Poisson statistics will be applicable. By inserting (8) into (6) one can simulate values of the Rayleigh power which includes not only Rayleigh fluctuations but also cosmic and gamma ray fluctuations. The random variable $\hat{x}$-s where s is calculated from (7) gives the significance of the ON-source counts relative to the OFF source counts after subtraction of the pulsed events and will be used to determine the validity of a periodic result.

A case study was undertaken with $\bar{n}_\gamma = 19$ and $\bar{n}_c + \bar{n}_\gamma = 447$ so that x =0.9σ. The distribution of $\hat{x}$-s is shown in Figure 1. A peak is seen at $\hat{x}$-s ≈0.9σ which shows that s can sometimes be equal to zero (i.e. $n\bar{R}^2 < 1$ for some samples) and the width of this peak indicates the range over which $\hat{x}$ alone can fluctuate (i.e. Poisson fluctuations of both the γ-ray signal and cosmic rays). The wide secondary peak around $\hat{x}$-s = 0 indicates the additional effect of Rayleigh fluctuations. We can thus see how significant the effect of Rayleigh fluctuations can be: One either do not see any periodic signal ($\hat{x}$-s≈0.9σ) or one can observe a periodic excess up to ≈$\hat{x}$+3. In the latter case there will be a 3σ hole in the ON-source data relative to the OFF-source data after the subtraction of the pulsed excess and the probability to obtain such a result from real periodic data can be determined as follows: For a certain ON-OFF comparison, one can simulate the distribution of $\hat{x}$-s to see whether the ON-source excess is compatible with the observed DC excess. If one obtains a hole in the ON-source data after subtraction (i.e. $\hat{x}$-s < 0), the area under the density function of $\hat{x}$-s from −∞ to the experimental value will give the probability of obtaining a hole larger than or equal to that obtained experimentally. The smaller this probability, the less confidence can one have in the significance of the periodic result obtained. Further, let us assume that a Rayleigh power of 9.7 was obtained for the 447 events ON source under discussion. The p-level or "chance probability for uniformity" equals exp(-9.7)=6x10$^{-5}$. From relation (7) we obtain s=5.9 so that $\hat{x}$-s=-5σ which corresponds to a hole of ≈5σ and the probability to obtain such a hole is calculated from Figure 1 to be less than 0.1%. This indicates that such a periodic effect must be either a statistical fluctuation or a systematic effect.

In general it seems as if a hole larger than two sigma can occur ≈7% of the times after subtraction of the pulsed events from the ON-source data while a hole larger than 3σ can occur only ≈2% of the times. This seems to be largely independent of the values of $\bar{n}_\gamma$, $\bar{n}_c$ or whether fluctuations are larger than Poisonian (the latter being due to variable atmospheric conditions in VHE γ-ray astronomy). Thus, a hole of approximately two sigma should not cause much alarm but we should be careful of holes larger than three sigma. These conclusions are so far only applicable to sinusoidal light curves and using the tools developed in Sections 2 and 3, one can determine the significance of such holes for other kinds of light curves also.

## 4. THE ESTIMATION OF SIGNAL STRENGTHS

It is important to estimate the total flux from a periodic source. If the geometry of radiation is unknown, then we have to use a non-parametric estimator to estimate the signal strength p from the observed light curve

$$f(\theta) = pf_s(\theta) + \frac{(1-p)}{2\pi}$$

25

where $f_s(\theta)$ is the unknown source function. The problem here is that we have one equation and two unknown quantities p and $f_s(\theta)$ which will be difficult to solve. If there is a relative large region on the interval $[0, 2\pi]$ whereby $f_s(\theta)=0$, then it becomes easier to solve for p. However, this is not always the case.

The von Mises density (Mardia, 1972) is considered by most statisticians as being one of the most important density functions on a circle

$$f(\theta) = \frac{1}{2\pi I_0(\kappa)} e^{\kappa \cos(\theta - \mu)}$$

De Jager (1987) has shown that this density function is applicable to periodic light curves in VHE $\gamma$-ray astronomy only if the concentration parameter $\kappa < 1$. Then we have the well known and frequently observed cosine or cardioid density function

$$f(\theta) \simeq \frac{(1 + \kappa \cos(\theta - \mu))}{2\pi} \tag{9}$$

where $\kappa \simeq p$ when comparing this with relation (4). Our main aim here is to estimate this p but the problem is that the resulting estimate of $\kappa$ is biased (Schou, 1978 and Batschelet, 1981) and we shall show to which degree this is true when considering different estimators of p. From this section one will see that this problem is serious when the signal is weak (i.e. $p \to 0$) and at present there seems to be no solution for this problem. First we shall discuss various estimators of p for relation (4) and their bias. Following that the variance of these estimators will be discussed together with the information inequality (which gives the asymptotic lower bound for the variance of any estimator). An attempt will also be made to propose the best estimator for p based on the mean squared error and bias properties.

## 4.1 Different Estimators of Signal Strength

(i) Formally the best way to estimate p in (4) is by means of the maximum likelihood method: Let $\underset{\sim}{\theta} = \theta_1, \theta_2, \ldots, \theta_n$ be a random sample from (4). The log - likelihood function is then

$$\log L = -n \log(2\pi) + \sum_{i=1}^{n} \log(1 + p \cos(\theta_i - \mu))$$

For weak signals one can develop the logarithmic term into a sum of linear combinations of $\hat{\alpha}_m$ and $\hat{\beta}_m$ until $p^m$ is negligibly small. Let $p_m$ be the maximum likelihood estimator (MLE) of p (i.e. that value of p which maximises Log L) for cosine light curves alone.

(ii) Another estimator will now be developed which is very simple to calculate: It can be shown that relations (1) and (2) are also related as follows (Hart, 1985)

$$E\left[\frac{n\bar{R}_m^2 - 1}{n - 1}\right] = \alpha_m^2 + \beta_m^2 \tag{10}$$

26

For sinusoidal light curves the right hand side of (10) equals $p^2/4$ when using relation (5). We can then define the "adapted moment estimator" of p as (De Jager, 1987).

$$p_d = 2\sqrt{\frac{n\overline{R}^2 - 1}{n - 1}} \quad \text{if} \quad n\overline{R}^2 > 1 \tag{11}$$

and $p_d = 0$ if $n\overline{R}^2 < 1$. This estimator is aimed at a reduction of the bias due to the subtraction of one from the Rayleigh power. This estimator is closely related to the one proposed by Middleditch and Nelson (1976) for a time series.

(iii) Schou (1978) proposed an estimator of $\kappa$ which is the unique positive solution of

$$nA(\kappa) = \overline{R}A(\kappa\overline{R}) \tag{12}$$

if $n\overline{R}^2 > 1$, giving $\kappa_s$ as the estimator of $\kappa$ but $\kappa_s = 0$ whenever $n\overline{R}^2 < 1$. Thus, we suggest $p_s = \kappa_s$ as the "Schou estimator" of the signal strength for cosine light curves.

## 4.2 The Bias of Signal Strength Estimators

For all these methods it is obvious that any estimator of p is always zero or positive and for uniformly distributed samples one can also obtain estimates larger than zero so that the expected value of any estimator of p is also larger than zero - thus these estimators are biased whenever p is very small. The "bias" b(p) (which may be a function of p) is defined as the expected value of the signal strength estimator minus the true p. Figure 2 gives the bias of the different estimators. It is obvious that the signal strength is always underestimated for $E(n\overline{R}^2) > 1.5$ and overestimated for $E(n\overline{R}^2) < 1.5$, but the MLE always overestimates p. However, all these estimators become unbiased $(b(p)\approx 0)$ whenever the signal becomes strong - this holds for Rayleigh powers larger than approximately fifteen.

## 4.3 The Information Inequality and the Variance of Estimators

The information inequality (Lehmann, 1983) gives one the lower bound for the variance of any estimator of some parameter. This means that no estimator will be found with a variance which is smaller than the given lower bound. We are interested in the estimation of the signal strength p of a cosine light curve. If b'(p) is the derivative of the bias with respect to p, then it can be shown that the lower bound for the variance of any esimator (say $\hat{p}$) of p is given by the following

$$\text{var}_p(\hat{p}) \geq \frac{2(1 + b'(p))^2}{n(1 + \frac{3}{4}p^2 + \frac{5}{8}p^4 + \cdots)} \tag{13}$$

According to Figure 2 all estimators of p are biased and $b'(p) < 0$ for $E(n\overline{R}^2) < 2.5$ while $b'(p) > 0$ for $E(n\overline{R}^2) > 2.5$. Consequently the variance

27

of an estimator of p can be smaller than $(2/n)$ for small values of the Rayleigh power (see Figure 3 for the variance of different estimators). The variance $(2/n)$ was also defined by Linsley (1975) as the "standard conventional" measure of error in signal strength and is shown by the horizontal line at the ordinate value of 2 in Figure 3. According to this figure the variance of the different estimators considered is already close to the information inequality and no other unbiased estimator of p will be found with a variance smaller than $\simeq(2/n)$.

## 4.4 Mean Squared Error properties of Signal Strength Estimators

Finally we should compare the mean squared errors (MSE) (which equals the sum of the variance and the square of the bias) to get a picture of the performance of the the signal strength estimators: In Figure 4 the performance of the MLE, adapted moment and Schou estimators are compared. The MLE has a very large MSE for $p\simeq0$ (which is due to its large bias there) and the smallest MSE for $3<E(nR^2)<12$. All estimators perform equally well for $E(nR^2) > 12$. However, the adapted moment and Schou estimators have on average the smallest MSE and seems therefor to be the most promising to use.

## 5. Conclusions

We have derived the general distribution of the Rayleigh power (relation (3)) for any harmonic and any light curve which can be evaluated numerically. Mardia (1972) derived the distribution of the Rayleigh power for two cases: (I) $\alpha_1^2 + \beta_1^2 > 0$ and (II) $\alpha_1^2 + \beta_1^2 = 0$. The distribution for case (II) does not follow from the distribution of case (I) when $\alpha_1^2 + \beta_1^2 \to 0$. Consequently Mardia's distributions will not be applicable for TeV $\gamma$-ray astronomy where the signals are very weak. However, relation (3) is valid for all possible values of $\alpha_m^2 + \beta_m^2$ (and also for different harmonics m = 1, 2, ...) and is in general more applicable than Mardia's relations.

The existence of a "clear DC excess" as a prerequisite for a periodic analysis has been acknowledged by most TeV $\gamma$-ray astronomers. From all the case studies for sinusoidal light curves it follows that after the subtraction of the periodic events, holes larger than two sigma and three sigma in the ON-source data can occur $\simeq$ 7% and $\simeq$ 2% of the times respectively if the DC excess is due to periodic $\gamma$-rays only. Thus, a two sigma hole still seems to be acceptable but any hole larger than three sigma must be treated with caution. This conclusion holds irrespective of the total counts ON-source, the signal strength of $\gamma$-rays and whether the fluctuations in the cosmic and $\gamma$-ray count rates are larger than Poissonian.

It was also evident that all signal strength estimators considered overestimate the signal strength on average if p=0 and for the adapted moment and Schou estimators, one will then obtain estimates which equal $\simeq0.7/\sqrt{n}$ on average while this average equals $1.8/\sqrt{n}$ for the MLE. Furthermore, the adapted moment and Schou estimators underestimate p with less than $0.4/\sqrt{n}$ if x>1 whereas the MLE overestimates p in this case. The adapted moment and Schou estimators are therefor conservative if x>1. However these three estimators are approximately unbiased if x>8 - i.e. for

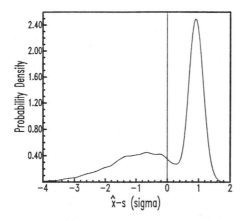

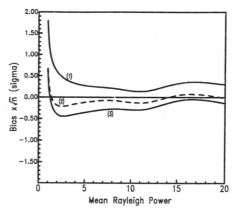

Figure 1. The distribution of the DC excess after the subtraction of the pulsed excess estimated from the Rayleigh statistic. The parameters for this Figure are $\bar{n}_\gamma=19$ and $\bar{n}_c=428$ so that the original DC excess x=0,9. All the $\gamma$-rays are assumed to be periodic.

Figure 2. The bias (expected value of the signal strength estimator minus the true signal strength) times $\sqrt{n}$ for (1) the MLE, (2) Schou and (3) the adapted moment estimators.

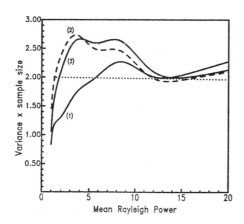

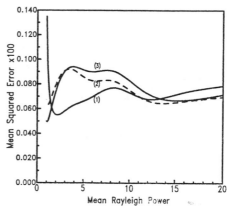

Figure 3. The variance of the signal strength estimators according to (1) the MLE, (2) Schou and (3) adapted moment.

Figure 4. The mean squared error of the signal strength estimators according to (1) the MLE, (2) Schou and (3) adapted moment.

relatively strong signals. To obtain the Schou estimator one should solve the non-linear equation (12), but the adapted moment estimator (relation 11) are the easiest to calculate. We therefor leave it to the researcher to decide between the adapted moment and Schou estimators. The variance of the adapted moment and Schou estimators are both $\approx 2/n$ except for weak signals where the bias is nonzero. Then we have the following distinction: If the mean Rayleigh power is less than two, then the variance is less than $2/n$, but the variance is larger than $2/n$ if the mean Rayleigh power is between two and thirteen. However, the quest should be to search for an unbiased estimator of the signal strength. Only then will one be able to quote reliable estimates of the signal strength for various observation cycles and the error of this estimator will then also be nearly independent of the signal strength.

## REFERENCES

Batschelet, E., 1981., Circular Statistics in Biology, Academic Press

Buccheri, R. 1985. (In Proceedings of the Workshop on Techniques in Ultra High Energy Γ-ray Astronomy. La Jolla, (USA)) 98-103.

Chadwick, P.M., Dowthwaite, J.C., Harrison, A.B., Kirkman, I.W., McComb, T.J.L., Orford, K.J., and Turver, K.E. 1985a. **Nature**, 317:236.

Chadwick, P.M., Dowthwaite, J.C., Harrison, A.B., Kirkman, I.W., McComb, T.J.L., Orford K.J. and Turver, K.E. 1985b., **Astron. Astrophys.**, 151 :L1.

Chadwick, P.M., Dipper, N.A., Dowthwaite, J.C., Gibson, A.I., Harrison, A.B., Kirkman, I.W., Lotts, A.P., Mcrae, H.J., McComb, T.J.L., Orford, K.J., Turver, K.E. and Walmsley, M. 1985c., **Nature**, 318:642.

Chardin, G. 1986, In Proceedings of the NATO Advanced Research Workshop, Durham (UK). D.Reidel. Dortrecht.

Chardin, G. and Gerbier, G., 1987, Proc. 20th ICRC, Moscow, 1:236.

De Jager, O.C. 1987. Ph.D. Thesis. Unisversity of Potchefstroom. South Africa.

Dowthwaite, J.C., Gibson, A.I., Harrison, A.B., Kirkman, I.W., Lotts, A.P., Mcrae, H.J., Orford, K.J., Turver, K.E., Walmsley, M. 1983. **Astron. Astr.**, 126:1.

Gerardi, G., Buccheri, R., Sacco, B. 1982. (In Proceedings of "COMPSTAT 82", Toulouse (France)), Preprint.

Hart, J.D. 1985, **J. Stat. Comp.**, 21:95.

Lamb, R.C. and Weekes, T.C., 1987, **Science**, 238:1483.

Lehmann, E.L., 1983, Theory of point estimation. New York: John Wiley and Sons.

Li, T-P. and Ma, Y-Q., 1983. **Astrophys. J.**, 272:317.

Linsley, J., 1975, Proc. 14th ICRC. München., 592.

Mardia, K.V., 1972, Statistics of directional data. New York : Academic Press.

Middleditch, J. and Nelson, J., 1976. **Astrophys J**, 208:567.

North, A.R., Raubenheimer, B.C., De Jager, O.C., Van Tonder, A.J. and Van Urk, G. 1987, **Nature** 326(6113):567.

Raubenheimer, B.C., North, A.R., De Jager, O.C., Van Urk, G. and Van Tonder, A.J., 1986, **Astrophys J**, 307 :L43.

Schou, G., 1978., **Biometrika**, 65(1):369.

# STABILITY IN FACTOR ANALYSES WITH RESPECT TO SMALL PERTURBATIONS OF THE DATA AND TO DEPARTURE FROM GAUSSIAN LAW OF THE VARIABLES

Mireille Ludivine Bougeard (1) (2)

(1) Univ. Paris X,IUT,1 chemin Desvallières, F-92410 Ville d'Avray

(2) UA 1125 CNRS, Observatoire Paris, F-75014 Paris, France

Abstract: In this paper, we consider the problem of approximations by sampling in the case of the principal component factor analysis. In a first part we recall the asymptotic results of Anderson, Arconte, Davis, Dauxois, Pousse and Romain about the distribution of the proper values of the operators. In a second step, we recall the principle of two non-parametric processes, namely jackknife and bootstrap. In a last stage, we compare the robustness of the asymptotic results mentioned earlier with the results obtained through non parametric processes by means of simulation and drawing of random samples in a given population.

## 1. INTRODUCTION

Principal Component Analysis (PCA) methods are of greast interest in many fields of astronomy and in particular in stellar statistics where they have been used for a long time with classification aims and to explain a large set of data in terms of a smaller number of dimensions (i.e. with a smaller number of characters called "underlying factors"), than one started with. As demonstrated in different works (see for instance: Tomassone and al 1983), PCA can be used not only as a descriptive tool but also as an aid in modelisation before applying regression techniques and in particular in astrometry (for application to astrolabe data, see for instance: Bougeard 1987,1988). In astrometry, observational errors are inherent to the data, so the variables are subject to random errors. If we have assumptions about the form of the covariance matrix from which the data come, parametric methods for analysing the stability of the principal components with respect to small perturbations (errors) on the data are of particular interest.

In this Paper, we intend to compare some parametric methods with non-parametric processes such as jackknife or bootstrap, by adding in a first step normal errors to the initial data and in a second step log-normal errors: the behaviour of the proper values ofthe covariance matrix (used in PCA) is analysed in each case. In Section 2, we shall recall the asymptotic results about the distributions of the proper values of the covariance matrix operator in the gaussian

31

case (Anderson 1963) and in the nongaussian case (Davis 1977), results which have been extented to the infinite dimensional case (Arconte and al 1980). In Section 3, we briefly recall the principle of some nonparametric processes, namely jackknife and the more recent bootstrap. In a last Section, we compare the robustness of the asymptotic results with the results obtained through a nonparametric method by means of simulation and drawing of random samples in a given population.

## 2. PROPER VALUES OF A COVARIANCE MATRIX: SOME ASYMPTOTIC RESULTS

### 2.1. The gaussian case

Assume that the data under analyse have a p-gaussian distribution, written as $N_p$ $(\mu, \Sigma)$ (mean $\mu$, covariance matrix $\Sigma$). Let S be the covariance matrix of a n sample, then the matrix nS is known to have a Wishart ([*]) distribution $W_p(n-1, \Sigma)$. Through a decomposition of this matrix - after an orthogonal transformation - as a sum of (n-1) Wishart matrices $W_p(1, \Lambda)$ and the use of the central limit theorem, Anderson (1963) proves especially

**Theorem (Anderson 1963).** Let $\Sigma = \Gamma \Lambda \Gamma'$ be a spectral decomposition ($\Lambda$ diagonal) of $\Sigma$, $\Sigma$ being supposed to have distinct proper values $\lambda_i$ and $\Sigma \geq 0$. Let $U = GLG'$ be a spectral decomposition (L diagonal) of a matrix U such that mU has a Wishart distribution $W_p(m, \Sigma)$, then the proper values $l_i$ of the matrix U are asymptotically gaussian, unbiased, independent with respective variances (2 $\lambda^2_i/m$).

**Corollary.** Let $l_i$ i=1,..p be the distinct proper values of the matrix (n/n-1)S, where S is a n-sample covariance matrix, then asymptotically $l_i$ is distributed as $N(\lambda_i, 2 \lambda^2_i/(n-1))$.

The above results can be applied asymptotically to the covariance matrix S of the sample and lead to asymptotic intervals of confidence

$$\log \lambda_i \in (\log l_i \pm t_\alpha \sqrt{2/(n-1)})$$

(1)

where $t_\alpha$ is the critical value for the chosen (normal) test level. Conversely, if we assume that the population parameters (especially $\lambda_i$) are known, then the corollary gives asymptotic intervals of confidence for the proper values $l_i$ of any q-sample drawn from this population, namely

$$l_i \in (\lambda_i(1 \pm t_\alpha \sqrt{2/(q-1)}))$$

(2)

where $l_i$ i=1,.. are the proper values in the sample.

### 2.2. The non-gaussian case

Anderson's asymptotic results hold even in the non-gaussian case provided that the four first moments of the distribution are taking into account:

---

[*] nS is often referred as the matrix of the centered products and cross-products.

**Proposition** (**Davis 1977 p209-210**). With the above notations, let $l_i$ be a proper value with multiplicity one of the covariance matrix S of a n-sample, then asymptotically, $\sqrt{n-1}\left(l_i - \lambda_i\right)$ is distributed as $N(0, 2\lambda_i^2 + k_{iiii})$ where k stands for a moment of order 4 of the variable $\Gamma'(X - \mu E)$.   **

Of course some difficulties can appear in the estimations of these moments. The above results have been extended to the infinite dimensional case by Arconte, Dauxois, Pousse and Romain (1980), through the use of properties of linear operators in Hilbert spaces (namely spectral analyses of Hilbert-Schmidt operators). Their results confirm the effect of the moments of order 4 in the non-gaussian case.

## 3. SOME NON PARAMETRIC TECHNIQUES

In this Section, we recall the principle of some non-parametric processes: jackknife, bootstrap. Historically, the idea was to perform a partition of the population under analyse in groups g and to evaluate a given parameter for each group in order to observe its variablity, a technique fluently applied around years 1950.

### 3.1. Quenouille-Tukey estimator or jackknife

First introduced by Quenouille (1954) with the aim of reduting estimator bias, this process has been studied by Tukey (1958) under the name of "jackknife" about the construction of approximated intervals of confidence in sampls in non-parametric contexts. For a detailed review, we refer for instance to (Miller 1974).

The method is the following: (i)-consider $X_1,...X_n$ a n random sample drawn from a population X ($X_1,..X_n$ are i.i.d. ie independently identically distributed) for which a parameter T has to be estimated through $\overrightarrow{T}$, (ii)-perform a partition of the population into say k subsets of same size. Then consider the new randon sample obtained by deleting the $i^{th}$ subset, i=1..k. For each new sample, estimate the related value of $\overrightarrow{T}$, denoted by $\overrightarrow{T}(i)$, (iii)- let $\left[J_i=k\overrightarrow{T} - (k-1)\ \overrightarrow{T}(i)\right]$ be a pseudo-value, then the jackknife estimate is defined by the arithmetic mean of these pseudo values:

$$J(\overrightarrow{T}) = (1/k)\ \Sigma_{i=1,.}\ \overrightarrow{J_i(T)} \tag{3}$$

It can easily be proved that if the bias of $\overrightarrow{T}$ is of order 1/n, then the bias of $J(\overrightarrow{T})$ is only of order $1/n^2$. At this stage, Tukey considers the statistic:

$$t= (J(\overrightarrow{T})-\overrightarrow{T})/s(J), \quad \text{with} \quad s^2(J)= \Sigma\ (J_i(\overrightarrow{T})-J(\overrightarrow{T}))^2/k(k-1) \tag{4}$$

and conjectures that it has a Student distribution with (k-1) degrees of freedom, conjecture which has appeared as satisfied for large classes of statistics.

### 3.2. Bootstrap

Introduced by Efron (1979), the bootstrap method appears as a "new look to jackknife" which is relevant to computer intensive methods. It is a resampling method which consists in resampling indepently in a given sample $(Y_1,..,Y_n)$ of

---

** X- $\mu E$ stands for the centered variables and $\Sigma = \Gamma\Lambda\Gamma'$

independent variables. In many cases, theoretical properties can be exactly proved and asymptotic results can be deduced (Bickel and Freedman 1981). For application in the domain of opinion pools, the reader is referred for instance to (Deville 1987).

## 4 STABILITY OF PARAMETRIC AND NON PARAMETRIC INTERVALS OF CONFIDENCE IN PCA.

Let us now apply the above methods in the context of PCA where the principal components are associated with the proper values of the variance covariance matrix of the data.

### 4.1. The chosen data

As an example, we shall deal here not with astrometric data, but with a sample of 525 student's examination performances. These students were judged on 11 subjects A1-A11 (namely: A1=general accountacy, A2=analytic accountacy, A3=tax science, A4=economy, A5=marketing, A6=computer science, A7=law, A8=mathematics, A9=french redaction, A10=psychology, A11=english). Principal component analyses were performed on covariance and correlation matrices and presented elsewhere (Bougeard 1981,1982). In this Paper, we restrict ourselves to the study of the stability of the principal components in the context of covariance-PCA.

We give on Tab.1 the results of some classical normality tests applied to the variables A1-A11, namely the Kolmogorov's D statistic, Fisher's G statistics for skewness and kurtosis which are (Fisher 1973) based on the first four cumulants: A1,2,5,7,9 are seen as nongaussian.

To observe a sensibility of the results to the first four moments, we have also considered the sample of the logarithms of the given variables $B1=\log_{10}A1$, $..B11=\log_{10}A11$. We can notice on Tab.1, that if there is no change in the "bad" level of significance of the Kolmogorov's D test (1%), the level of some G1 and G2 tests are drastically modified : none of the variable B1-B11 is gaussian.

### 4.2. Sampling and Anderson-Davis's intervals of confidence

When the data under analyses are assumed to be normally distributed, then applying the asymptotic results of Anderson, Davis and Romain, we obtain intervals of confidence as given in formula (2).

12 samples of size 50 have been drawn at random in the complete population A1-A11 (sampling rate 50/525=9,5%). The proper values of the related covariances matrices have been computed and compared with the a priori intervals of confidence (2); Tab.2 shows a good agreement with the theorical intervals, except for the last proper values (n°10,11).

Drawing, at this stage, at random 12 samples of size 50 of the population B1-B11, and comparing the proper values of the covariances matrices to Anderson's intervals (Tab.2), we observe not so good agreement especially as far the first proper value - that is the first component- is concerned. This gives a numerical confirmation of the sensibility of these parametric intervals to the moments of order 4 of the distributions.

## 4.3. Stability with respect to normal and log-normal perturbations

Since observational errors are inherent in many astronomical data, it was of greast interest to analyse the stability of the parametric and the non-parametric intervals of confidence, when perturbations are present.

For this purpose, two kinds of perturbations have been applied to the data A1-A11:
i) gaussian perturbations with zero mean and standard error proportional to the standard error, $\sigma(j)$, on each variable Aj, j=1,..,11

$$v(k,j) = (0,05 \ k \ \sigma(j)) \times N(0,1) \qquad k=1,2,..10$$

ii) log-normal perturbations given by

$$\varepsilon(k,j) = (0,05 \ k \ \sigma(j)) \times \exp(N(0,1)) \qquad k=1,2,..10$$

In each case, the proper values of the covariance matrix have been computed for k=1,..,10 and compared with parametric and nonparametric intervals of same level of confidence: the highest value of k such tant the interval is valid has been noticed (MGP(k) or MLGP(k) on Tab.3). ([***])

To obtain parametric intervals, we have considered the studied population A1-A11 of size 525, as drawn from a more general population and Anderson's formula (1) has been applied. The results are on Tab.3 : it can be seen a good stability of the first two eigenvalues in case of the gaussian perturbations, but in case of log-normal perturbations the maximum value "admitted" for k is 5 for the 90% and the 95% intervals of confidence. As far as the other proper values are concerned, we notice a large lack of robustness in case of log-normal perturbations ("long-tailed errors").

To save computer time, the non-parametric intervals which will be constructed here are only the jackknife intervals of confidence. For this purpose, the population P of 525 students has been partitioned in separate subsets of size 25 each (S1-S21). Then, we have successively computed:
a) for each subpopulation (P-$S_m$) of size 500, the proper values of the covariance matrix in decreasing order: $1_1(m),..,1_{11}(m)$ (they were non nul and non equal), m=1,..21,
b) the pseudo-values and the Quenouille-Tukey estimate, given by formula (3), for each proper value of rank i.

Then, by using the Tukey statistic (4) (Student with 20 degrees of freedom), non-parametric intervals of confidence have been obtained. Tab.3 col.4-9 gives the comparisons with the proper values of the covariance matrices when the data are under a k gaussain or log-normal perturbation. It can be remarked that these intervals are, of course, larger than the above Anderson's parametric intervals. It can also be noticed that these intervals are less robust in case of log-normal perturbations. Nevertheless, as far as the first and the second proper values (of the PCA) are concerned, they appear to be much more robust that the parametric intervals with respect to both types of perturbations on the studied data. This superiority fails as far as the eigenvalues n°3,4 are concerned (Tab.3 col12-15).

---

[***] Values of k superior or equal to 5 are often considered as "gross" errors in astrometric data.

# 5. CONCLUSION

The analysis presented here in the context of our students'examination performances can be considered as a "school-study" which can be easily transposed in any other context, especially in the field of astrometry where observational errors are inherent to the data. Such errors can often be assumed as gaussian or at least log-normal.

When the data under analysis can be supposed as approximately normally distributed, Anderson's asymptotic results about the eigenvalues of a sample covariance matrix (eigenvalues important in PCA to define the "underlying factors") have appeared as robust with respect to small perturbations on the data when the errors are gaussian. Nevertheless, they are very sensitive to log-normal perturbations (long tailed errors) and to the moments of order four in the case of distributions showing large departures from a gaussian law as shown in Section 4.2.

In comparison, nonparametric intervals of confidence constructed with the jackknife technique— which is, like bootstrap, relevant of computer intensive methods— have been seen as larger and so a little more robust with respect to gaussian and log-normal perturbations but only, here, for the first two eigenvalues of the studied covariance matrix. This partially justifies the popularity of these nonparametric methods even if they do not save computer time and money. Nevertheless, we have observed in Sect. 4.3. that this a priori superiority in terms of robustness may fail for higher ranks of eigenvalues. Furthermore, as seen in Sect. 4.2, it is clear that the form under which the variables are analysed (use of A=exp(B) instead of B,...) appears as a factor of importance in the final robustness of the results.

**ACKNOWLEDGEMENTS.** We acknowledge G. Saporta (CNAM, Paris) who directed the study (Bougeard 1981) and J.M. Grosbras (ENSAE, F-92241 Malakoff) for fruitful computational suggestions. The computations were performed with S.A.S. at ENSAE and C.I.R.C.E. (Orsay, France).

Table 1. Some classical normality tests applied to the variables A1–A11 and B1–B11.

| VARIABLE | N | D-MAX | PROB | SKEWNESS G1 | G1/SERG1 G1 | P-LEVEL G1 | KURTOSIS G2 | G2/SERG2 G2 | P-LEVEL G2 | MEAN | ST DEV |
|---|---|---|---|---|---|---|---|---|---|---|---|
| A1 | 525 | 0.0742 | .01 | -0.425 | -3.984 | 0.000 | 0.473 | 2.225 | 0.026 | 123.5238 | 26.7367 |
| A2 | 525 | 0.0935 | .01 | -0.677 | -6.350 | 0.000 | 1.061 | 4.987 | 0.000 | 122.5333 | 28.7736 |
| A3 | 525 | 0.2485 | .01 | -0.423 | -3.969 | 0.000 | 0.267 | 1.253 | 0.210 | 113.0381 | 23.1811 |
| A4 | 525 | 0.3470 | .01 | 0.143 | 1.339 | 0.181 | 0.799 | 3.754 | 0.000 | 116.1810 | 12.2254 |
| A5 | 525 | 0.3361 | .01 | -0.439 | -4.122 | 0.000 | -0.513 | 2.410 | 0.016 | 115.2476 | 13.7673 |
| A6 | 525 | 0.2536 | .01 | -0.009 | -0.088 | 0.930 | -0.304 | 1.430 | 0.153 | 114.2361 | 20.6477 |
| A7 | 525 | 0.3910 | .01 | -0.269 | -2.523 | 0.012 | 1.457 | 6.846 | 0.000 | 111.1714 | 13.8596 |
| A8 | 525 | 0.1192 | .01 | -0.413 | -3.878 | 0.000 | 0.109 | 0.514 | 0.607 | 130.2571 | 25.5060 |
| A9 | 525 | 0.3684 | .01 | -0.411 | -3.858 | 0.000 | 1.047 | 4.920 | 0.000 | 116.1905 | 11.3355 |
| A10 | 525 | 0.3333 | .01 | 0.151 | 1.420 | 0.156 | 0.359 | 1.687 | 0.092 | 115.3524 | 13.8324 |
| A11 | 525 | 0.1639 | .01 | 0.012 | 0.117 | 0.907 | -0.212 | -0.997 | 0.319 | 129.0095 | 17.5196 |
| VARIABLE | | | | | | | | | | | |
| B1 | 525 | 0.1255 | .01 | -1.877 | -17.608 | 0.000 | 7.841 | 36.845 | 0.000 | 2.0795 | 0.1104 |
| B2 | 523 | 0.1369 | .01 | -1.923 | -18.003 | 0.000 | 8.340 | 39.119 | 0.000 | 2.0763 | 0.1168 |
| B3 | 525 | 0.2726 | .01 | -1.270 | -11.909 | 0.000 | 2.571 | 12.080 | 0.000 | 2.0427 | 0.1002 |
| B4 | 525 | 0.3421 | .01 | -0.398 | -3.736 | 0.000 | 2.314 | 10.875 | 0.000 | 2.0627 | 0.0463 |
| B5 | 525 | 0.3240 | .01 | -0.948 | -8.891 | 0.000 | 2.032 | 9.547 | 0.000 | 2.0583 | 0.0546 |
| B6 | 525 | 0.2681 | .01 | -0.547 | -5.128 | 0.000 | 0.492 | 2.313 | 0.021 | 2.0504 | 0.0818 |
| B7 | 525 | 0.3731 | .01 | -1.062 | -9.966 | 0.000 | 3.961 | 18.612 | 0.000 | 2.0424 | 0.0571 |
| B8 | 525 | 0.1051 | .01 | -1.196 | -11.217 | 0.000 | 2.419 | 11.369 | 0.000 | 2.1053 | 0.0945 |
| B9 | 525 | 0.3563 | .01 | -0.911 | -8.547 | 0.000 | 2.549 | 11.977 | 0.000 | 2.0630 | 0.0440 |
| B10 | 525 | 0.3301 | .01 | -0.301 | -2.824 | 0.005 | 0.776 | 3.647 | 0.000 | 2.0589 | 0.0527 |
| B11 | 525 | 0.1691 | .01 | -0.393 | -3.690 | 0.000 | 0.309 | 1.453 | 0.146 | 2.1065 | 0.0603 |

Table 2. (Nsam = number of random samples in agreement with the given confidence interval.)

| N° | Proper value | Percent | Anderson's conf. intervals (2) level 90% | | level 95% | | Nsam. 90% | Nsam. 95% | |
|---|---|---|---|---|---|---|---|---|---|
| 1 | 1673.2 | 38.9 | 1117.2; | 2229.2 | 1010.6; | 2335.8 | 10 | 12 | V |
| 2 | 642.4 | 14.9 | 428.9 | 855.8 | 388.0 | 896.7 | 11 | 11 | A |
| 3 | 386.6 | 9.0 | 258.1 | 515.1 | 233.5 | 539.7 | 11 | 11 | R |
| 4 | 317.2 | 7.4 | 211.8 | 422.6 | 191.6 | 442.8 | 11 | 11 | . |
| 5 | 296.6 | 6.9 | 198.0 | 395.1 | 179.1 | 414.0 | 11 | 12 | A1 |
| 6 | 277.0 | 6.4 | 184.9 | 369.0 | 167.3 | 386.6 | 12 | 12 | |
| 7 | 223.2 | 5.2 | 149.0 | 297.3 | 134.8 | 311.6 | 12 | 12 | t o |
| 8 | 154.8 | 3.6 | 103.4 | 206.3 | 93.5 | 216.2 | 11 | 12 | A11 |
| 9 | 131.2 | 3.0 | 87.6 | 174.9 | 79.3 | 183.2 | 12 | 12 | |
| 10 | 109.0 | 2.5 | 72.8 | 145.2 | 65.8 | 152.1 | 9 | 10 | |
| 11 | 96.4 | 2.2 | 64.4 | 128.5 | 58.3 | 134.6 | 2 | 5 | |
| 1 | 40.54 | | 27.07; | 54.01 | 24.49; | 56.60 | 4 | 8 | V |
| 2 | 13.01 | | 8.68 | 17.33 | 7.86 | 18.16 | 8 | 7 | a |
| 3 | 6.10 | | 4.07 | 8.13 | 3.60 | 8.52 | 12 | 12 | r |
| 4 | 5.39 | | 3.60 | 7.19 | 3.26 | 7.53 | 12 | 12 | i |
| 5 | 4.68 | | 3.13 | 6.24 | 2.83 | 6.54 | 11 | 12 | e s |
| 6 | 4.37 | | 2.92 | 5.83 | 2.64 | 6.11 | 10 | 12 | B1 |
| 7 | 2.95 | | 1.97 | 3.93 | 1.78 | 4.12 | 11 | 11 | t |
| 8 | 2.23 | | 1.49 | 2.98 | 1.35 | 3.12 | 12 | 12 | o |
| 9 | 2.07 | | 1.38 | 2.76 | 1.25 | 2.90 | 7 | 9 | B11 |
| 10 | 1.67 | | 1.11 | 2.22 | 1.01 | 2.33 | 5 | 9 | |
| 11 | 1.37 | | 0.92 | 1.83 | 0.83 | 1.92 | 3 | 6 | |
| | | multiply by 10**-3 | | | | | | | |

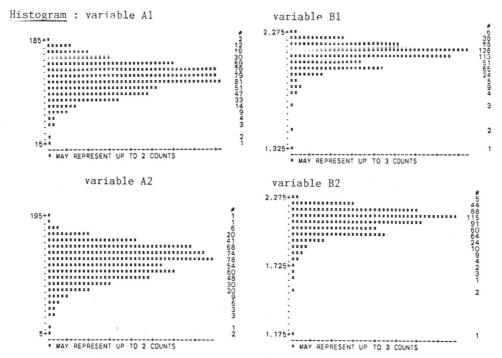

Histogram : variable A1    variable B1

variable A2    variable B2

Fig. 1. Histogram of some of the variables A1–A11 studied here and for graphical comparison of the variables B1 = $\log_{10}A1$, B2 = $\log_{10}A2$,...

37

## Table 3. Variables A1-A11

Col. 2 = Quenouille estimator J of the proper value 1; col. 3 = Tukey standard error for J; col. 10-11 = Anderson's 90% interval of confidence; MGP = maximum value of k, for the k gaussian perturbation; MLGP = maximum value for the k log-normal perturbation (see text).

| Proper value 1: N° | Quenouille-Tukey estimates | | Tukey's : level 90% interval | MGP(k) | MLGP(k) | Tukey's:level 95% MGP(k) | MLGP(k) | Anderson's:level 90% interval | MGP(k) | MLGP(k) | And.:level 95% MGP(k) | MLGP(k) |
|---|---|---|---|---|---|---|---|---|---|---|---|---|
| | J(1) | s(J(1)) | | | | | | | | | | |
| 1 | 1678.2 | 241.8 | 1261.0;2095.4 | 10 | 8 | 10 | 8 | 1511.5;1852.2 | 9 | 5 | 10 | 5 |
| 2 | 647.6 | 59.6 | 544.8 750.6 | 7 | 4 | 8 | 4 | 580.3 711.1 | 5 | 3 | 6 | 4 |
| 3 | 382.2 | 20.4 | 347.1 417.3 | 4 | 1 | 4 | 2 | 349.3 428.0 | 4 | 2 | 5 | 2 |
| 4 | 301.3 | 12.3 | 280.1 322.4 | 1 | 1 | 2 | 1 | 287.0 351.6 | 4 | 2 | 5 | 2 |
| 5 | 289.3 | 23.2 | 249.3 329.3 | 5 | 2 | 5 | 2 | 267.9 328.3 | 5 | 2 | 5 | 2 |
| 6 | 303.2 | 29.8 | 251.7 354.7 | 8 | 3 | 9 | 4 | 250.2 306.6 | 4 | 2 | 5 | 2 |
| 7 | 235.3 | 15.5 | 208.5 263.1 | 8 | 3 | 8 | 3 | 201.6 247.0 | 6 | 2 | 7 | 2 |
| 8 | 156.8 | 11.2 | 137.6 176.1 | 5 | 2 | 6 | 2 | 139.9 171.4 | 5 | 2 | 5 | 2 |
| 9 | 135.4 | 11.5 | 115.5 155.2 | 7 | 2 | 8 | 2 | 118.6 145.3 | 5 | 1 | 6 | 2 |
| 10 | 113.3 | 7.9 | 100.0 126.9 | 7 | 6 | 7 | 3 | 98.5 120.6 | 6 | 1 | 6 | 3 |
| 11 | 103.1 | 6.2 | 92.5 113.7 | 6 | 2 | 6 | 3 | 87.1 106.8 | 4 | 2 | 5 | 2 |

## 6. REFERENCES

Anderson T.W. 1963. "Asymptotic theory for PCA", Ann. Math. Sta., 34, p. 121

Arconte A. Dauxois J. Pousse A. Romain A. 1980. CRAS Paris, t 2918, p. 319-322

Bickel P.J. Freedman D.A. 1981. Ann. Stat., 9(6), P.1196-1217.

Bougeard M.L. 1981. "Tests et validation en analyses factorielles" mémoire E.N.S.A.E.

Bougeard M.L. 1982. "Les étudiants en DUT gestion: une étude statistique", in: Mathématiques dans les enseignements de gestion, Colloque Dijon, ed. Bougeard (IUT Sceaux-Ville d'Avray).

Bougeard M.L. 1987. Astron. Astroph., 173 p. 191-203

Bougeard M.L. 1988. "Analyse de l'effet temps-magnitude-couleur dans des données d'astrométrie", Cahiers Mathématiques Paris X, n°15.

Davis A.W. 1977. "Asymptotic theory for PCA", Austr. J. Stat., 19(3), p.206-212.

Deville J.C. 1987. "Replications d'échantillons", in "Les sondages", ed. Fichet-Tassi, Economica Paris, p.158-171.

Efron B. 1979. "Bootstrap: another look to the jackknife", Ann. Stat.,7, 1-26

Fisher R.A. 1973. "Statistical Methods for Research Workers",3rd ed., Hafner Press.

Miller R. 1974. "The jackknife: a review", Biometrika, p1-15

Tomassone R. Lesquoy E. Miller C. 1980. "La régression: nouveaux regards sur une ancienne méthode statistique", ed. Masson Paris.

# MOVING REGRESSION: A TOOL FOR THE STATISTICAL AND ASTROMETRICAL ANALYSES OF SOLAR OBSERVATIONS

M.L. Bougeard (1)(2), J.M. Grosbras (3), and F. Laclare (4)

(1) Univ. Paris X,IUT,1 chemin Desvallières, F-92410 Ville d'Avray

(2) UA 1125 CNRS, Observatoire de Paris, F-75014 Paris

(3) ENSAE, 3 avenue Pierre Larousse, F-92241

(4) CERGA, avenue Copernic, F-06130 Grasse-France

Abstract: In this Paper, we perform regressions on the solar observations made at the CERGA Observatory, through a moving window the size of which is chosen taking into account the physical constraints. After some statistical considerations, the results obtained with different sizes of window arc analysed. Graphical comparisons are carried out on the estimators obtained by the usual method and by the new one for the 1978, 1979 and 1980 compaigns. They show that the latter is much more adequate to exhibit some periodic terms confirmed in the case of the sun demi-diameter through another analysis of one o the authors (Laclare, 1985). Furthermore, the increased accuracy of the estimations in the parameters of the relative solar orbit (Bougeard and al 1983) shows that this method is relevant.

## 1. INTRODUCTION AND STATISTICAL BACKGROUND

Since 1974, solar observations are performed with the solar Astrolabe at CERGA (Grasse). The observational data can be expressed as linear fonctions of differential corrections in zenithal distance (z) and in solar semi-diameter (d), right ascension $(\alpha)$ and declination $(\delta)$. A "complete observation" is the measure on the same day, of the time of passage, east and west (subscript: e,w), of the upper and lower edges (subscript: u, 1) of the sun disk croosing the same almucantar, which gives the following equations, written with the usual notations

$$\begin{bmatrix} R_{ue}=(1, \ -15| \sin Z| \cos\phi, \ 1)\beta & (1_{ue}) \\ R_{1e}=(1, \ -15| \sin Z| \cos\phi, -1)\beta & (1_{1e}) \end{bmatrix} \begin{bmatrix} R_{1w}=(1, \ 15 \ | \sin Z| \cos \phi, -1)\beta & (1_{1w}) \\ R_{uw}=(1, \ 15 \ | \sin Z| \cos\phi, \ 1)\beta & (1_{uw}) \end{bmatrix}$$

where, with the usual notation, R, Z, $\phi$ , respectively stand for observational data (observed-computed), the sun azimuth, the latitude. $\beta$ is the 3x1 vector of unknown parameters $\beta= (\cos S_i \Delta\delta+\Delta z_i, \Delta\delta ,\Delta d)'$ where (') stands for transposition, S for the star angle, and the subscript i for the chosen observational zenith distance (30°, 45°, 60°, ..).

A direct solution of system (1) can be given as

$$\left\{ \begin{array}{l} 4(\cos S_i \Delta\delta+\Delta zi)= \quad R_{ue}+R_{1e}+R_{1w}+R_{uw}, \quad 4 \Delta d= R_{ue}-R_{1e}-R_{1w}+R_{uw} \quad \text{(in arcsec)} \\ \qquad \Delta\alpha = (-R_{ue}-R_{1e}+R_{1w}+R_{uw})/(60 \sin Z \cos \phi) \qquad \text{(in time sec.).} \end{array} \right.$$

When the data are divided into morning and afternoon observations, determinations of the solar diameters can be obtained respectively through equations $\{1_{ue}-1_{1e}\}$ and $\{1_{uw}-1_{1w}\}$: the related series have been studied elsewhere (Laclare 1983; Delache and al 1985).

In fact, the primary goal of these solar observations is to estimate the orbital elements of the Earth around the Sun (Bougeard and al 1983; Bougeard and Laclare 1984). The first reduction process of system (1) (Chollet 1981) required complete observations, the system of the 4 equations being solved by a least Squares (LS) fit. First, it produced high fluctuations in the estimated error variances (one degree of freedom) and very large intervals of confidence for $\Delta\alpha$. Furthermore, in the case of CERGA sun observations, west data can be missing for several consecutive observational, days (Fig.1) even in summer, in relation with meteorological conditions. In such a case, system (1) is not invertible. There was a loss of information due to the incomplete observations being given up, which introduced, in a second step, difficulties in the calculation of accurate determinations of the orbital elements of the Earth. This led us to introduce a reduction method which we called "moving regression". This method consists in performing regressions through a moving window applied on the data and grouping a given number of observations: this number is chosen taking into account the physical constraints.

Window estimates have been used for a long time in astrometric time series analyses. Several theoritical and statistical results have been obtained in this field, especially in terms of non parametric prediction for a stationary process, autoregressive or ARMA process (see: Collomb 1984 and its references).

By using moving-window regressions in the context of the sun astrolabe observations, our aim was not to attempt time series modelisation of the variables $\Delta\alpha, \Omega = (\cos S \Delta\delta + \Delta z)$ and $\Delta d$; our purpose was to ascertain the LS determinations of these parameters (by using more than 4 equations), to obtain LS estimates with comparable error variances (same degrees of freedom) and to use incomplete observations.

In Section 2, we recall some statistical results in the context of moving regressions. Graphical comparisons are carried out on the estimators obtained by the usual method and by the new one. They show that the latter is much more adequate to exhibit some periodic terms confirmed through another analysis (Laclare 1985). Furthermore, the increased accuracy of the estimations in the parameters of the (relative) solar orbit (Bougeard and al 1983) shows that this method is relevant.

## 2. MOVING WINDOW REGRESSIONS

For expositional convenience, let us write the model in statistical and matrix algebra notation

$$Y = X \beta + \varepsilon$$

where Y is the (T,1) observed vector, X is the (T,3) regressor matrix and $\varepsilon$ the (T,1) vector of unknown errors. Assuming a common variance of the errors, $\sigma^2$, the LS fit produces a well-known estimate b of $\beta$

$$b = (X'X)^{-1}X'Y$$

$\sigma^2$ being estimated by the mean square error $s^2 = e'e/(T-3)$ where $e = Y - Xb$ is the usual (T,1) vector of the LS residuals. We denote by S the sum of the square residuals : $S = e'e$.

In pooling T solar astrolabe observations, we have to pay attention to the fact that the observations are made at the same zenith distance, that they belong to an interval of time less than two weeks in order to neglect time variations of the unknown parameters and that, in case of visual observations, it has to be the same observer (personal equation). Furthermore, to ensure X'X is of full rank, observations of each solar edge in each transit (east, west) have to be taken into account.

### 2.1. Deleting one observation

Let us suppose that in a first step, a LS regression has been performed on T observations $(y_k, x'_k)$ $k=1,..T$ with

$$Y = (y_1,...,y_T)' \qquad X = (x'_1,..x'_T)'$$

to obtain with obvious notations:

$$b(T) = (X'X)^{-1}X'Y \quad , \quad S(T) = (Y - Xb(T))'(Y - Xb(T)). \qquad (2)$$

If, we want to delete the observation j of the T sample, by introducing $Y = (Y_{T-1}, y_j)'$ and $X = (X_{T-1}, x'_j)'$, we obtain following a well-known[*] Rao's lemma (Rao and Miller 1971), subscript j being omitted:

$$b(T-1) = (X'_{T-1}X_{T-1})^{-1}X'_{T-1}Y_{T-1} = (X'X - xx')^{-1}(X'Y - yx)$$

$$\begin{cases} b(T-1) = b(T) - (y - x'b(T))(X'X)^{-1}x/D & \text{with } D = 1 - x'(X'X)^{-1}x \\ S(T-1) = S(T) - (y - x'b(T))^2/D \end{cases}$$

The above relations are useful for iterative estimations in jackknifing regressions or to evaluate the influence of a particular observation on the parameter estimates.

### 2.2. Adding and deleting one observation

Let us suppose now that n data are available in chronological order of observation. Let T be the size of the window. A sample of T consecutive data $((y_t, x'_t),$

---

[*] Let A(n,n), u(n,1) and v(n,1) : $(A+uv')^{-1} = A^{-1} - A^{-1}uv'A/(1+u'A^{-1}v)$ when the operators exist.

$(y_{t+1}, x'_{t+1}), ..., (y_{t+T-1}, x'_{t+T-1}))$ t=1,2... leads to an estimate $b_t$ such that (2) holds with

$$Y_t = (y_t, Y_*)' \qquad X_t = (x'_t, X'_*)'$$

while $b_{t+1}$ is solution of a LS fit with

$$Y_{t+1} = (Y_*, y_{t+T})' \text{ and } X_{t+1} = (X'_*, x'_{t+T})'$$

So : $\quad b_{t+1} = (X'_t X_t - x_t x'_t + x_{t+T} x'_{t+T})^{-1} (X'_t Y_t - x_t y_t + x_{t+T} y_{t+T})$

which yields, performing some algebra (Bougeard 1982), subscript t omitted

$$\begin{cases} b_{t+1} = b + (y_T - x'_T b)(X'X)^{-1}(dx + cx_T)/m \ + \ (y - x'b)(X'X)^{-1}(dx_T - ax)/m \\ S_{t+1} = S + (y_T - x'_T b)^2/a \ - \ ((y_T - x'_T b) - (y - x'b)d/a)^2/(c + d^2/a) \end{cases}$$

$$\text{where:} \begin{cases} a = 1 + x'_T(X'X)^{-1}x_T & d = x'(X'X)^{-1}x_T \\ c = 1 - x'(X'X)^{-1}x & m = ac + d^2 \end{cases} \qquad (3)$$

From (3), the correlation between $b_t$ and $b_{t+1}$ can be deduced.

When we simultaneously add and delete not one but a given number of observations, formulae similar to (3) can be written using partitioned matrices.

## 2.3. Application to the solar astrolabe data

In reference to the facts that the astrolabe observations of the sun are not obtained at equidistant dates, and that under favourable atmospherical conditions there are 4 observations per day, we have chosen the size T of the moving LS regression, MLS(T), as a multiple of four. Furthermore, since the estimates $b_t$ have to be used through a second LS fit to obtain corrections to the main parameters of the Earth orbit, the LS determinations are moved from 4 to 4 taking into account the physical constraints given at the beginning of this Section. For example, for MLS(8), successive regressions are a priori performed on the samples (t=1-8), (t=5-12), (t=9-16),(t=13-20).. and for MLS(16) on (t=1-16), (t=5-20), (t=9-24),(t=13-28)..., the related estimates being associated to a reference julian date which is the mean of the julian date of the observations in the chosen sample.
We can note that, since the size of the window is constant in all these MLS regressions, the successive estimates have comparable error variances (same degrees of freedom). Furthermore, it is clear that the density of the estimations $b_t$ will not be so high in MLS(16) as in MLS(8) owing to the physical constraints (see Fig.1). In MLS(8), a vector of estimates is a priori only correlated with its precedent and its successor; things get more complicated in the case of MLS(16).

## 3. ANALYSIS OF THE RESULTS OF DIFFERENT MOVING REGRESSIONS

In a first step, we have analysed, through MLS(8), MLS(16) and MLS(20) regressions, the observations of the 1978 campaign made at the same instrumental zenith distance 45°, mainly by F. Laclare. The results are graphically given in Fig.2,3,4 for the right ascension $\Delta\alpha$, the demi-diameter $\Delta d$ and $\Omega = \cos S \, \Delta\delta + \Delta z$. They are compared with the estimations obtained by the method of "Complete Observations" (COC) (Chollet 1981). It can be seen on Fig.3 that the MLS(8) estimations in demi-diameter show four points nonhomogeneous with the others around the julian date 43700: they correspond to observations made by an

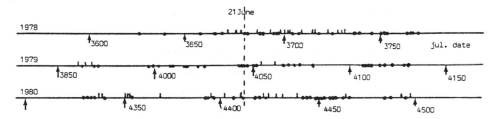

Figure 1. Cerga Solar Astrolabe (z=**45°** zenith distance)
● (resp ı ) = complete (resp. incomplete observation),
jul. date = julian date-40000.

observer who is not F. Laclare and these MLS determinations clearly give evidence of an observer's personal effect on the measures.

As far as Fig. 2 and 4 are concerned, by successive consideration of the COC, MLS(8), MLS(16) and MLS(20) estimates, it can be seen that periodic effects are more and more underlined in the different curves which appear as if they were deduced one from the other by smoothing.

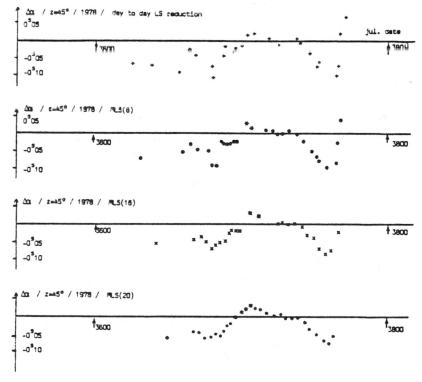

Figure 2 .

43

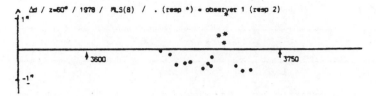

Figure 3.

In a second step, MLS(8) regressions have been systematically performed on the observations of the 1978, 1979 and 1980 compaigns. The related corrections to some parameters of the earth orbit can be found in (Bougeard and al 1983) where the increased accuracy of these estimations shows that the method is relevant. We present here the graphs related to the different MLS(8) estimates per julian date and for the same instrumental zenith distance 45°. A Vondrak's smoothing (ε=0.0001) (Vondrak 1969; Huang and Zhou 1981) has been applied to bring possible periodicities into evidence.

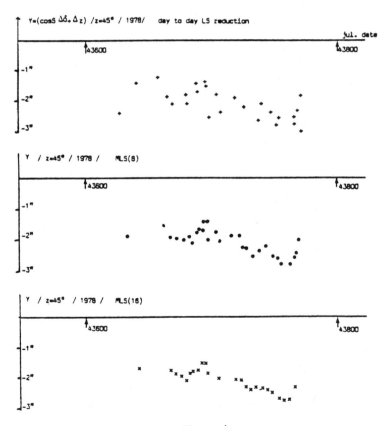

Figure 4.

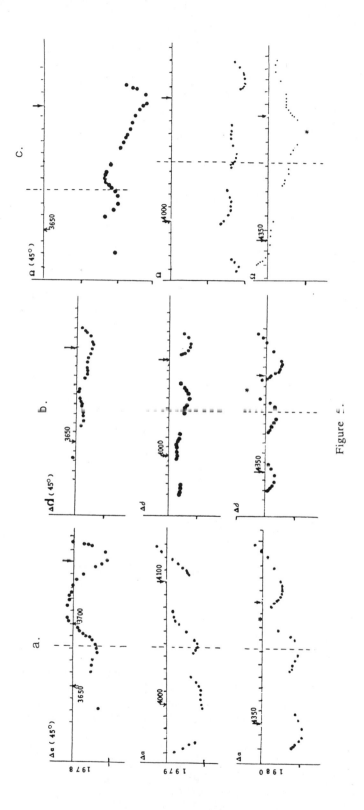

Figure 5.

45

As already remarked in (Bougeard 1982), althrough there are many missing values in the time series of the studied years, periodicities appear not only in the estimations of the demi-diameter (Fig. 5b)) - periodicities confirmed through east and west determinations of this demi diameter in other works of one of the authors (Laclare 1983; Delache and al 1985) - but also in the estimations in right ascension (Fig. 5a) of the sun and in declination or more precisely in the term cos S $\Delta\delta$ +$\Delta$z (Fig. 5c). These effects were not detected in the COC estimations (Chollet 1981) since there were too high fluctuations in the estimated quantities. Furthermore, it can be observed on Fig. 5c, that the mean value of the parameter (cos S $\Delta\delta$ + $\Delta$z) depends on the year, which is a consequence of different instrumental changes in the studied observational years.

## 4. CONCLUSION

As proved in this Paper, moving regressions through a moving window of T points, denoted by MLS(T), appear as a very good tool for the statistical and astrometrical analysis of the solar observations made at the CERGA OBSERVATORY astrolabe. We have shown that the MLS(8) regressions, especially studied here, were adequate to detect some observer effect in the sun demi-diameter measures and some instrumental effects. They are also adequate to exhibit periodicities not only in the demi-diameter (confirmed through east and west determinations in other works of one of the authors) but also in the estimations in right ascension and declination, although there were many missing data in the studied compaigns. At present, the instrument is being becoming impersonal (through the use of a C.C.D. camera for the measurements) with an improvement of the refraction corrections obtained by adequate filters. So, the periodicities detected here through the moving regressions will be underlined more easily in the future campaigns; in particular, the different periodic components could be determined with more accuracy as soon as the whole (apparent) sun orbit will be covered.

## 5. REFERENCES

Bougeard M.L. 1982. "Analyse des observations du Soleil à l'astrolabe du Cerga, mémoire Observatoire de Paris.

Bougeard M.L., Chollet F., Laclare F. 1983. Astron. Astroph., 126, 161-169

Bougeard M.L., Laclare F. 1984. "Mesures du diamètre solaire et des éléments orbitaux de la Terre, Symposium GS5 Paris, éd. Requième, 87-91.

Chollet F. 1981. Doctor thesis, Paris Observatory.

Collomb G. 1986. "Analyse d'une série temporelle et prédictions non-paramétriques", in Asymptotic Theory for non iid Process, ed. Florens, Pub, Fac. Univ. Saint-Louis, Bruxelles, p.37-75.

Delache P. Laclare F. Sadsaoud H. 1985. "Long period oscillations in solar diameter measurements", Nature, 317, p. 416-418.

Huang K.I. Zhou X. 1981. "The Whittaker-Vondrak method of data smoothing as a numerical filter", Chin. Astron. Astroph., 5, p.449-454

Laclare F. 1983. Astron. Astroph., 125, p.200-205.

Rao P. Miller R.L. 1971. "Applied econometrics", Wadsworth.

# MULTIVARIATE LINEAR REGRESSION

# WITH SELF-ESTIMATION OF ERRORS

*Rino Bandiera*

Osservatorio Astrofisico di Arcetri
Largo E. Fermi 5
I-50125 Firenze  (Italy)

*Leslie Hunt*

C.A.I.S.M.I.–C.N.R.
Largo E. Fermi 5
I-50125 Firenze  (Italy)

## INTRODUCTION

A common approach to the analysis of a sample of data is to search for linear correlations between the variables that describe the sample. Most of the fitting techniques used in that context derive from a statistically rigorous method of estimation, the maximum likelihood method. In its simplest and most widely used form, the method is applied to the case of two variables, only one of which is affected by error; while in the more general case all variables under consideration are beset by errors. Moreover, what we call generically "errors" can actually have various origins: they can either be real measurement errors, or rather originate from a scatter intrinsic in the sample, appearing when the variables included in the analysis are not sufficient to completely describe the physical attributes of the objects. The former case is easier to analyze because measurement errors can, in principle, be estimated and standard multivariate regression techniques can be adopted; the latter, instead, is more difficult to deal with as one must extract simultaneously from the data both the direction of the main correlation and the properties of the scatter. Principal components analysis is usually adopted for this purpose but by this technique arbitrary structure is introduced in the residuals.

We have developed a specific technique for the case described above, in which several measured mutually correlated quantities are available and all variables are affected by errors; we want to fit the data to a single correlation and determine the properties of the residual scatter. Our method is based on the separation of all the correlations into two terms: one contains only the primary correlation (described by a single parameter) and the other represents residual scatter and is treated as a statistical dispersion. Some reasonable constraints are then put on the properties of the scatter: in the simplest version, residual scatters along the various axes are assumed to be mutually uncorrelated.

## ASSUMPTIONS AND DEFINITIONS

Let us consider a sample of $N$ objects, and assume that a set of $M$ different quantities is available for each object; each set of quantities can be represented by a point ($x_k$, where $k = 1, M$) in a $M$-dimensional space. Let every quantity be linearly correlated with every other quantity; in the absence of scatter, all the points must lie on a straight line in $M$-space: this line can be represented by $M$ parametric equations, $x_k = \alpha_k + \beta_k \tau$. This is equivalent to assuming that one parameter, namely $\tau$, is sufficient to replace the $M$ different quantities that are originally used to describe the sample. However, in any real situation the $M$ measured quantities do not perfectly fit on the line but are scattered about it; each data point can then be expressed as $x_k = \alpha_k + \beta_k \tau + \epsilon_k$, where $\epsilon_k$ is the error about the regression line in the direction of the $k^{th}$ variable.

According to the standard problem, we shall assume that the measured quantities are affected by errors which follow a normal distribution, with zero mean ($\langle \epsilon_k \rangle = 0$), constant along the regression line (the error matrix $\mathcal{E}_{hk} = \langle \epsilon_h \epsilon_k \rangle - \langle \epsilon_h \rangle \langle \epsilon_k \rangle$ is constant). The $k^{th}$ component of the standard deviation about the fitted line is proportional to the square root of the diagonal element $\mathcal{E}_{kk}$ and, if the errors are mutually uncorrelated, $\mathcal{E}_{hk}$ will be a diagonal matrix. In the more general case we are addressing here, $\mathcal{E}_{hk}$ will be positive-definite and symmetric.

In order to simplify the notation, we do not use any index to refer to a specific object in the sample; when summing up over the $N$ points (objects) we shall simply use the symbol $\sum_p$. Therefore, we define $A_{hk} = \sum_p x_h x_k / N$ and $B_k = \sum_p x_k / N$ so that the covariance matrix can be expressed as $S_{hk} = A_{hk} - B_h B_k$.

## THE METHOD

We want to derive the best-fit line from a sample of data. When the error matrix is known *a priori*, the standard maximum likelihood technique simplifies to the minimization of the sum of the square deviations, in a metric defined by the inverse matrix of $\mathcal{E}_{hk}$; in the simple case of a diagonal $\mathcal{E}_{hk}$, the cartesian distance is used, but with coordinates scaled with the respective scatters. Our method, instead, works under the assumption that the error matrix is not known.

By direct substitution of $x_k = \alpha_k + \beta_k \tau + \epsilon_k$ in $A_{hk}$ and $B_k$, the covariance matrix can be written as:

$$S_{hk} = \lambda \beta_h \beta_k + \mathcal{E}_{hk}, \quad \text{where } \lambda = \sum_p \tau^2 / N - \left( \sum_p \tau / N \right)^2; \tag{1}$$

we used also the identities $\sum_p \tau \epsilon_k / N - \sum_p \tau \sum_p \epsilon_k / N^2 = 0$ (true if errors do not depend on the position) and $\sum_p \epsilon_h \epsilon_k / N - \sum_p \epsilon_h \sum_p \epsilon_k / N^2 = \mathcal{E}_{hk}$ (from the definition of $\mathcal{E}_{hk}$).

The statistics of data can be inferred as follows: if we define $u_k = \sqrt{\mathcal{E}_{kk}}$ (the magnitude of the $k$ component of the scatter about the regression line), $E_{hk} = \mathcal{E}_{hk} / u_h u_k$ (the error matrix normalized on its diagonal) and $V_k = \sqrt{\lambda} \beta_k / u_k$, the equation for $S_{hk}$ becomes:

$$S_{hk} = (V_h V_k + E_{hk}) u_h u_k. \tag{2}$$

When $\sum_k |V_k|^2 \gg 1$ the covariance matrix is dominated by correlation; when $\sum_k |V_k|^2 \ll 1$ scatter plays the major role. Incidentally, an estimate of scatter can be directly obtained from the diagonal elements of $S_{hk}$ only when the correlation in the sample is low; otherwise these elements depend also on the direction of the correlation.

Solving for $u_k$ one gets $u_k = \sqrt{S_{kk}/(1 + V_k V_k)}$, that, when substituted in Eqn.2, allows one to evaluate:

$$E_{hk} = \sqrt{(1 + V_h V_h)(1 + V_k V_k)} S_{hk} - V_h V_k, \qquad (3)$$

where $S_{hk} = S_{hk}/\sqrt{S_{hh} S_{kk}}$ is the correlation matrix (normalized on its diagonal). In order to determine $V_k$ an additional constraint must be placed on $E_{hk}$: a natural one is to assume that correlations between errors on different axes are as small as possible (the terms on the diagonal of $E_{hk}$ are excluded, because they are already set to unity). In other words, $V_k$ can be derived minimizing:

$$F = \sum_{h \neq k} |E_{hk}|^2. \qquad (4)$$

All the other best-fit parameters follow: $u_k$ directly from its definition, $\lambda$ and $\beta_k$ from the definition of $V_k$ (with the additional constraint that $\beta_k$ is a unit vector); furthermore $\alpha_k = B_k$, if we fix the arbitrarity of the choice by the condition $\alpha_k \beta_k = B_k \beta_k$.

Standard multivariate regression techniques adopt as the best-fit line the direction of the eigenvector of $S_{hk}$ with the largest eigenvalue (see, e.g., Murtagh and Heck 1987); the same direction is derived by our method when $E_{hk} = S_{hk} - V_h V_k$, namely when $\sum_k |V_k|^2 \ll 1$: therefore it reduces to the standard method in the limit of large scatter. It should be noted, however, that while with our method the normalized error matrix $E_{hk}$ evaluated at the minimum of $F$ resembles the unit matrix, with standard regression techniques the residual matrix is a projection onto a hyperplane orthogonal to the direction of the best-fit line. In other words, structure is introduced into the residual errors by standard fitting techniques; this condition is much less reasonable than the one we assumed above, namely that the sum of the square of the off-diagonal error matrix elements is a minimum.

## A TEST ON THE QUALITY OF THE FIT

The number of independent quantities in $S_{hk}$ is $M(M-1)/2$, being $S_{hk}$ a symmetric matrix with all the elements on its diagonal equal to unity; conversely, $M$ components of $V_k$ must be derived: therefore one needs $M(M-1)/2 \geq M$. When $M = 2$ the problem is underdetermined, and no self-estimation of errors can be performed (see, e.g., Kendall and Stuart 1963). At $M = 3$ the number of unknown variables is equal to that of equations: a solution is reached giving $F = 0$. The criterion of minimum can be properly used only for $M \geq 4$, when the number of equations in eccess over the unknown variables is $M(M-3)/2$.

Therefore for $M \geq 4$ one could in principle determine the statistics for $F$, and then use it to estimate how good the fit was. In the general case the distribution of $F$ is very complicated, and can be determined only numerically; however a good analytical approximation can be derived in the limit of a large number of points, provided that scatter is small and that $M$ is large enough (say $\geq 5$). In fact for large $N$ each term $\mathcal{E}_{hk}$ (with $h \neq k$) follows a normal distribution with zero mean and $\sqrt{\langle\epsilon_h^2\rangle\langle\epsilon_k^2\rangle/2N}$ standard deviation. $\mathcal{E}_{kk}$, instead, follows a normal distribution with $\langle\epsilon_k^2\rangle$ mean and $\sqrt{2/N}\langle\epsilon_k^2\rangle$ standard deviation: for large $N$, $\mathcal{E}_{kk}$ can be directly approximated by its mean value. Therefore the distribution of each term $E_{hk}$ (with $h \neq k$) is a gaussian with zero average and $1/\sqrt{2N}$ standard deviation; as a result $NF$ follows a $\chi^2$ distribution with $M(M-1)/2$ degrees of freedom. More specifically, the average value of $F$ is $M(M-1)/2N$, while its variance is $M(M-1)/N^2$.

The possibility of performing a $\chi^2$ test is very important in order to apply our technique, because it can tell us how confident we can be about the resulting fit. In real data, due to various effects, our basic assumptions may not be completely fulfilled: the most common are a non linear dependence between data, a statistics of scatter depending on the position in the parameter space, and the presence of secondary correlations between data. A successful $\chi^2$ test allows one to put upper limits on those effects, the larger the number of objects available, the tighter the limits.


NUMERICAL RESULTS

A routine has been developed that solves the criterion of minimum expressed in Eqn.4; the algorithm uses, as a first guess for $\lambda$ and $\beta_i$, the maximum eigenvalue and the corresponding eigenvector of matrix $S_{hk}$, and then proceeds by a minimization technique. The performances of this routine have been tested on simulated data. For each set of data a primary correlation is defined, with an arbitrary direction in the $M$-dimensional parameter space; then $N$ points are chosen randomly on the correlation line, over a distance $D$; finally a scatter is assigned to each point, following a normal distribution with zero mean, and in general with different standard deviations along different axes.

The results of simulations are displayed in Figg. 1–5. Each figure is divided vertically into three parts, corresponding to $M$ equal to 4, 6, 8; in each part the lines for $N$ equal to 20, 50, 100, 200 are shown, labelled with letters from $(a)$ to $(d)$, respectively; each line connects results for different values of the scatter. The variable $\epsilon_{\mathrm{act}}$, along the x-axis, represents the scatter in simulated data, or the maximum scatter when different values are used for different axes: we performed simulations for $\epsilon_{\mathrm{act}}$ ranging from $10^{-2}D$ to $D$, with 10 steps per decade. A few hundred simulations were required for each set of parameters in order to accurately trace the average behaviour of that routine.

We started simulating data with the same scatter along all the axes. The values estimated from the fitting routine follow a distribution with mean value $\epsilon_{\mathrm{est}}$ (see Fig. 1) and spread $\Delta\epsilon_{\mathrm{est}}$ (see in Fig. 2). In general $\epsilon_{\mathrm{est}}$ reproduces the actual scatter rather well; for $N \geq 100$ and $\epsilon_{\mathrm{act}} \leq 0.1D$ the agreement is within $\approx 1\%$; however, for a smaller number of points, or for a larger scatter, $\epsilon_{\mathrm{act}}$ tends to be slightly underestimated, because structures created by the random noise can be partly ascribed to the primary correlation; moreover, the results improve for larger values of $M$. Also the spread around $\epsilon_{\mathrm{act}}$ depends on $M$, $N$ and $\epsilon_{\mathrm{act}}$; for $M \geq 6$, $N \geq 100$ and $\epsilon_{\mathrm{act}} \leq 0.1D$ each estimated scatter agrees within 10% with the actual one.

Then we tested the ability of the routine to recover the original scatters when they are different along different axes; we basically repeated the simulation described above, but decreasing the scatter along one axis (axis 1) by a factor 10 ($0.1\,\epsilon_{\mathrm{act}}$) with respect to the others. The ratio between the mean estimated scatter along axis 1 ($\epsilon_1$) and that along the other axes ($\epsilon_{est}$) is plotted in Fig. 4; the presence of a smaller scatter along axis 1 is always recognized, but the estimated ratio is smaller than 0.15 $\epsilon_{est}$ only for $M \geq 6$ and $N \geq 50$, with a negligible dependence on the amount of scatter; for smaller $M$ or $N$ the tendency is to level the estimated scatters along different axes.

Also the angle between the actual and the estimated direction of the primary correlation has been computed; its mean value, $\theta$, is compared with that obtained by standard multivariate regression ($\theta_{std}$); the case with equal scatters along all the axes is shown in Fig. 3, while that with a smaller scatter along axis 1 is shown in Fig. 5: our routine usually gives better results, particularly when the number of points is rather large; in the case of the same scatter along all the axes the standard routines seem to give slightly better results when $\epsilon_{act} \geq 0.3D$, but in the case of different scatters our routine gives better results for a wider range of scatters.

To summarize, our method successfully reconstructs both the direction of the primary correlation and the residual scatters, whenever the number of points is not too small and the scatters are not too large. Otherwise the scatters on the various axes tend to be leveled at smaller values than the original ones, and the information on scatter is lost; but even in that case a reasonable attempt of reconstruct the correlation direction is obtained, with an accuracy usually higher than that of standard multivariate regression.

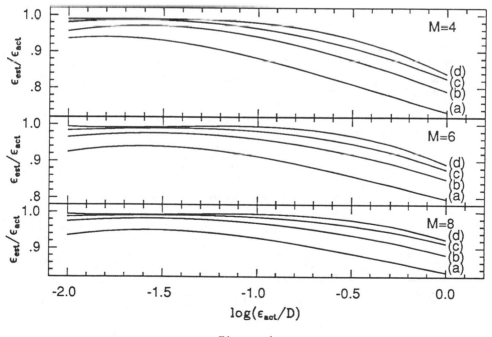

Figure 1.

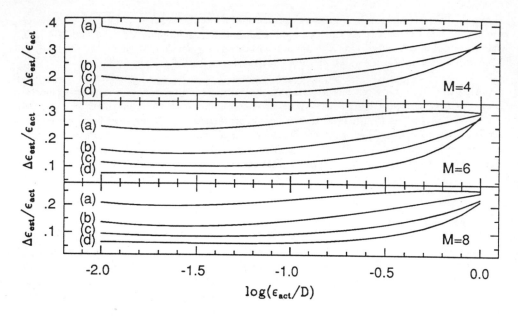

Figure 2.

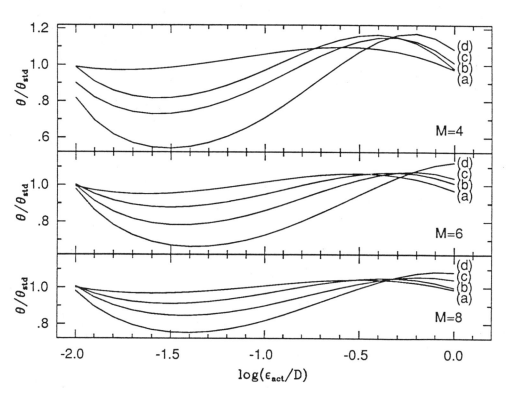

Figure 3.

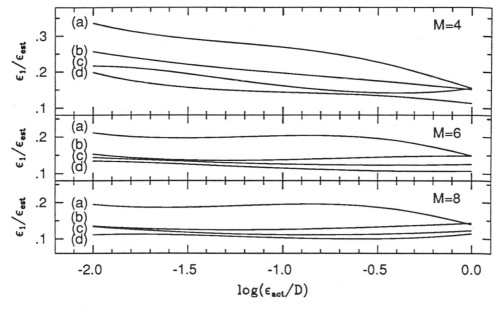

Figure 4.

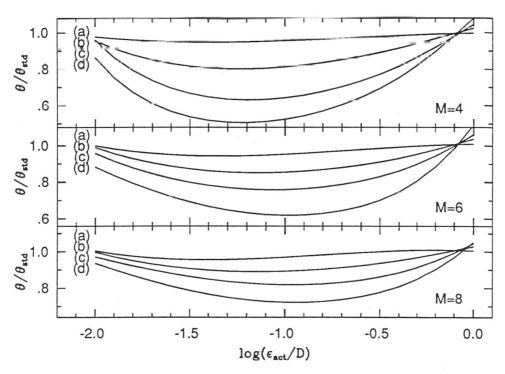

Figure 5.

## CONCLUSIONS

The method of linear regression presented here can be used for the analysis of samples with the following properties: 1) various kinds of measurements are available for each object; 2) there is a clear correlation between each pair of quantities; 3) an intrinsic residual scatter is present, that cannot be estimated *a priori*.

For these problems our method is more powerful than two dimensional linear regressions applied to all pairs of quantities, because only by treating the $M$-dimensional problem can all the statistical information derived from data be properly used. It is also more powerful than standard multivariate regression techniques, because they rely on an estimate of scatter which turns out to be arbitrary in all cases presenting good correlations; while a procedure that self-estimates scatter not only gives a more accurate direction of the correlation, but also provides information on what quantities are most affected by fluctuations.

Our technique assumes that, apart from the general correlation, all residuals are statistically uncorrelated; however it can be generalized to also account for secondary correlations, the maximum number of which only depends on the number of different measurements processed simultaneously. But we believe that a treatment of secondary correlations can be performed only in those rare cases (at least for the moment) where the number of objects in the sample is very large; for the procedure presented here, instead, the requirements on the sample match rather well the quality of data commonly available.

## REFERENCES

Kendall, M.G. and Stuart, A.S. 1963, *The Advanced Theory of Statistics; Vol.II* (Griffing & C., London)

Murtagh, F. and Heck, A. 1987, *Multivariate Data Analysis* (Reidel, Dordrecht)

# MAXIMUM ENTROPY IMAGE PROCESSING

# IN GAMMA-RAY ASTRONOMY

A.W. Strong and R. Diehl
Max-Planck Institut für Extraterrestrische Physik
D-8046 Garching,
W. Germany

## INTRODUCTION

Imaging $\gamma$-ray astronomy is a slowly evolving field because experiments are expensive, few and far between. For energies above 100 MeV the only satellite experiments to date with real imaging capabilities have been NASA's SAS-2 and ESA's COS-B. At lower energies (1-30 MeV) only balloon experiments have so far had imaging capabilities (e.g. the MPE Compton Telescope). Much of the scientific analysis of such data can be done without actually generating images at all - for example source searches can be done by cross-correlation or similar methods, and quantitative studies of the diffuse Galactic emission are best done by model fitting using radio surveys of Galactic gas as a basis. However visual presentation is also essential as a means for understanding the data, comparing it with other wavelengths, and looking for the unexpected. Because of the particular problems associated with $\gamma$-ray observations, it is logical that considerable investment be made at this time in studying methods to obtain the best possible images for scientific analysis, especially since the forthcoming launch of NASA's Gamma-Ray Observatory will increase the amount of $\gamma$-ray data by two orders of magnitude.

The maximum entropy method is an obvious choice for imaging because of its characteristic treatment of the relation between image and data; it requires only that the detector response to any given image can be computed, and does not require any explicit relation between individual points in image and data space. In this way it differs fundamentally from the simple 'binning and smoothing' normally adopted. The advantage is particularly clear when the statistics are small so that there are few events per angular resolution width. In the normal approach a larger

55

binning must then be adopted which degrades the intrinsic resolution of the instrument. Even more compelling however is the case where the response is such that no unique relation exists between individual image and data bins, so that there is no 'conventional' method which can produce an image at all - such is the case for the Compton telescope.

Here we give examples of studies of maximum entropy applied to both spark-chamber (COS - B) and Compton (GRO-COMPTEL) telescopes. Since the principles of the maximum entropy method are well described elsewhere (e.g. Gull and Skilling 1984) they will not be further discussed here.

## SPARK-CHAMBER TELESCOPES

This type of instrument uses pair production as the primary conversion process, and this means in practice that the energy range is restricted to >30 MeV. The $\gamma$-ray arrival direction is estimated from the tracks of the electron-positron pair produced in the first interaction. The errors inherent in this procedure are responsible for the rather wide point-spread-function (PSF) characteristic of these $\gamma$-ray energies.

The characteristics of COS-B are summarized below:

| | |
|---|---|
| Energy range | 50–5000 MeV |
| Field of view | $40^{\circ}$ |
| Energy resolution | 50 % |
| Angular resolution | $1-3^{\circ}$ |
| Counts/resolution element | 1-100 |
| Mission duration | 6.7 years |

Skilling et al. (1979) applied a maximum entropy method to some COS-B data in the region of the Galactic anti-centre. Although acceptable images were obtained the investigation did not lead to any general adoption of the method in future COS-B work, perhaps because of the novelty of the method but also because of the problem of knowing just how to use and interpret a maximum entropy image.

A signficant problem in the application of maximum entropy is that there are typically only a few events per bin so that the usual $\chi^2$ statistic is not applicable. A more appropriate statistic is the likelihood calculated from Poisson statistics, which remains valid even for empty bins. A much extended application was later made to the entire Galactic plane using the full COS-B database and the MEMSYS maximum entropy package from MEDC Ltd, modified to use the Poisson likelihood statistic as an alternative to $\chi^2$. A sample map for the energy range 150-5000 MeV is shown in Fig 1.

An interesting finding from this map is the source apparent south of the plane in the Cygnus region ($l=76^o$, $b=-8^o$). This is confirmed by more detailed study of the region in comparison with the predictions of a model of the emission from cosmic ray /gas interactions. Also of interest is the considerable emission between the Crab pulsar ($l=184^o$) and Geminga ($l=195^o$) - this evidently orginates in clouds of molecular hydrogen in the Galactic anticentre (see Strong et al. 1988).

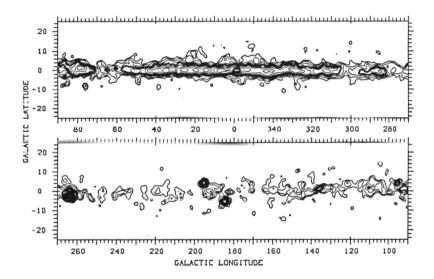

**FIG 1.** Maximum entropy map of the Galactic plane using COS-B data in the energy range 150-5000 MeV

A certain amount of deconvolution is achieved for the strong sources as can be seen in a comparison of the raw binned data with the maximum entropy image in a longitude plot (Fig 2). For example the peak around longitude $80^o$ lies above the data points because the deconvolved intensity is peaked within the latitude band ($-5.5^o < b < +5.5^o$) used in this plot. Deconvolution in latitude is illustrated in Fig 3 which shows the same region ($70^o < l < 90^o$) as a latitude profile.

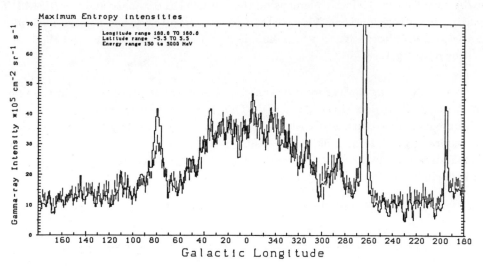

**FIG 2.** Longitude profile corresponding to Fig 1. The histogram shows the maximum entropy image, the vertical bars the original binned data, both averaged over the latitude range. $-5.5° < b < +5.5°$.

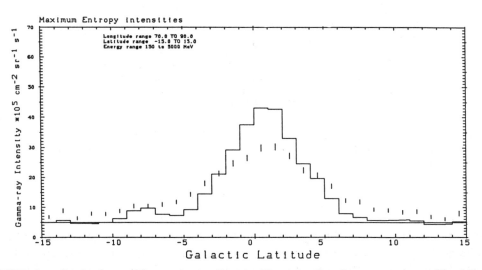

**FIG 3.** Latitude profile corresponding to Fig 1 in the Cygnus region. The histogram shows the maximum entropy image, the vertical bars the original binned data, both averaged over the longitude range $70° < l < 90°$.

# COMPTON TELESCOPES

## Imaging in the MeV Range

In the MeV-regime of the $\gamma$-ray spectrum, Compton telescopes have been applied for imaging (Graser and Schönfelder 1982, von Ballmoos et al. 1987a,b). These instruments employ a rather different detection principle due to the penetrating power of MeV $\gamma$-rays : photon interactions with matter are of particle rather than wave nature, which prevents construction of focusing devices; the photons are however not sufficiently energetic to produce a large number of particle interactions, which would generate a track of secondary particles along their trajectory within the instrument. Therefore the detection of an MeV $\gamma$-ray is based on measurement of many parameters for a few interaction processes within the instrument. A Compton telescope detects a Compton scattering interaction of the incoming photon in an (upper) plane of detector elements, followed by a total absorption of the Compton scattered photon in another (lower) plane of highly absorbing detector material (see figure 4).

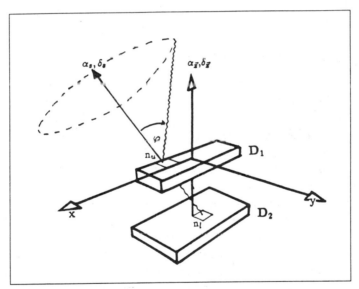

FIG 4. Schematic diagram of COMPTEL

The characteristics of GRO-COMPTEL are summarized below:

| | |
|---|---|
| Energy range | 1-30 MeV |
| Field of view | $60^{\circ}$ |
| Sensitive area | 20–40 cm$^2$ |
| Energy resolution | 10 % |
| Angular resolution | $0.7$–$2.1^{\circ}$ |
| Mission duration | 2–5 years (expected) |

Early attempts at imaging made use of special raw data filtering to select detector events with minimum distortions from the 'ideal' Compton scatter event. The simplest way to generate an 'image' from Compton telescope data is the so-called 'event-circle' method. The projection of measured scatter direction and scatter angle results in an 'event circle', which contains the true photon arrival direction in the ideal case. Regarding the circles as measures of the probability distribution on the sky for each event, summation of circles for many events results in an 'image'. The first skymap of the Galactic anticentre in the MeV range was generated with this method (Graser and Schönfelder 1982). v. Ballmoos et al. (1987a,b) improved the event filtering techniques, applied systematic corrections (derived from Monte-Carlo calculations) to the 'event circle' parameters, and combined the individual event's information into an image via a likelihood product. Thus they were able to derive images of the Galactic Centre region and the nearby active galaxy Cen A. The event circle method does produce a kind of image, but one which does not satisfy the key criterion, that folded with the instrumental reponse it should be consistent with the observed data. In addition, the filtering effort becomes particularly complex when instrumental parameters (gains, thresholds) vary over the exposure, and minor filtering inconsistencies may lead to false features in the skymaps(*). Therefore application of a least biased image construction technique was investigated for the satellite version of the Compton telescope, the COMPTEL instrument to be launched in 1990 aboard NASA's Gamma Ray Observatory.

## Characteristics of COMPTEL Data

The prime measurement quantities of an event in a Compton telescope are the energy deposits in the two detector planes, and the position of the interactions in both planes. These detector planes are separated by a distance of order one metre. In the COMPTEL telescope there are 7 upper detector modules of liquid scintillator and 14 lower detector modules of NaI, the two planes being separated by 1.5 m. In both detector module types the positional measurement is made by the Anger camera principle.

The Compton scatter process in the upper detector plane determines the particular directional information which is measured by the instrument for each photon; because of the probabilistic nature of this scatter process a wide range of scattering angles is possible for photons incident on the instrument with the same angles and energies. Therefore each pixel in image-space is translated into a multi-parameter distribution in data-space. (From this it is evident that an event-by-event backward projection of Compton telescope data cannot provide adequate images).

The measurement process in a Compton telescope can be described by

$$D(\vec{x}_1, \vec{x}_2, E_1, E_2) = R(E_\gamma, \alpha, \delta, \vec{x}_1, \vec{x}_2, E_1, E_2) I(E_\gamma, \alpha, \delta)$$

---

(*) v. Ballmoos et al. (1987) were able to exclude such effects via extensive realistic Monte-Carlo calculations of the balloon flight; this possibility is unrealistic for larger datasets compared to their 4 hour exposure

where D represents the data-space distribution resulting from an image intensity distribution I, R is the response function, $\vec{x}_1, \vec{x}_2$ are the coordinates of the interactions, and $E_1, E_2$ the energy deposits in the upper and lower detectors, $(E_\gamma, \alpha, \delta)$ represents the energy and direction of the incoming photon. The azimuthal symmetry of the Compton interaction (for unpolarized photons), and the fact that $\vec{x}_1 - \vec{x}_2$ defines the scatter direction $(\chi, \psi)$, allows the number of independent dataspace parameters to be reduced from 6 to 4:

$$D(E_T, \bar\varphi, \chi, \psi) = R(E_\gamma, \alpha, \delta; E_T, \bar\varphi, \chi, \psi) I(E_\gamma, \alpha, \delta)$$

where

$$E_T = E_1 + E_2$$

and

$$\bar\varphi = arccos(1 - \frac{mc^2}{E_2} + \frac{mc^2}{E_T})$$

( $mc^2$=.511 MeV), from the physics of Compton scattering. The total measured energy deposit can be separated from the directional variables to first order, given the energy resolution of about 10% for COMPTEL. Thus for intervals of input photon energies $E_\gamma$, the angular response of the Compton telescope can be vizualized in the 3-dimensional dataspace of Compton scatter angle and scatter direction $(\bar\varphi, \chi, \psi)$ (see figure 5).

The idealized measurement of a Compton scatter interaction, combined with total absorption of the scatted photon in the lower detector plane, results in a pattern of allowed datapoints per image pixel which lies on a cone in $(\bar\varphi, \chi, \psi)$ space, where the cone apex is at $\chi_0, \psi_0$ corresponding to $\alpha, \delta$ and the cone semi-angle is $45^\circ$. The response density along the cone is given by the variation of the Klein-Nishina cross-section for Compton scattering, which yields e.g. a falloff by 0.14 between $20^\circ$ and $40^\circ$ scattering for a 5 MeV photon. This idealized 'cone mantle' response is blurred by measurement inadequacies in the scintillator detectors. In addition, detector trigger thresholds and the finite extent of the detector modules in both planes result in a significant deviation from a smooth distribution along the cone by cutoffs and complex exposure coefficients per dataspace element. The blurring function in dataspace, however, is relatively insensitive to the direction of the incoming photon, and can be parameterized as a function of the energy deposits in the two detector planes, $E_1$ and $E_2$, or equivalently, by $E_T$ and the Compton scatter angle $\bar\varphi$. Therefore the angular response of a Compton telescope can be factorized to first order into

$$R = g(\bar\varphi, \chi, \psi) \cdot f(\bar\varphi, \chi - \chi_0, \psi - \psi_0)$$

This implies that the convolution of an input sky model into the Compton telescope dataspace can be done in two steps:

1) application of the 'PSF' part (the ringlike pattern in each $\bar\varphi$ plane), and

61

2) multiplication with instrumental geometrical constraints g depending on the shape and layout of the individual detectors.

If, for simplicity, the PSF part is assumed to be independent of the infalling direction, then the same PSF representation can be applied over the whole field of view. This simplifies the convolution of the input sky model with the PSF sufficiently to allow use of a Fast Fourier convolution method, at least for restricted image sizes.

### Application of maximum entropy to COMPTEL data

The maximum entropy imaging approach is to convolve a (2-dimensional) image with the full response of the telescope in the 3-dimensional dataspace $(\bar{\varphi}, \chi, \psi)$, and to try and match the 'mock data' from the trial image to the measured data in the 3-dimensional dataspace. The trial image yielding a statistically acceptable match to the measured data, and at the same time fulfilling the entropy criterion, is defined to be the solution of maximum entropy imaging. An example of simulated data for a single 'mini-telescope' consisting of a single upper and lower detector pair is shown in Fig 5. There are 20 bins of $1^{\circ}$ in $(\chi$ and $\psi)$ and 5 of $4^{\circ}$ in $\bar{\varphi}$. The (Gaussian) half-width of the PSF cone was taken as $1^{\circ}$, to simulate a possible COMPTEL response. The test image is an array of 7 point sources separated by $3^{\circ}$. The whole image is $20^{\circ}$ square.

Using this response, we investigated the deconvolution of various test images using the maximum entropy method (Fig 6). The program is a prototype version of that to be used for COMPTEL, based on the MEMSYS package. In the COMPTEL case the statistics are much larger than for COS-B, so that it is possible to use the $\chi^2$ statistic. Examples of results for various background intensities are shown, varying from zero background to an intensity per $(1^{\circ})$ pixel corresponding to 1/10 the flux from one of the point sources. The expected $\chi^2$ is 2000; therefore we choose the iteration having the smallest $\chi^2$ above this value.

For case 1, the deconvolution is practically perfect, as would be expected. Case 2, with approximately equal source and background counts still resolves all sources; the 'halo' in the region between the sources indicates the problems induced by background data. Case 3 again shows all the sources, so that such an image would provide still a good basis for identification of regions of interest; the unreal structures, with up to one third of the intensity of a single source, demonstrate the limitation caused by low signal-to-noise.

### CONCLUSIONS

The maximum entropy method provides an promising method for producing images from $\gamma$-ray telescopes. It has proved its potential scientific usefulness for COS-B data, and tests on simulated Compton telescope data indicate that it will

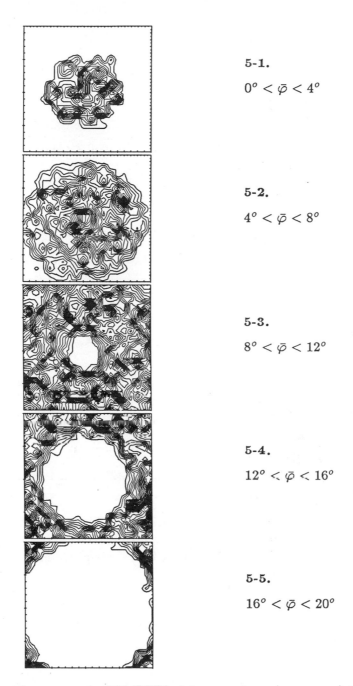

5-1.

$0^o < \bar{\varphi} < 4^o$

5-2.

$4^o < \bar{\varphi} < 8^o$

5-3.

$8^o < \bar{\varphi} < 12^o$

5-4.

$12^o < \bar{\varphi} < 16^o$

5-5.

$16^o < \bar{\varphi} < 20^o$

FIG 5. Sample response in COMPTEL data space to an image consisting of seven point sources. The corresponding maximum entropy image is shown in Fig 6-1.

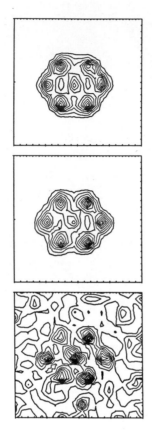

**6-1.**

Total source counts : 10000

Background counts: 0

Iterations: 17

$\chi^2$: 2004

**6-2.**

Total source counts : 10000

Background counts: 5400

Iterations: 20

$\chi^2$: 2007

**6-3.**

Total source counts : 10000

Background counts: 54000

Iterations: 40

$\chi^2$: 2041

**FIG 6.** Maximum entropy image from data generated using a pattern of seven point sources, with varying amounts of background. The parameters of the data and maximum entropy runs are shown next to the corresponding Figures.

provide imaging even for regions with complex structure; it can be demonstrated that closely spaced multiple sources can be resolved in the imaging process, unlike the more simple methods used for skymap generation earlier.

Some caution must be applied in interpretation of maximum entropy $\gamma$-ray images: in principle of course a maximum entropy image shows only what is 'significant'; however this does not mean that every slight wiggle on the contour presentation has to be taken as a real feature! The correct view seems to be to regard maximum entropy images as visual presentations of the data to be used as a guide to further analysis; for example in the search for localized sources, the image shows which points are worth further investigation by other methods; any region which does not show an excess in the maximum entropy map is certainly *not* worth further investigation on the basis of the available data. The method therefore acts as a highly effective filter, highlighting only the places where something interesting may be present.

The power of the method depends critically on the prior knowledge about response and data; our investigations indicated that the background contributions in dataspace are most critical. Realistic simulations of the in-flight situation will require more knowledge of the actual background situation, which will not be known until flight data from the NASA Gamma Ray Observatory are in hand.

REFERENCES

Graser U., Schönfelder V., (1982), Astrophys. Journal , **263**, 677

Gull S. F., Skilling J., (1984), IEE Proceedings , **131**, 646-659

Skilling J., Strong A. W., Bennett K. (1979), Mon. Not. R. astr. Soc., **187**, 145-152

Strong A. W., Bloemen J.B.G.M., Dame T.M., Grenier I.A., Hermsen W., Lebrun F., Nyman L.-A., Pollock A.M.T, Thaddeus P., (1988), Astron. Astrophys , in press

v. Ballmoos P., Diehl R., Schönfelder V., (1987a), Astrophys. Journal , **312**, 134

v. Ballmoos P., Diehl R., Schönfelder V., (1987b), Astrophys. Journal , **318**, 654

# SEARCH FOR BURSTS OF GAMMA-RAY EMISSION FROM THE CRAB PULSAR IN THE COS-B DATA BY ADAPTIVE CLUSTER ANALYSIS

R. Buccheri,  M. Busetta,  and  C.M. Carollo

Istituto di Fisica Cosmica e Applicazioni dell'Informatica
Consiglio Nazionale delle Ricerche, Palermo, Italy

ABSTRACT. The photon arrival times detected by COS-B around the direction of the Crab pulsar have been analyzed by an adaptive cluster method in order to search for short bursts of gamma-ray emission from the source. As a result, two clusters of 3 and 6 photon arrival times within 264 s and 1813 s respectively were found with the corresponding residual phases having close correlation with the peaks of the Crab pulsed emission. The comparison with a similar analysis performed by using  photon arrival times not correlated with the Crab direction shows that the events have chance occurrence probabilities of about $10^{-3}$ and therefore may be consistent with (rare) random fluctuations of the background. Finally, 3 $\sigma$ upper limits on the existence of bursts of emission from the Crab are given.

## 1. INTRODUCTION

Gibson et al.,1982 and Bhat et al.,1986 have reported  detection of transient pulsed emission from the Crab pulsar in the TeV energy range.  According to their observations the flux from the Crab pulsar increased occasionally by a factor of about 25 from the low, persistent pulsed flux level. Recent  observations from the Haleakala observatory (Fry, 1988) seem to confirm that the VHE gamma radiation from the Crab pulsar may appear in short bursts of up to 1000 s duration.

From the theoretical point of view, VHE transient phenomena observable in pulsars have been predicted as derived from collisions of primary gamma rays (produced by curvature radiation of relativistic electrons interacting with the strong magnetic field at the surface of the neutron star) with the soft tertiary photons produced in the outer gaps of Cheng et al., 1986.

67

We present in this paper the results of an analysis for the search of gamma-ray bursts of emission from the Crab pulsar in the range 50 MeV to 5 GeV using the COS-B data base.

## 2. ANALYSIS OF THE COS-B DATA

COS-B observed the Crab pulsar 6 times along the 6.7 years of the satellite lifetime (see Scarsi et al., 1977 for the experimental characteristics of COS-B). Table I shows the relevant parameters of the COS-B observations used for the present analysis (observation no. 44 was excluded because the collecting photon statistics was too poor; see Clear et al., 1987 for more detail). The selection of the photon arrival directions was done by using an energy dependent cone of acceptance which optimizes the signal-to-noise in the case of an $E_\gamma^{-2}$ spectrum; the semi-angle $\theta_{max}$ of the cone is given by

$$\theta_{max} = 12.5 \, E_\gamma^{-0.16} \tag{1}$$

where $E_\gamma$ is the photon energy in MeV (see Buccheri et al., 1983).

TABLE I

| Obs. no. | year | useful time T (days) | sensitive area (cm$^2$) | photons N |
|----------|------|----------------------|-------------------------|-----------|
| 00 | 1975 | 18.81 | 23 | 505 |
| 14 | 1976 | 20.61 | 11 | 235 |
| 39 | 1978 | 16.75 | 18 | 292 |
| 54 | 1980 | 12.02 | 20 | 262 |
| 64 | 1982 | 19.20 | 19 | 365 |

For each observation, the selected arrival times were searched for accumulations in short time intervals by applying the cluster method described in Buccheri et al.,1988. The method is self adaptive and consists in subdividing the list of times in clusters within which the interdistance between each consecutive pair of times is less than a pre-defined threshold $\alpha$. For all the clusters so obtained, we computed the probability for chance occurrence $P_1$ by means of the Erlangian function

$$P_1 = 1 - \exp(-Nt/T) \sum_1^{n-1} (Nt/T)^{n-i-1}/(n-i-1)! \tag{2}$$

(where t is the width of the cluster and n the number of photon times within it) and decided to consider as due to random fluctuations all clusters for which the product $P_1 M_\alpha$ (where $M_\alpha$ is the number of clusters analyzed at the given threshold $\alpha$) was greater than 0.01. The COS-B satellite had an orbital period of 1.5 days, in about 20% of which it was switched-off due to the presence of the Van Allen Belts. As a consequence we are in presence of a systematic clusterization of the selected photons within every 1.2 days. The average cluster resulting from the orbit switch-offs contains 25 photons and the probability $P_1$ for chance occurrence is about 0.1, therefore not considered significant by the method; however, due to the statistics, to variability of the sensitivity and to occasional malfunctions of the instrument, the real clusterizations may sometimes be sufficiently high to be detected by the method. In order to get rid of these orbital effects and also to limit the investigation to short clustering intervals, we used for the threshold $\alpha$ values smaller than 10000 s.

The application of the cluster analysis procedure to obs. 14, 39, 54 and 64 did not reveal any "significant" flares; two such events were instead found in the data of obs.00 the parameters of which are given in table II. Fig. 1 shows the result of the application of the method to the data of obs. 00;

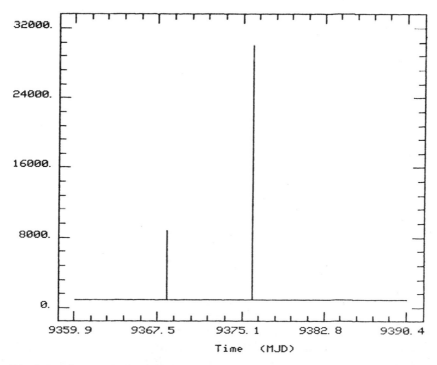

Fig.1 - *Photon density (arbitrary units) along the COS-B observation 00 as derived by our cluster analysis. The two peaks refer to clusters for which the probability for chance occurrence is lower than 0.01.*

## TABLE II

|   | time (MJD) of center | interval (s) t | photons n | probability $M_\alpha P_1$ |
|---|---|---|---|---|
| 1 | 9368.32120 | 1813 | 6 | $30*3.\ 10^{-4}$ |
| 2 | 9376.01448 | 264 | 3 | $2*3.\ 10^{-3}$ |

These two events are interesting for our purposes, their lenghts and intensities corresponding to possible bursts of fluxes approximately equal respectively to $1.10^{-4}$ and $2.5\ 10^{-4}$ ph/cm$^2$ s, i.e. 18.5 and 46 times the steady pulsed flux from the Crab. We have therefore investigated the effects in more details by performing a phase analysis of the corresponding arrival times. Fig. 2 shows the light curve of the Crab pulsar valid for obs. 00 in the form of a phase histogram with 66 bins with superimposed the positions of the phase residuals of the two events of table II. It is interesting to note that the most of the residual phases of the two events coincide with the position of the two main peaks of the pulsar.

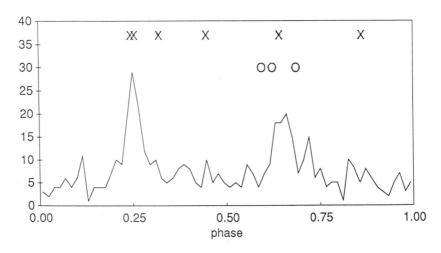

Fig 2 - *Phase histogram of the Crab light curve during observation 00. Points marked with X and O refer to the residual phases corresponding to the times of event 1 and 2 respectively.*

70

We want to stress that the probabilities $P_1$ refer to chance occurrence from uniformly distributed photon arrival times in a single trial and have been calculated by using eq. 2 which considers for the two clusters only 5 and 2 photons respectively (instead of 6 and 3); this aims to include a statistical time lag before the first time and after the last time of each cluster but results in a conservative value for the probability especially in the case of small n (in our case for the event 2). On the other side the final probabilities must take into account the total number of independent clusters detected in the complete process of search. For a calibration of the results found above, the cluster analysis procedure was applied to the arrival times selected from acceptance cones around several directions of the sky, with the exclusion of the directions of the Crab pulsar and of Geminga, the latter being a strong source of gamma radiation and therefore a possible candidate burster (see fig. 3). In order to have similar statistical conditions we summed for each of the 5 observations the cones that provided roughly the same number of photons. Table III gives the cones added for each of the Crab observations, the numbers being referred to fig. 3; we obtained a total of 10 calibration cones, three for obs. 14, two for observations 00, 39 and 64 and only one for observation 54. As a result of the analysis, three clusters were found with probability below $10^{-2}$ in the observations 00, 39 and 54 having widths of 103, 905 and 1176 s and photon content of 3, 4 and 4 respectively. The comparison between the total number of "significant" clusters found on-source (2 in 5 observations) and off-source (3 in 10 observations) does not favour a clearly positive interpretation for the on-source events and more extensive Montecarlo simulations are needed.

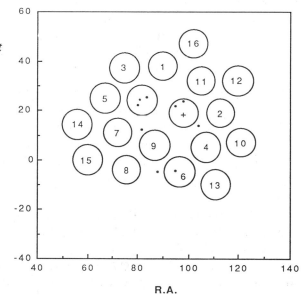

*Fig. 3 - Positions ( in Right Ascention and Declination ) and widths of the acceptance cones used to select off-source photons for background analysis. The two circles marked with a cross and an asterisk refer to Geminga and Crab resp. The points refer to the the positions of some radio pulsars in the COS-B field of view.*

TABLE III

| Obs 00 | off 1 | 1+3+9+14 |
| | off 2 | 2+4+5+6+7+8+11+12+14+15+16 |
| Obs 14 | off 1 | 2+10+11+13 |
| | off 2 | 3+7+9 |
| | off 3 | 4+6+12 |
| Obs 39 | off 1 | 1+2+3+5+6+12 |
| | off 2 | 4+7+8+9+11+14+16 |
| Obs 54 | off 1 | 1+2+3+5+7+8+9 |
| Obs 64 | off 1 | 1+2+3+4+12+16 |
| | off 2 | 5+9+11 |

## 3. DERIVATION OF UPPER LIMITS AND CONCLUSIONS

Let us estimate the sensitivity of our method in detecting a cluster of any width in the Crab data of Obs. 00. From the table I,we see that the useful observing time was T= 1625561 s = 18.814 days, the effective area A = 23 cm$^2$ and the number of photons with energies between 50 and 5000 MeV within the optimal acceptance cone was N=505. The Crab pulsed flux in this energy range, measured for obs. 00, is $\phi$=5.54 10$^{-6}$ ph/cm$^2$ s equivalent to 206 "pulsed" photons; a flux R times higher in a burst of duration t equals to have m = R $\phi$ A t photons to discriminate against the background inside the time interval t. In our case the observation of, for example, 4 photons in 1500 s, means to have observed a flux enhancement of a factor R=21 during the burst; the corresponding value of $P_1$ according to the Erlangian probability distribution would be 1.3 10$^{-3}$. Fig. 4 shows the value of R versus the width t for two values of the probability $P_1$ as derived from eq. 2 and using the parameters of the COS-B observation 00. The two plots refer to two different intervals of the width which have been chosen such to include the two events detected in our on-source analysis. The probabilities $P_1$ $M_\alpha$ of the two evetns are also shown in the plots .

According to De Jager (1986) the gamma-ray luminosity of pulsars, as derived from the model of Cheng et al. (1986), depends on the square of the density n of the tertiary infrared and optical photons in the VHE range but is linear with n below 3 GeV. In this framework, possible bursts of gamma-ray emission in the flux of the Crab pulsar are expected to be about 5 times larger than the steady flux, the square root of the increase in the TeV range. Although the sensitivity of the present method is not such to make feasible the investigation down to values of R=5, the possibly negative result of the analysis is consistent with the model of Cheng et al. (1986) and with the TeV results. On the other side, the two events detected in the on-source analysis are near the visibility limit and deserve further investigation especially because of the coincidence in phase with the pulsar peaks. We plan to improve the method by a better optimization of the cone for the selection of the photon arrival times and by a better estimate of the probability for chance occurrence via extensive Montecarlo simulations of the experiment.

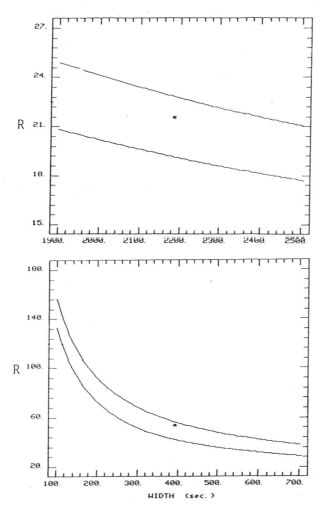

*Fig. 4 - Upper limits of the burst flux (referred to the steady pulsed Crab flux) at the specified confidence level for two ranges of the burst lenght. The two ranges were chosen in such a way to include the lenghts of the two events found in the data of observation no. 00.*

# REFERENCES

- Bhat P.N., Ramana Murthy P.V., Sreekantan B.V., Vishwanath P.R. ; Nature, 319, 127, 1986
- Buccheri R., Bennett K., Bignami G.F., Bloemen J.B.G.M., Boriakoff V., Caraveo P.A., Hermsen W., Kanbach G., Manchester R.N., Masnou J.L., Mayer-Hasselwander H.A., Ozel M.E., Paul J.P., Sacco B., Scarsi L., Strong A.W. ; Astron. Astrophys., 128, 245, 1983
- Buccheri R., Di Gesu' V., Maccarone M.C., Sacco B. ; Astron. Astrophys., 201, 194, 1988
- Clear J., Bennett K., Buccheri R., Grenier I.A., Hermsen W., Mayer-Hasselwander H.A., Sacco B. ; Astron. Astrophys., 174, 85, 1987.
- De Jager O.C.; PhD thesis, Univ. of Potchefstroom, 1986
- Fry W.F. ; private communication
- Gibson A.I., Harrison A.B., Kirkman I.W., Lotts A.P., McCrae H.J., Orford K.J., Turver K.E., Walmsley M. ; Nature, 296, 833, 1982
- Scarsi L., Bennett K., Bignami G.F., Buccheri R., Hermsen W., Koch L., Mayer-Hasselwander H.A., Paul J.A., Pfeffermann E., Stiglitz R., Swanenburg B.N., Taylor B.G., Wills R.D. ; Proc. of the 12th ESLAB Symp., ESA SP-124, p.3, 1977

# STATISTICAL RESTORATION OF ASTRONOMICAL IMAGES

R. Molina [1], N. Perez de la Blanca [1], and B.D. Ripley [2]

(1) Dept. de Ciencias de la Computacion e Intelligencia Artificial
Univ. Granada and Inst. de Astrofisica de Andalucia, Spain
(2) Dept. of Mathematics, Univ. of Strathclyde, Glasgow, Scotland

## INTRODUCTION

It is well known that the theoretical intensity at a point (x,y) from an astronomical image is given

$$\widetilde{Z}(x,y) - \int \int S(x-x',y-y') \, h(x',y') \, dx' \, dy' \tag{1}$$

where S(,) represents the true underlying intensity and h(,) is the point spread function (psf). However, when we consider discrete samples on a rectangular grid, the intensity measured at each pixel (i,j) is given by

$$Z_{ij} = \int_{i-\frac{1}{2}}^{i+\frac{1}{2}} \int_{j-\frac{1}{2}}^{j+\frac{1}{2}} \widetilde{Z}(x,y) \, dx \, dy + \varepsilon_{ij} \tag{2}$$

where $\varepsilon_{ij}$ represents the external contribution, in our case background and random noise. The main problem in the restoration of astronomical images is to estimate S(,) on a grid of points from the Z values, knowledge of the psf and adequate assumptions about the noise process.

This paper is divided in two parts. In the first we will consider the problem of fitting the psf and in the second we propose methods to estimate S(, ).

75

# THE PSF FUNCTION

The psf fitted on each particular image must be based on two main goals, (a) to estimate its functional form, (b) to estimate the flux emitted from each star in the image. An estimation procedure meeting both these conditions will allow us not only to do stellar photometry but will also provide a function h at (1) to estimate S.

In order to estimate $h(,)$, we consider a point source at $(p,q)$, so $S(x,y)=k$ if $(x,y)=(p,q)$ and $S(x,y) = 0$ otherwise. Then from (1) and (2) we obtain

$$Z_{ij} = k \int_{i-\frac{1}{2}}^{i+\frac{1}{2}} \int_{j-\frac{1}{2}}^{j+\frac{1}{2}} h(x-p,y-q)\, dx\, dy + \varepsilon_{ij} \tag{3}$$

To solve (3) we will need an hypothesis about $\varepsilon$. The most common assumption is $\varepsilon_{ij} = b_{ij} + n_{ij}$ where b is a function given by the background values and n represents the noise. Buonanno et al. (1983) and Stetson (1987) considered $b_{ij} = Ai+Bj+C$ and $n_{ij}$ independent Gaussian random noise $N(0,\sigma^2)$. In our experience there is a need to consider other hypotheses; we will also consider $n_{ij}$ with variance proportional to $h*S$. In those situations where there is a single isolated star we will show that $b = $ constant suffices.

The psf is defined by two main components, atmospheric turbulence and instrumental aberrations. Some efforts to model these effects to find a theoretical analytical expression for the psf have been made by Moffat (1969), but the tail behaviour of his theoretical function does not agree with the observational data. The following heuristic analytical expression have been proposed (Buonanno et al., 1983, Moffat, 1969):

$$h_1(r) = \left(\beta/\pi R^2\right) / \left[1 + \left(r/R\right)^2\right]^\beta \tag{4}$$

where R and $\beta$ are the parameters to be estimated and r represents the distance from the pixel with highest function value.

We also considered an alternative expression given by

$$h_2(r) = \left(\alpha e^{R\alpha}/2\pi\right) \exp\left(-\alpha S\right)/S, \qquad S = \left[1 + (r/R)^{1/2}\right] \tag{5}$$

This expression is a particular case of the generalized two-dimensional hyperbolic distribution (Barndoff-Nielsen, 1978, Blaesild, 1981).

Penny and Dickens (1986) proposed a combination of a two-dimensional elliptical Lorentzian with a wide low circular modified Gaussian base, but this approach needs a very high number of parameters and is not considered here.

If we suppose the coordinates (p,q) of the star are <u>known</u>, expressions (4) and (5) present the same advantages for fitting since both depend on only two parameters and both have slowly decaying tails.

The first step of our experiment was to extract the star from image. To do this we select a closed curve around each star containing those pixels with intensities greater or equal than a given value. That value could be an estimate of the background intensity or some values near this. To those pixels inside the curve we will fit $Z_{ij}$ as given by (3). We considered several alternatives for h and n. Let us denote by $Z^*_{ij}$ the convolution of the rue image with the psf. We present here 2 cases:

Case 1:      h evaluated at the centre of each pixel, $n_{ij} \sim N(O,\sigma^2)$.

Case 2:      h integrated over the pixel and $n_{ij} \sim N(O,k_n Z^*_{ij})$.

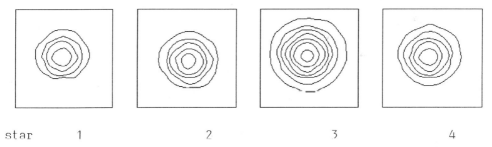

star      1           2           3           4

Figure 1. Contour plots of four stars. Contours are $100 \times 2^n$, $n \geq 1$.

Table 1.   Fluxes from fitted psfs

|  | observed | $h_1$ | | $h_2$ | | R |
|---|---|---|---|---|---|---|
|  |  | case 1 | case 2 | case 1 | case 2 |  |
| Star 1 | 27798 | 28918 | 27799 | 28861 | 28485 | 6 |
| Star 2 | 103929 | 108304 | 104067 | 95888 | 104756 | 6 |
| Star 3 | 225698 | 227081 | 226572 | 225003 | 225906 | 10 |
| Star 4 | 380071 | 381263 | 381489 | 380560 | 381722 | 10 |

For the moment we will consider b as a constant on each star. We present a pilot experiment in which we took four different stars (fig.1). The stars have been measured with a CCD detector, the first two with 512 x 340 resolution, the others with 1024 x 656 resolution, so for the same sky field the last two stars have four times as many pixels than the first two. Table 1 presents the estimated fluxes obtained. For $h_1$ we found that the best value of $\beta$ was 3. Figure 2 shows the fitted psf functions. The main conclusions were that integration over the pixels gave much better residual displays, and that for stars 1 and 2 (but not 3 and 4, sampled at higher rate) $h_1$ fits better than $h_2$.

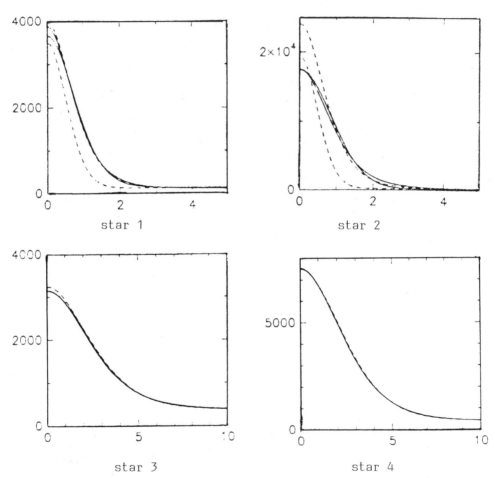

Figure 2. Fitted radial psf functions. Solid line $h_1$, dashed line $h_2$, higher curve case 1.

# THE DECONVOLUTION PROBLEM

Let us assume that the observed image Z is derived from S' by the expression

$$Z_{ij} = S'_{ij} + \varepsilon_{ij} \tag{6}$$

where $\varepsilon_{ij} \sim N(0, k_n S'_{ij})$, S'=HS with blurring matrix H, S the true image and i $\in \{1,2,..,M\}$, $j \in \{1,2,..,L\}$. For our deconvolution problem we will consider only sections of the image without point sources. Thus $S_{ij}$ is expected to vary slowly. Images with extent and point objects will be considered elsewhere.

A mathematical model which has been found to work well to obtain "smooth solutions" is the conditional autoregressive or CAR process (Ripley, 1981). Such models have been used in smoothing problems for agricultural field trials (Besag and Kempton, 1986) and in smoothing problems in astronomy (Ripley, 1988) and were proposed for image analysis by Besag (1986). A CAR process is defined by

$$E\left(S_{ij}\middle| S \text{ at all other pixels}\right) = \sum_{kl \neq ij} C_{kl,ij} S_{kl}$$

$$\text{Var}\left(S_{ij}\middle| S \text{ elsewhere}\right) = k_s$$

for a symmetric matrix C such that I-C is positive definite. Then S has covariance matrix $k_s(I-C)^{-1}$. (Obviously, we are considering S as a one dimensional vector with ML components). Let $N_{kl,ij}$ indicate neighbours (that is, pixels at distance one from (i,j)) with one and be zero otherwise. Then the simplest choice for C is $\varphi N$ with $\varphi$ just less than 1/4. However, this causes difficulties at the edges, since the conditional mean at those points will have a smaller sum and will be pulled toward zero. To overcome this problem we used the following modification of the CAR process. Let r denote the number of neighbours of a pixel and define P = diag {( # nhbrs of i)/r}. The modified model is

$$E\left(S_{ij}\middle| S \text{ elsewhere}\right) = \left(r\varphi\right)\left(\text{average of S over nhbrs of ij}\right)$$

$$\text{Var}\left(S_{ij}\middle| S \text{ elsewhere}\right) = k_s / p_{ij}$$

Then the covariance matrix of S is $k_s(P-\varphi N)^{-1}$. This model for S is viewed as a priori probability in a Bayesian analysis. It reflects our belief in a smooth underlying image.

The simplest analysis comes from combining our prior model with the noise model. Then the posterior density of S given Z is given by

$$-2\ln P(S|Z) = \text{const} + S^t(P-\varphi N)S/k_s + \Sigma[(Z-HS)^2_{ij}/HS_{ij}]/k_n \tag{7}$$

giving us a probability distribution over all possible images. We can summarize it by taking its mode, the so-called MAP estimator. The maximization is done by solving iteratively the equation for a zero derivative of (7); details will be given elsewhere.

## PARAMETER ESTIMATION

Our method depends on the three parameters $\varphi$, $k_n$ and $k_s$. Since $\varphi \approx 1/4$, the exact value affects the restoration only marginally. It is easy to find a reliable estimator $k_n^*$ of $k_n$ by comparing the signal with the average of its neighbours in regions where the real image is almost flat. The main problem is to estimate $k_s$, which should perhaps be done locally in the image. Since

$$E[Z^t(P-\varphi N)Z|S] = \text{tr } (P-\varphi N)HSS^tH^t + k_n\text{tr } (PHS)$$

we used

$$\widehat{k}_s = \left[Z^t(P-\varphi N)Z - k_n^* \text{ tr } (PHS)\right]/ML \tag{8}$$

changing the value every few iterations, using the current estimator of S. There is a potential problem, since (8) can give negative estimates. To avoid this we actually used

$$k_s^* = \max\left(\widehat{k}_s, \ S^t(P-\varphi N) S / ML\right)$$

again using the current value of S.

## TEST EXAMPLES

We will use for these examples $\varphi=1/4$ and $r=4$ neighbours.
Our first example is a 40×40 simulated image, a quadratic function (fig.3). The convolution function is given by $h_{kl}\alpha[1-(k^2+l^2)/50]$ where k and l $\in \{0,1,...,5\}$, and we added multiplicative noise with $k_n=9$ (fig.4). The restored image is shown in figure 5.

We also applied the method to an actual image of two galaxies, a 156×156 image (fig.6). The blurring function was estimated by $h_1$ with $\beta=3$ and $R=6.18$, by fitting to the stars that appeared in the whole image. Figure 7 shows the result, re-convolved with h in figure 8 for comparison with the original image.

Our methods have the advantage of being very rapid to compute compared to procedures such as maximum entropy, and on being based on firm statistical theory.

## ACKNOWLEDGMENT

We wish to thank the Extragalactic Astrophysics group of the IAA for allowing us to use the data of figure 6 and for the use of their computer facilities.

80

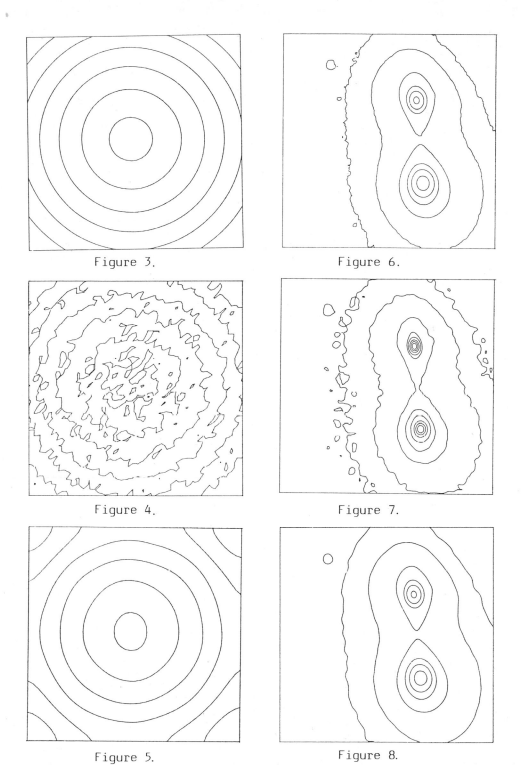

Figure 3.

Figure 6.

Figure 4.

Figure 7.

Figure 5.

Figure 8.

# REFERENCES

Barndorff-Nielsen, O., 1979, Models for non-Gaussian variation with applications to turbulence, <u>Proc. R. Soc. London A</u>, 368:501.

Besag, J., 1986, On the statistical analysis of dirty pictures, <u>J. Roy. Statist. Soc. B</u>, 48:259.

Besag, J., and Kempton, R., 1986, Statistical analysis of field experiments using neighbouring plots, <u>Biometrics</u>, 42:231.

Blaesild, P., 1981, The two dimensional hyperbolic distribution and related distributions with an application to Johannsen's bean data, <u>Biometrika</u>, 68:251.

Buonanno, R., Buscema, G., Corsi, C.E., Ferraro, I., and Iannicola, G., 1983, Automated photographic photometry of stars in globular clusters, <u>Astron. Astrophys.</u>, 126:278.

Mofatt, A.F.J., 1969, A theoretical investigation of focal stellar images in the photographic emulsion and application to photographic photometry, <u>Astron. Astroph.</u>, 3:455.

Penny, A.J., and Dickens, R.J., 1986, CCD photometry of the globular cluster NGC 6752, <u>Mon. Not. R. Astr. Soc.</u>, 220:845.

Ripley, B.D., 1981, "Spatial Statistics", Wiley, New York.

Ripley, B.D., 1988, "Statistical Inference for Spatial Processes", Cambridge University Press, Cambridge.

Stetson, P.B., 1987, Daofot: A computer program for crowded-field stellar photometry, <u>Publ. Astr. Soc. Pacific</u>, 99:191.

# MULTIDIMENSIONAL DISCRETE SIGNALS DESCRIPTION USING

# ROTATION AND SCALE INVARIANT PATTERN SPECTRUM

M. Binaghi(-), V. Cappellini(*), and C. Raspollini(+)

(-)Dipartimento di Ingegneria Elettronica, University of Firenze, Via S.Marta 3, Firenze, Italy
(*)Dipartimento di Ingegneria Elettronica, University of Firenze and IROE - C.N.R., Via Panciatichi, 64, Firenze, Italy
(+)IBM Rome Scientific Center, Via Giorgione 159, Roma, Italy

## INTRODUCTION

Mathematical morphology was developed in the mid 1960's by G.Matheron and J.Serra as a methodology for continuous and discrete multidimensional signal analysis.
The basic idea underlaying this methodology is to trasform the original signal into a simpler and more expressive one, by interacting with a structuring element, called kernel, strategically chosen by the observer. A morphological operation is then constitued by a transformation followed by some measurement on the transformed signal. The measurement on the transformed signal can be its lenght, area, volume, etc., depending on the dimension of the signal.
In the theory of mathematical morphology signals are represented as closed sets inside a M-dimensional (MD) Euclidean space. The four basic morphological operations are erosion, dilation, opening and closing.
The erosion of the set X by the kernel B can be defined as the locus of points of X to which the origin of the structuring element can be traslated such that B is entirely contained within X:

$$X \ominus B = \{z : B_z \subset X\}$$

The dilation of X by B can be defined as the locus of points to which the origin of the structuring element B can be translated such that B intersects X:

$$X \oplus B = \{z : B_z \cap X \neq 0\}$$

The dilation is the dual operation of erosion with respect to complementation; in other words, dilating X is equivalent to eroding its complement.
The opening of X by B is obtained by first eroding X by B and then dilating the result, always by B:

$$X_B = (X \ominus B) \oplus B$$

It represents the union of all the translations $B_y$ of B that can be inscribed in X:

$$X_B = \{x : \exists B_y \ni B_y \subset X\}$$

The closing of X by B is performed by first dilating X by B and then eroding the result by B:

$$X^B = (X \oplus B) \ominus B$$

Geometrically, the closing is the union of all z such that all translations of B, containing z, hit X:

$$X_B = \{z : \forall B_y \supset z \ni B_y \cap X \neq 0\}$$

The opening smooths the contours of X, removing details smaller than B from the signal; the effect of the opening operation is similar to that of a non-linear low-pass filter, which reduces the sharp variations of the signal.
By increasing the size of B and computing the measures of the corresponding openings, a monotonically decreasing function, called **pattern spectrum** (p.s.) is obtained. It represents the amount of detail removed, step by step, by the structuring element after each successive opening operation.
The p.s. of a discrete MD signal is defined as:

$$P(n) = \frac{[Mes(X_{nB}) - Mes(X_{(n+1)B})]}{Mes(X)} \qquad n = 0,1,2,....+\infty$$

where

- n is a positive integer

84

- B denotes a compact convex structuring element

- $X_{nB}$ denotes the morphological opening of X by the structuring element nB (nB can be thought as a magnified version of B, with a scale factor of n)

- Mes(X) denotes the measure of X and it is related to the definition space of X (length, area, volume, etc.). The measure of X is equivalent to the total number of points belonging to the set X.

Besides, if the "size of a point x belonging to the set X" is the quantity:

$$\lambda_X(x) = \sup\{n \ni \exists y \ni x \in (nB)_y \subset X\}$$

then the p.s. is the probability density function of $\lambda_X(x)$ .

The main properties of the p.s. are:

- translation invariance

- rotation invariance if the structuring element is isotropic

- size dependence

These properties make the p.s. a good tool to quantify the geometrical structure of continuous and discrete multidimensional signals. In particular, the p.s. has been investigated to characterize the shape and the size of bidimensional discrete patterns (Maragos, 1987; Bronskill et al., 1987).

In the present work the p.s. is employed to recognize objects in discrete binary images. The p.s. is made size and rotation invariant using two new procedures. The first concerns a suitable normalization of the structuring elements succession; in the second solution two orthogonal directions, which are univocally defined starting from the objects, are evaluated and the kernel is aligned with them. A minimum distance classifier allows to recognize the objects.


ROTATION AND SIZE INVARIANCE

The aim of shape description is the quantification of the concept of shape in a concise, yet simple manner; to achieve this goal we use the pattern spectrum. A good shape descriptor must be translation, rotation and size invariant. These attributes are important, since in most of the practical applications (such as object recognition), we will not know a priori exactly where in the space the shape is located, how it is oriented, or how large or small it will appear to a sensor. The pattern spectrum is intrinsically translation invariant;

the rotation invariance is ensured only using perfectly isotropic kernels, but the approximation introduced by the discretization of a isotropic shape make necessary the use of suitable expedients to compute the p.s. for digital applications. As far as the scale invariance is concerned, it must be noticed that the p.s. is a shape and size descriptor and then a normalization technique must precede the computation of the p.s..

The choice of the kernel shape is closely conditioned by the geometrical properties of the object, which have to be emphasized. In particular, if we have to choose among several shapes with the same characteristics, but to different degrees, we should retain the most extreme of this choices. In the technique presented in this paper, to investigate the morphological properties of the objects along two orthogonal directions, a totally anisotropic not convex shape is preferred. The basic structuring element used is obtained by the union of two segments $B_1$ and $B_2$ :

$$B_1 = \bullet \, o \, \bullet \qquad B_2 = o$$

and the kernel nB is obtained as:

$$nB = nB_1 \cup nB_2 = (B_1 \oplus B_1 \, ... \, \oplus B_n \cup (B_2 \oplus B_2 \, ... \, \oplus B_n)$$

Using the structuring elements defined in this way, the result of the opening operations $X_{nB}$, closely depends on the orientation of X except for 90° multiple integer rotations. To overcome this limitation, two orthogonal directions, which are univocally defined from the objects, are evaluated and the kernel is aligned with them. In this way, the relative position between the kernel and the object is independent of its original orientation. For this purpose we have chosen the principal axes of inertia as basic directions.

To obtain the size invariance the structuring elements succession is properly normalized. A standard succession of kernels of increasing size is established; then this succession is multiplied by a scale factor defined as the ratio between the dimension of the maximum kernel enclosed in the object and the dimension of the last element in the standard succession. In this way, the amount of detail removed after each successive openings is always comparable, even if the objects have very different sizes. In conclusion, to obtain a size and rotation invariant p.s, is necessary, at first, to rotate the object so that its axes of inertia are aligned with those of the kernel; then the maximum kernel enclosed in the object is computed and the standard succession of structuring elements is normalized.

The unknown object that is being analyzed can be classified on the basis of the distance from its p.s. to the p.s. of the reference shapes; the minimum euclidean distance between the vectors containing the p.s. values of object and models respectively is computed and, if it is less than a fixed threshold, the object is recognized.

A shape description is also obtained computing some probabilistic parameters which are derived starting from the p.s. values. They are defined as:

$$PAR_1 = \sum_{n=1}^{N} \frac{P(n)}{n}$$

$$PAR_2 = P_N$$

$$PAR_3 = \sum_{n=1}^{N} P(n)n$$

$$PAR_4 = - \sum_{n=1}^{N} P(n) \log(P(n))$$

with:

$$P(n) = 0, \forall n > N$$

The parameters $PAR_1$ and $PAR_2$ can be derived from the "roundness" and "circularity" factors defined in the theory of isotropic kernels (Serra, 1987). $PAR_3$ can be defined as the average size $\bar{n}$ of X relative to $B_j$ and, using concepts from information theory, $PAR_4$ is the average uncertainty (entropy) of the random variable $\lambda$. The maximum value of $PAR_4$ is obtained whenever the p.s. is flat, otherwise its minimum value is obtained when the p.s. contains just an impulse.

The classification is performed computing four minimum distances between the probabilistic parameters of the object and those of the models; the object is then recognized if at least three minimum distances are obtained for the same model.

## RESULTS AND CONCLUSION

Experimental tests are performed using six sinthetic shapes (letter E, letter L, disk, annulus, rectangle, square, triangle) and two real shapes representing two mechanical parts called object A and B.
The models are created by averaging the p.s. values computed starting from a set of rotated and scaled shapes.
Table 1 shows the classification results using the original shapes, and table 2, using rotated (30°, 45°, 60° and 90°) and scaled shapes (their size varying from 1/3 to 1/10 of the entire image size).
Table 3 shows the classification result for the object B using the probabilistic parameters.

The aforementioned techniques can be straightforwardly applied to sets in the 3D Euclidean space allowing an efficient recognition of 3D objects. The drawback is the high computational cost but it can be avoided implementing the repetitive simple operations on special-purpose hardware.

Table 1

|  | LETT.E | LETT.L | DISK | RING | RECT. | SQUARE | TRIAN. | OBJ.A | OBJ.B |
|---|---|---|---|---|---|---|---|---|---|
| LETT.E | 2.13 | 13.32 | 70.08 | 99.28 | 100.37 | 69.26 | 75.05 | 61.06 | 77.73 |
| LETT.L | 15.44 | 0.69 | 61.69 | 91.27 | 95.74 | 59.27 | 66.96 | 51.09 | 70.11 |
| DISK | 72.27 | 62.19 | 0.65 | 34.74 | 49.34 | 11.34 | 11.76 | 23.74 | 24.17 |
| RING | 100.64 | 91.07 | 34.67 | 1.35 | 62.56 | 38.74 | 39.02 | 56.09 | 46.77 |
| RECT. | 102.67 | 96.54 | 50.15 | 64.56 | 0.88 | 56.89 | 40.32 | 56.53 | 37.73 |
| SQUARE | 70.38 | 58.56 | 11.02 | 39.96 | 55.94 | 0.98 | 16.48 | 18.46 | 25.52 |
| TRIAN. | 77.14 | 67.02 | 10.11 | 38.29 | 41.42 | 14.84 | 2.33 | 23.29 | 16.04 |
| OBJ.A | 63.98 | 52.49 | 23.88 | 57.52 | 54.37 | 20.62 | 23.23 | 2.26 | 30.21 |
| OBJ.B | 80.10 | 70.73 | 24.67 | 48.50 | 35.11 | 26.76 | 14.70 | 30.76 | 2.02 |

Table 2

|  | LETT.E | LETT.L | DISK | RING | RECT. | SQUARE | TRIAN. | OBJ.A | OBJ.B |
|---|---|---|---|---|---|---|---|---|---|
| LETT.E | 1.21 | 13.04 | 70.46 | 99.07 | 101.38 | 68.34 | 75.79 | 62.56 | 78.54 |
| LETT.L | 12.49 | 0.59 | 62.27 | 91.12 | 96.46 | 58.26 | 67.46 | 53.02 | 71.12 |
| DISK | 69.46 | 62.51 | 1.11 | 34.02 | 49.98 | 11.99 | 10.34 | 25.56 | 25.04 |
| RING | 98.12 | 91.45 | 34.25 | 0.21 | 62.95 | 39.71 | 36.91 | 58.01 | 47.74 |
| RECT. | 100.07 | 96.78 | 49.28 | 63.04 | 0.35 | 57.61 | 42.11 | 53.81 | 35.18 |
| SQUARE | 67.38 | 58.93 | 12.13 | 39.48 | 56.57 | 0.96 | 14.97 | 21.38 | 26.95 |
| TRIAN. | 74.20 | 67.36 | 9.81 | 37.30 | 42.05 | 15.65 | 0.62 | 23.69 | 16.37 |
| OBJ.A | 60.94 | 52.66 | 24.73 | 56.86 | 55.02 | 19.93 | 23.86 | 2.14 | 30.34 |
| OBJ.B | 77.06 | 71.18 | 24.14 | 47.35 | 35.71 | 27.42 | 16.18 | 29.87 | 0.75 |

Table 3

| | PAR 1 | PAR 2 | PAR 3 | PAR 4 |
|---|---|---|---|---|
| LETT.E | 28.17 | 393.1 | 52.38 | 61.36 |
| LETT.L | 8.24 | 426.4 | 73.37 | 27.61 |
| DISK | 84.27 | 116.8 | 21.88 | 1.45 |
| RING | 165.0 | 83.04 | 70.55 | 63.55 |
| RECT. | 153.4 | 306.2 | 90.47 | 67.86 |
| SQUARE | 11.07 | 197.1 | 19.04 | 20.96 |
| TRIAN. | 46.60 | 64.49 | 13.88 | 6.37 |
| OBJ.A | 2.04 | 171.3 | 34.09 | 19.49 |
| OBJ.B | 0.14 | 5.22 | 0.82 | 0.15 |

REFERENCES

Serra, J., 1982, "Image Analysis and Mathematical Morphology", Academic Press, New York.

Maragos, P., 1987, Pattern Spectrum of Images and Morphological Shape-Size Complexity, Proceedings of IEEE Int. Conf. on Acoustics, Speech and Signal Processing.

Bronskill, J.F., and Venetsanopoulos, A.N., 1987 Multidimensional Shape Recognition using Mathematical Morphology, in: "Time-Varying Image Processing and Moving Object Recognition", V. Cappellini, ed., North-Holland, Amsterdam.

# APPLICATIONS OF DECOMPOSITION ALGORITHM IN CLUSTERING METHODS

Zaki Ahmed Azmi

Dept. of Statistics, Faculty of Economics and Political Science,
Cairo University, Cairo, Egypt

## ABSTRACT

The aim of clustering is to partition a given collection of objects into a number of classes in such a way that objects within each class are strongly similar each another, while objects of different classes are appreciably less similar.
MM (Minimax) Decomposition algorithm is of special importance to study properties of non-fuzzy partitions derived from a number of algorithms providing convex decomposition of matrix U.
We suggest the possibility of the application of decomposition algorithm in clustering methods in two approaches:
(1)     The application of non-metric similarity/dissimilarity measures which may be of value when the data do not admit the geometrical properties.
(2)     The application of fuzzy numbers instead of non-fuzzy numbers.

Keywords: decomposition algorithm - clustering - fuzzy numbers.

## 1. INTRODUCTION

The study of Bezdek's Fuzzy ISODATA [§] was performed under the assumption that all measurements are correct similarly all final decisions, although producing soft classification, were based on this assumption too. The simplest formulation of the problem of set of objects $P = \{p_1, ..., p_n\}$ may be stated as follows:

Let $p_i, p_j \in P$ and $x_i, x_j$ denote their mathematical representations. The aim is to find a transparent algorithm $A_{tr}$ such that :

$$A_{tr}[m(p_i),m(p_j)] = A_{op}(p_i,q_j) \quad i,j = 1,...,n; \quad p,q \in P \qquad (1)$$

------------------------------------

[§] Bezdek,J."Pattern Recognition with Fuzzy Objective Functions", Plen. Pr. (1981)

where $$A_{tr}(x_i, y_j) = r(x_i, y_j) = r_{ij} \qquad (2)$$

is interpreted as a grade of similarity between patterns $x_i$ and $x_j$.

This way we obtain a mapping $r: X*X \rightarrow [0,1]$ called a fuzzy relation (Zadeh, 1971). The immediate implementation of this idea leads to the use of similarity relation, i.e. the fuzzy relation that is reflexive, symmetric and transitive. Similarity relation should provide the inference of the form:

$$\text{IF "x is similar to y" AND "y is similar to z"} \qquad (3)$$
$$\text{THEN "it is reasonable to assume that x is similar to z"}$$

Applying the fuzzy logic rules (Zadeh, 1971) to this statement we obtain the concept of (max-min) transitivity, i.e. a fuzzy relation is transitive if

$$r_{ij} \geq \max [\min (r_{ij}, r_{kj})] \qquad 1 \leq k \leq n \qquad (4)$$

Such defined similarity relation is an equivalence relation and it possesses the property: all its $\alpha$-cuts are equivalence relations, where by $\alpha$-cut we mean a non-fuzzy subset of $X*X$ defined as:

$$r_\alpha = \{ (x,y) \in X*X \mid r(x,y) \geq \alpha \} \qquad (5)$$

$\alpha$ means here the minimal strength of relationship between the elements of $r_\alpha$.

Moreover $r_{\alpha 1} > r_{\alpha 2}$ if $\alpha_1 < \alpha_2$.

To cluster a given data collection we start from a reflexive and symmetric relation $s$ defined on $X*X$. Denote by $S$ a matrix build from the values $s_{ij}$, i.e. $S = [s_{ij}]$. These values are determined subjectively or by using some similarity index (e.g. cosine function, Tanimoto measure, etc.). To find a similarity matrix $R$ we use one of the four algorithms: "multiplying" (using max-min operator, defined by (4) ) matrix $S$ q times ($q \leq n-1$) (Tamura et al., 1971); a column-row scanning algorithm (Kandel, Yelowitz, 1971); Prim's minimal spanning tree algorithm (Dunn, 1974); or using a heuristic method given by Watada et al., (1981). Thresholding resulting $R$ (in the sense of $\alpha$-cuts) we build a nested hierarchy of fuzzy partitions. Dunn (1974) showed that because (max-min) transitivity is equivalent to the ultra metric inequality, the resultant hierarchies are in fact a subset of single-linkage hierarchies. Hence this method may be placed among graph theoretic method. The method was used by Negoita (1975) in information retrieval system.

Ruspini (1981) has observed that the application of Aleph-I Lukasiewicz's logic to statement (3) leads to the concept (max-T) transitivity defined as:

$$r_{ij} \geq \max [\max (r_{ik} + r_{kj} - 1, 0)] \qquad (6)$$

Reflexive, symmetric and (max-T) transitive relation is called a likeness relation. Its complement is a bounded pseudometric in $X$. Likeness relations form the largest class of equivalence relations (similarity relations are very sparse among them) and they were studied extremely by Bezdek and Harris (1978). Ruspini (1977) showed that a necessary and sufficient condition for the existence of fuzzy clusters is that a relation R defined on $X*X$ be a likeness relation. All fuzzy sets that are elements of the fuzzy quotient $X/R$ are clusters and can therefore be chosen as components of a clustering of $X$. This observation provide a bases for new clustering techniques (see also Ruspini, 1981).

Ruspini (1969) was the first who suggested the use of objective function method for fuzzy clustering. Some generalization of this method was proposed by Roubens (1987) and Bezdek (1981) applied this methodology in a very constructive way introducing the infinite family of fuzzy ISODATA algorithms. The motivation was that the hard ISODATA always yield some partition even when compact and well separated clusters do not exist. Hence, when it is not known in advance that such clusters are actually present, inferences drawn from hard ISODATA partitions can be very dangerous (Dunn, 1974). Following Ruspini a fuzzy non-degenerate partition of a set $X$ is determined by specifying a set of maps $u: X \rightarrow [0,1]$ satisfying the conditions:

$$\sum_{i=1}^{c} u_i(x) = 1 , \qquad \sum_{x \in X} u_i(x) > 0 \qquad (7)$$

The first of them insures that each $x \in X$ must have a total membership in $X$ of unity (this membership may vary arbitrary among the fuzzy subset partitioning $X$) while the second condition means that each of $c$ clusters is non-empty. It is convenient to represent a fuzzy partition as a matrix $U = [ u_{ik} ]$ where $u_{ik} = u_i(x_k)$ . The set of all such defined partitions forms a non-degenerate partition space denoted $M_{fc}$ . In particular it contains a space of all hard partitions $M_c$ (they are defined by the characteristic function $h_i : x \rightarrow (0,1)$ ). The $M_{fc}$ provides a nice mathematical structure for classification models: it is closed, compact and convex subset in the positive orthant of $V_{cn}$ (the vector space of all $(c*u)$ matrices) with cardinality card $(M_{fc}) = u(c-1)$. This last property is especially attractive: for example in the case of $M_c$ and $n=25, c=10$ we have card $(M_c) \approx 1018$ while card $(M_{fc}) = 225$. Studying connection between $M_{fc}$ and $R_T$ (the space of all likeness relations defined on $X*X$) Bezdek and Harris (1978) showed that the composition $U^T(sum,min)U$ induces a unique likeness relation that enables one to convert fuzzy membership of patterns into metrical fuzzy relationship between these patterns.

## 2. EXTENSIONS OF FUZZY ISODATA

### 2.1 - Classification of fuzzy objects

In this subsection we will assume that: using precise observations, we have computed the vector $V$ of centroids defined in the equation:

$$v = \sum_{k=1}^{n} u_{ik}^{m} x_k \sum_{k=1}^{n} u_{-k}^{m} \qquad 1 \le i \le c \qquad (8)$$

Further we assume that a new observation $x$ was made and this $x$ is an s_dimensional vector whose elements are fuzzy numbers.
We are interested in the membership degree

$$\mu_i(x) = \sum_{j=1}^{c} \left[ \|x-v_i\| / \|x-v_j\| \right]^{2/(1-m)} , \quad m>1 \qquad (9)$$

Of course this $\mu_i(x)$ becomes now a fuzzy number. To get an insight into the way of computing $\mu_i(x)$ let us rewrite (9) in the form:

$$\mu_i(x) = \sum_{j=1}^{c} d(x,v_i) / d(x,v_j) \tag{10}$$

where

$$d(x,v_j) = \| x - v_j \|^{2/(m-1)} \tag{11}$$

Proposition 2.1: Let $x$ be vector composed from TFN (Triangular Fuzzy Number), i.e. $x = [x^1, x^2, ..., x^s]$ and each $x^l$ is of the form $x^l = (x^l_d, x^l_m, x^l_g)$. Then the generalized distance $d(x,v_i)$ can be computed as the distance between $v_i$ and the $\alpha$-cuts of the measurement $x$.

To prove this proposition, assume for simplicity that $s=2$ i.e. we have two-dimensional observations and let us consider fig. 2.1.

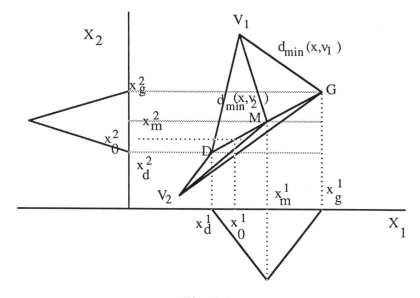

Fig. 2.1.

Our first observation is that to compute the distance between a centroid $v$ and the fuzzy observation $x$ it sufficies to compute distances between $v$ and the points from the thickened line D-M-G. Indeed, let us fix a point $x^1_0$. As degree of membership of the observation $(x^1_0, x^2_0)$ equals to the minimum of $\mu_{x1}(x^1_0)$ and $\mu_{x2}(x^2_0)$, then for any $x^2 \in (x^2_d, x^2_0)$ this degree will be less than

94

$\mu_{x1}(x^1{}_0)$ and for any $x^2 \in (x^2{}_d, x^2{}_0)$ this degree will be not greater than $\mu_{x1}(x^1{}_0)$. Thus the distance $d(x_0,v)$, where $x_0 = (x^1{}_0, x^2{}_0)$, gets the membership degree of $\mu_{x1}(x^1{}_0) = \mu_{x2}(x^2{}_0)$ and it is the maximal value that can be assigned to the number $d(x_0,v)$.

Proposition 2.2: Let $A_1 \ldots A^2$ be a collection of TFN's.
Denote A the sum $1/A_1 + \ldots + 1/A_k$ . Then this A is characterized by the membership function:

$$\mu_A(r) = \begin{cases} L_A(r) = \dfrac{r-\alpha}{\beta-\alpha} & \text{for } \alpha \le r \le \beta \\[2mm] R\ (r) = \dfrac{\gamma-r}{\gamma-\beta} & \text{for } \beta \le r \le \gamma \\[2mm] 0 & \text{otherwise} \end{cases} \tag{12}$$

where:

$$\alpha = \left[ \sum_{i=1}^{k} a_d^i \right] / \prod_{i=1}^{k} a_g^i$$

$$\beta = \left| \sum_{i=1}^{k} a^i \right| / \prod_{i=1}^{k} a^i$$

$$\gamma = \left[ \sum_{i=1}^{k} a_g^i \right] / \prod_{i=1}^{k} a_d^i$$

To prove this proposition suppose first that k=2. In this case $A = (A_1+A_2)/A_1*A_2$. The numerator of this expression is a TFN, $N = [a^1{}_d + a^2{}_d, a^1 + a^2, a^1{}_g + a^2{}_g]$ and the denominator can be approximated by the TFN, $D = [a^1{}_d a^2{}_d, a^1 a^2, a^1{}_g a^2{}_g,]$. Now using the equation (13) we find the membership function of the fuzzy number B = N/D:

$$\mu\ (r) = \begin{cases} L(r) = \dfrac{r - a_d/b_g}{a/b - a_d/b_g} & \text{for } a_d/b_g \le r \le a/b \\[2mm] R(r) = \dfrac{a_g/b_d - r}{a_g/b_d - a/b} & \text{for } a/b \le r \le a_g/b_d \\[2mm] 0 & \text{otherwise} \end{cases} \tag{13}$$

$$B(r) = \begin{cases} L_B(r) = \dfrac{r - \left[a_d^1 + a_d^2\right] / a_g^1 a_g^2}{\left[a^1 + a^2\right] / a^1 a^2 - \left[a_d^1 + a_d^2\right] / a_g^1 a_g^2} \\[2em] R_B(r) = \dfrac{\left[a_g^1 + a_g^2\right] / a_d^1 a_{d^-}^2 \, r}{\left[a_g^1 + a_g^2\right] / a_d^1 a_{d^-}^2 \left[a^1 + a^2\right] / a^1 a^2} \\[2em] 0 \hspace{4cm} \text{otherwise} \end{cases} \qquad (14)$$

Extending this result to the general case, k-2, we obtain the formulas (12).

Observe now that the equation (10) is of the form:

$$\mu_i(x) = \dfrac{1}{d(x,v_i) \sum\limits_{j=1}^{c} d^{-1}(x,v_j)} \qquad (15)$$

Applying Propositions 2.1 and 2.2 we obtain the next algorithm which enables to compute $\mu_i(x)$:

STEP 1.   Compute the fuzzy distance $d(x,v_j)$, $j=1,...,c$.
STEP 2.   Find the fuzzy number $M = \Sigma_j \, d^{-1}(x,v_j)$.
STEP 3.   Compute the denominator of (15),i.e. a fuzzy number $D = d(x,v_j)M$.
          Again this number can be approximated to a triangular fuzzy number.
STEP 4.   Using the equation (16) we obtain $\mu_i(x) = 1/D$.

$$\mu_{1/A}(r) = \begin{cases} L_{1/A}(r) = R_A(1/r) = \dfrac{a_g r - 1}{r(a_g - a)} & \text{for } 1/a_g \le r \le 1/a \\[1.5em] R_{1/A}(r) = L_A(1/r) = \dfrac{1 - a_g r}{r(a - a_d)} & \text{for } 1/a \le r \le 1/a_d \\[1.5em] 0 & \text{otherwise} \end{cases} \qquad (16)$$

This procedure is much more complicated in comparison with the original Bezdek's algorithm. Notice however that it may be radically simplified if we agree to make final decisions by comparing the membership degrees immediately. Examining the expression (15) we see that x belongs to class i when:

$$\mu_i(x) = \min_{1 \le j \le c} \, d(x,v_j) \qquad (17)$$

Since $d(x,v_j)$'s are fuzzy numbers. This justifies the next:

Proposition 2.3: Let $x$ be a collection of fuzzy measurements. The $\alpha$-cuts of fuzzy centroids are found by solving a sufficient number of equations:

$$(v_i)_\alpha = \sum_{k=1}^{n} u_{ik}^m (x_k) \, / \, \sum_{k=1}^{n} u_{ik}^m$$

$$u_{ik}^{-1} = \sum \left[ \left\| (x_k)_\alpha - (v_i)_\alpha \right\| \, / \, \left\| (x_k) - (v_i) \right\| \right]^{2/m-1}$$

Here $(x_k)_\alpha$, $(v_i)_\alpha$ stands for the $\alpha$-cut of the corresponding quantity.

Note that if all $x$ are sufficiently precise (in the sense mentioned earlier), we need only 3 runs of the program. In the first run we use $x$ the collection $x_*$ composed from the lower bounds of the measurements and in the third run $x$ is the collection of the upper bounds of the measurements.

## 3. NUMERICAL EXAMPLE

Consider a sample of 25 two-dimensional points consisting "touching clusters". The main values of the measurements are shown in Figure 3.1 and the full measurements are displayed in Table 3.1. Tables 3.2 - 3.4 present computer results.

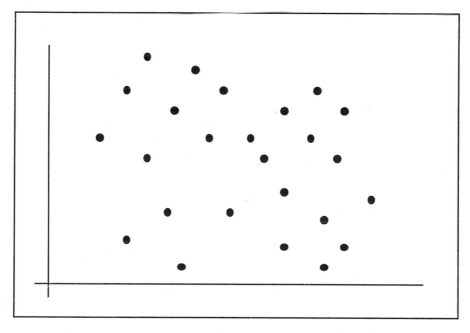

Fig. 3.1   Touching clusters used in the study

Table 3.1 - Input data.

| | | | | | | | |
|---|---|---|---|---|---|---|---|
| 1 = (( 1.0, | 2.0, | 3.0) | (5.5, 11.0, 16.5 )) | 14 = (( 5.0, 10.0, 15.0) | (5.0, 10.0, 15.0 )) |
| 2 = (( 1.5, | 3.0, | 4.5) | (3.5, 7.0, 10.5 )) | 15 = (( 6.0, 12.0, 18.0) | (3.5, 7.0, 10.5 )) |
| 3 = (( 1.5, | 3.0, | 4.5) | (6.5, 13.0, 19.5 )) | 16 = (( 6.0, 12.0, 18.0) | (4.5, 9.0, 13.5 )) |
| 4 = (( 2.0, | 4.0, | 6.0) | (7.5, 15.0, 22.5 )) | 17 = (( 6.0, 12.0, 18.0) | (6.0, 12.0, 18.0 )) |
| 5 = (( 2.0, | 4.0, | 6.0) | (5.0, 10.0, 15.0 )) | 18 = (( 7.0, 14.0, 21.0) | (3.0, 6.0, 9.0 )) |
| 6 = (( 2.5, | 5.0, | 7.5) | (4.0, 8.0, 12.0 )) | 19 = (( 7.0, 14.0, 21.0) | (4.0, 8.0, 12.0 )) |
| 7 = (( 2.5, | 5.0, | 7.5) | (6.0, 12.0, 18.0 )) | 20 = (( 6.5, 13.0, 19.5) | (5.5, 11.0, 16.5 )) |
| 8 = (( 3.0, | 6.0, | 9.0) | (7.0, 14.0, 21.0 )) | 21 = (( 7.0, 14.0, 21.0) | (6.5, 13.0, 19.5 )) |
| 9 = (( 3.0, | 6.0, | 9.0) | (3.0, 6.0, 9.0 )) | 22 = (( 8.0, 16.0, 24.0) | (3.5, 7.0, 10.5 )) |
| 10 = (( 3.5, | 7.0, | 10.5) | (5.5, 11.0, 16.5 )) | 23 = (( 8.0, 16.0, 24.0) | (5.0, 10.0, 15.0 )) |
| 11 = (( 4.0, | 8.0, | 12.0) | (4.0, 8.0, 12.0 )) | 24 = (( 8.5, 17.0, 23.5) | (6.0, 12.0, 18.0 )) |
| 12 = (( 4.0, | 8.0, | 12.0) | (6.5, 13.0, 19.5 )) | 25 = (( 9.0, 18.0, 27.0) | (4.5, 9.0, 13.5 )) |
| 13 = (( 4.5, | 9.0, | 13.5) | (5.5, 11.0, 16.5 )) | | |

Table 3.2

## Results for alpha = 0.0 (left reference function)

Data file: dane00.txt
Features file: pola.txt
control parameter = 1.50
Precision = 0.0001
Distance kind: Euclidean distance
Initialization variant: Suggested initialization
Relevant features: 37 38

Result centroids vectors

| 1 | 2.6835 | 7.0534 |
|---|---|---|
| 2 | 5.3650 | 4.7346 |

Frequency table

| 13 | 13.0 | 0.0 |
|---|---|---|
| 12 | 0.0 | 12.0 |

Statistical information

| 13 | 100.0 | 0.0 |
|---|---|---|
| 12 | 0.0 | 100.0 |

Classification entropy = 0.039948
Partition coefficient = 0.931584

Result fuzzy partition matrix

| | | |
|---|---|---|
| 1 | 0.9942 | 0.0058 |
| 2 | 0.9778 | 0.0222 |
| 3 | 0.9938 | 0.0062 |
| 4 | 0.9776 | 0.0224 |
| 5 | 0.9995 | 0.0005 |
| 6 | 0.9921 | 0.0079 |
| 7 | 0.9996 | 0.0004 |
| 8 | 0.9837 | 0.0163 |
| 9 | 0.9210 | 0.0790 |
| 10 | 0.9973 | 0.0027 |
| 11 | 0.8827 | 0.1173 |
| 12 | 0.9443 | 0.0557 |
| 13 | 0.8210 | 0.1790 |
| 14 | 0.3780 | 0.6220 |
| 15 | 0.0320 | 0.9680 |
| 16 | 0.0097 | 0.9903 |
| 17 | 0.0535 | 0.9465 |
| 18 | 0.0152 | 0.9848 |
| 19 | 0.0007 | 0.9993 |
| 20 | 0.0037 | 0.9963 |
| 21 | 0.0239 | 0.9761 |
| 22 | 0.0058 | 0.9942 |
| 23 | 0.0012 | 0.9988 |
| 24 | 0.0115 | 0.9885 |
| 25 | 0.0089 | 0.9911 |

Table 3.3

**Results for alpha = 1.0**

| Data file: dane.txt | Result fuzzy partition matrix | | |
| --- | --- | --- | --- |
| Features file: pola.txt | | | |
| control parameter = 1.50 | 1 | 0.9942 | 0.0058 |
| Precision = 0.0001 | 2 | 0.9778 | 0.0222 |
| Distance kind: Euclidean distance | 3 | 0.9938 | 0.0062 |
| Initialization variant: Suggested initialization | 4 | 0.9776 | 0.0224 |
| Relevant features: 37 38 | 5 | 0.9995 | 0.0005 |
| | 6 | 0.9921 | 0.0079 |
| Result centroids vectors | 7 | 0.9996 | 0.0004 |
| | 8 | 0.9837 | 0.0163 |

Result centroids vectors

| 1 | 5.3671 | 14.1069 |
| --- | --- | --- |
| 2 | 10.7299 | 9.4692 |

| 9 | 0.9210 | 0.0790 |
| 10 | 0.9973 | 0.0027 |
| 11 | 0.8827 | 0.1173 |

Frequency table

| 13 | 13.0 | 0.0 |
| --- | --- | --- |
| 12 | 0.0 | 12.0 |

| 12 | 0.9443 | 0.0557 |
| 13 | 0.8210 | 0.1790 |
| 14 | 0.3780 | 0.6220 |
| 15 | 0.0320 | 0.9680 |
| 16 | 0.0097 | 0.9903 |

Statistical information

| 13 | 100.0 | 0.0 |
| --- | --- | --- |
| 12 | 0.0 | 100.0 |

| 17 | 0.0535 | 0.9465 |
| 18 | 0.0152 | 0.9848 |
| 19 | 0.0007 | 0.9993 |
| 20 | 0.0037 | 0.9963 |
| 21 | 0.0239 | 0.9761 |
| 22 | 0.0058 | 0.9942 |

Classification entropy = 0.039948

Partition coefficient = 0.931584

| 23 | 0.0012 | 0.9988 |
| 24 | 0.0115 | 0.9885 |
| 25 | 0.0089 | 0.9911 |

The resulting centroids $v_i = (v_{ix}, v_{iy})$, i=1,2, are

$$v_{1x} = (2.6835, \ 5.3671, \ 8.0506)$$
$$v_{1y} = (7.0534, 14.1069, 21.1603)$$
$$v_{2x} = (5.3650, 10.7299, 16.0949)$$
$$v_{2y} = (4.7346, \ 9.4692, 14.2038)$$

Consider now the fuzzy point $x = (x_x, x_y)$ whose coordinates are TFN's:

$$x_x = (8, \ 9, \ 10); \qquad x_y = (9, 10, 11)$$

The precise determination of the shape of the distance $d(x,v_i)$ is complicated as we must refer to the Extension Principle by which

$$\mu_{d(x,v_i)}(r) = \sup_{r=d(u,z)} \min \ (\mu_x(u), \mu_{v_i}(z))$$

Table 3.4

**Results for alpha = 0.0 ( right reference function )**

Data file: data.txt
Features file: pola.txt
control parameter = 1.50
Precision = 0.0001
Distance kind: Euclidean distance
Initialization variant: Suggested initialization
Relevant features: 37 38

Result centroids vectors

| 1 | 8.0506 | 21.1603 |
| 2 | 16.0949 | 14.2038 |

Frequency table

| 13 | 13.0 | 0.0 |
| 12 | 0.0 | 12.0 |

Statistical information

| 13 | 100.0 | 0.0 |
| 12 | 0.0 | 100.0 |

Classification entropy = 0.039948
Partition coefficient   = 0.931584

Result fuzzy partition matrix

| 1 | 0.9942 | 0.0058 |
|---|--------|--------|
| 2 | 0.9778 | 0.0222 |
| 3 | 0.9938 | 0.0062 |
| 4 | 0.9776 | 0.0224 |
| 5 | 0.9995 | 0.0005 |
| 6 | 0.9921 | 0.0079 |
| 7 | 0.9996 | 0.0004 |
| 8 | 0.9837 | 0.0163 |
| 9 | 0.9210 | 0.0790 |
| 10 | 0.9973 | 0.0027 |
| 11 | 0.8827 | 0.1173 |
| 12 | 0.9443 | 0.0557 |
| 13 | 0.8210 | 0.1790 |
| 14 | 0.3780 | 0.6220 |
| 15 | 0.0320 | 0.9680 |
| 16 | 0.0097 | 0.9903 |
| 17 | 0.0535 | 0.9465 |
| 18 | 0.0152 | 0.9848 |
| 19 | 0.0007 | 0.9993 |
| 20 | 0.0037 | 0.9963 |
| 21 | 0.0239 | 0.9761 |
| 22 | 0.0058 | 0.9942 |
| 23 | 0.0012 | 0.9988 |
| 24 | 0.0115 | 0.9885 |
| 25 | 0.0089 | 0.9911 |

(Remember that $u$ and $z$ are s-dimensional vectors in general). We can use the triangular estimation, however. That is we compute distances $d(x_d,v_i)$, $d(x_m,v_i)$ and $d(x_g,v_i)$. Next we define:

$$d_* = \min [ d(x_d;v_c), d(x_g,v_i) ]$$
$$d^* = \max [ d(x_d;v_c), d(x_g,v_i) ]$$

If $d_* \leq d(x_m,v_i)$ then we take as the distance between the fuzzy point $x$ and the centroid $d_i$ the TFN $d(x,v_i) = (d_*, d(x_m,v_i), d^*)$. When $d(x_m,v_i) < d_*$ then $d(x,v_i)$ can be approximated by the fuzzy number $(d(x_m,v_i), d(x_m,v_i), d^*)$.
In our case (we recall that we were using m=1.5) we obtain the next approximations of fuzzy distances:

$$d(x,v_1) = (31.36, 31.36, 96.49)$$
$$d(x,v_2) = ( 3.24, 3.24, 50.84)$$

Let us conclude our results. It is quite easy to extend Bezdek's algorithm to the case of fuzzy measurements. We simply reiterate the program for different values of $\alpha$ in

100

$\alpha$ -cuts of the measurements. It is important that when data are modeled by TFN's it suffices to run the program only three times.

The only inconvenience concerns the computation of fuzzy distances among new observation and the fuzzy centroids. The solution of this problem requires a special procedure. For fast and rough decisions it suffices to approximate the distances by TFN as we already described.

## REFERENCES

[1]   J.M. Adamo, "Fuzzy decision trees", Fuzzy Sets and Syst., 4: 207-219 (1980)

[2]   S.M. Baas and H. Kwakernaak, "Rating and ranking of multiple-aspect alternatives using fuzzy sets", Automatica, 13:47-58 (1977)

[3]   J.F. Baldwin and N.C.F. Guild, "Comparison of fuzzy sets on the same decision space", Fuzzy Sets and Systems, 2:213-233 (1979)

[4]   G. Bortolan and R. Degani, "Ranking of fuzzy alternatives in electrocardiography", in Preprints IFAC Conf. on Fuzzy Informtion, Knowledge Representation and Decision Analysis, 394-402 (1983)

[5]   G. Bortolan and R. Degani, "A review of some methods for ranking fuzzy subset", Fuzzy Sets and Systems, 15:1-19 (1985)

[6]   J.J. Buckley, "The multiple judge, multiple criteria ranking problem: a fuzzy set approach", Fuzzy Sets and Systems, 13:25-37 (1984)

[7]   J.J. Buckley, "Ranking alternatives using fuzzy numbers", Fuzzy Sets and Systems, 15:21-31 (1985)

[8]   W. Chang, "Ranking of fuzzy utilities with triangular membership functions", in Proc. Int. Conf. on policy Anal. and Inf. Systems, 263-272 (1981)

[9]   R. Degani and G. Pacini, "Linguistic pattern recognition algorithms for computer analysis of ECG", in Proc. BIOSIGMA '78, 18-26 (1978)

[10] D. Dubois and H. Prade, "Fuzzy Sets and Systems: Theory and Applications", Academic Press (1985)

[11] D. Dubois and H. Prade, "Ranking of fuzzy numbers in the setting of possibility theory", Inform. Sci. 30:183-224 (1983)

[12] D. Dubois and H. Prade, "Fuzzy-set-theoretic differences and inclusions and their use in the analysis of fuzzy equations", Control and Cybernetics, 13:129-146 (1984)

[13] A.N.S. Freeling, "Fuzzy sets and decision analysis", IEEE Trans. Systems Man Cybernet.,10:341-354 (1980)

[14] L.J. Kohout and W. Bandler, "Knowledge representation in Medicine and Clinical Behavioural Science", Abacus Press (1987)

[15] H. Kwakernaak, "Fuzzy random variables. Part I: Definitions and Theorems", Inform. Sci., 15:1-15 (1978)

[16] H. Kwakernaak, "Fuzzy random variables. Part II: Algorithms and examples for the discrete case", Inform. Sci., 17:253-278 (1979)

[17] R. Kruse, K.D. Meyer, "Statistics and Vague Data", Reidel Publ.Co. (1987)

[18] M. Mizumoto and K. Tanaka, "Some properties of fuzzy numbers" in 'Advances in Fuzzy Set Theory and Applications' (M.M. Gupta, R.K. Ragade, R.R. Yager, Eds.) North-Holland, 153-165 (1979)

[19] S. Nahmias, "Fuzzy variables", Fuzzy Sets and Systems, 1:97-110 (1978)

[20] S. Nahmias, "Fuzzy variables in a random experiment" in 'Advances in Fuzzy Set Theory ad Applications' (M.M. Gupta, R.K. Ragade, R.R. Yager, Eds.) North-Holland, 165-180 (1979)

[21] S.T. Wierzchon, "Linear programming with fuzzy sets: A general approach", Mathematical Modelling, 9:447-460 (1987)

[22] S.T. Wierzchon, "On fuzzy stochastic programming" in 'Combining Fuzzy Imprecion with Probabilistic Uncertainty in Decision Making' (J. Kacprzyk and M. Fedrizzi, Eds.), Springer Verlag (1988)

[23] L.A. Zadeh, "Fuzzy sets as a basis for a theory of possibility", Fuzzy Sets and Systems, 1:3-29 (1978)

# CLUSTERING ANALYSIS AS A METHOD OF AUTO-CLASSIFICATION AND ITS APPLICATION TO THE SEARCH FOR GALAXY CHAINS IN TWO DIMENSIONS

Z.G. Deng [1,2], X.Y. Xia [3], H. Arp [1], and G. Börner [1]

1. Max Planck Institut für Astrophysik, 8046 Garching bei Munchen, FRG
2. Dept. Physics, Graduate Sch., Academia Sinica, P.O. Box 3908 Beijing, PRC
3. Dept. Physics, Tianjin Normal University, Tianjin, PRC

## ABSTRACT

Filaments and lines of galaxies on the sky have been discussed in recent years. The present paper outlines a method for quantitatively testing the connected nature of these features. Comparison with Monte Carlo simulations give probability values that the features are real.

Two large scale alignments of galaxies in the sky are tested. In the region $7^h$ < R.A. < $11^h$, $30°$ < Dec. < $65°$ filaments of galaxies of $3500 < cz_0 < 5500$ kms$^{-1}$ and $4200 < cz_0 < 5200$ kms$^{-1}$ are shown to be significant. The second field at $12^h30^m$ < R.A. < $15^h30^m$, $35°$ < Dec. < $70°$ shows a string like structure recognized here for the first time for galaxies with redshifts $2300 < cz_0 < 2800$ kms$^{-1}$.

## 1. INTRODUCTION

Clustering analysis is extensively applied in solid state physics for investigating percolation problems. Zeldovich et al. (1982) introduced this method to astrophysics. They used this method to examine whether the distribution of galaxies contains string or sheet like structures as the pancake model has predicted. Zhou et al. (Zhou, et al., 1986) have used the clustering analysis method to investigate whether string like structures exist in the distribution of quasars. Geller and Huchra (1983) have used the clustering analysis method with minor modifications to sort out double galaxies and galaxy groups from the CfA catalog.

Classification of objects is an important task in astronomy. A good classification can supply astronomers with a lot of information about the features and evolution of

103

these objects. The H-R diagram is one of the brilliant examples, which played an important role in the development of stellar astrophysics.

The rapid development of computer techniques permits us to develop methods by which the classification of a huge number of objects can be carried out automatically. Applications of the auto-classification programs hold the promise for scientists of obtaining more information on a larger sample in shorter time.

In this work, we present the preliminary results of developing an auto-clasification method based on clustering analysis and its application to the search for galaxy chains in a given field.

## 2. METHOD

The basic idea of a clustering analysis is the division of objects into different classes related to a given length scale r. Any two objects with relative distance less than r belong to the same class. Any object which is closer than r to any member of a class belongs also to that class.

For a very small value of r all objects will be isolated. If r is increased gradually, at first a few objects will be connected to one class. If progressively adjacent objects are found, then with r getting larger this defines connected structures of increasing size.

To examine the characteristics of structures we compute the maximum value $R_{max}$ of the distance between two objects for all groups found at a certain value of r. $R_{max}$ is obviously a function of r which increases when r increases. The way in which $R_{max}$ increases depends on characteristic features in the distribution of objects. For example, if there are string or sheet-like structures in the distribution, then $R_{max}$ will increase very fast when r arrives at a certain critical value.

In the case of hierarchical clustering the function $R_{max}$ will show a step-function type increase: As long as r is less than the distance between connected structures of one type, $R_{max}$ will be approximately constant. But as r approaches the mean distance between this type of structures $R_{max}$ will increase rapidly. Thus, each type of connected structure will show up in a rapid increase of $R_{max}$, followed by a plateau of constant $R_{max}$ for some range of values of r.

Besides the parameter $R_{max}$ we may also use the abundance spectrum of the distribution to show clustering behaviour in the distribution of objects. The abundance spectrum shows the number of objects in various clusters. The abundance of level m is defined to be the relative number of objects which are in groups with $2^m - 1$ to $2^m$ objects. For example, m = 0,1,2,3,..., correspond to groups with 1, 2, 3-4, 5-8, ... galaxies. The abundancies at these levels are the total number of objects in the sample. The abundance spectrum gives additional informatin about the distribution and clustering characteristics of our sample.

To see if the structure and clustering characteristics of the distribution are significant we compare the results obtained from the analysis with the mean results given by Monte Carlo sampling. If the value of $R_{max}$ given by our sample is different from the Monte Carlo mean by an amount larger than 3 σ we conclude that the differences are significant. Otherwise we shall consider that the differences are produced only accidentally. The same criterion can be used for the comparison of the abundance spectrum with the mean spectrum obtained from Monte Carlo sampling.

## 3. STUDY OF THE GALAXY DISTRIBUTION IN TWO DIMENSIONS

As an example, we use the method described in the previous section to study the behaviour of the galaxy distribution in three fields. We have only carried out the analysis in two dimensions. All galaxies are sorted out from the ZCAT catalog (J. Huchra, Private Communication) in which the observed redshift has been given. We have also used the redshift of galaxies as a parameter in the choice of our sample galaxies. Before sorting out the samples we have corrected the effect of rotation of our own Galaxy on the redshifts by the following formula (Arp, 1986)

$$V_{sol} = 251 \sin l \cos b - 20 \cos l \cos b - 5 \sin b \quad (kms^{-1}) \tag{1}$$

where, l and b are galactic latitude and longitude.

In Table 1, we give the fields, ranges of the corrected redshift and the number of galaxies in our three samples. The calculated results and preliminary conclusions are given in Figures 1-6 and their captions.

Table 1. The fields, ranges of redshifts and the numbers of our samples. The mean angular distances for each sample is also given in this table.

| sample | R.A. | DEC. | range of $z_0$ | number of galaxies | $n^{-1/2}$ |
|--------|------|------|----------------|--------------------|-----------|
| $A_1$ | $7^h$ - $11^h$ | 30° - 65° | 3500 - 5500 | 129 | 3.39 |
| $A_2$ | $7^h$ - $11^h$ | 30° - 65° | 4200 - 5200 | 77 | 4.26 |
| B | $12^h30^m$ - $15^h30^m$ | 35° - 70° | 1500 - 2500 | 80 | 3.44 |
| C | same field as B | | 2300 - 2800 | 64 | 3.84 |

Definitions of quantities used are:    $n^{-1/2}$ = ave density/square degree
$\theta$ = angular distance on sky
$\Theta_{max}$ = largest angular dimension of all connected points

## 4. DISCUSSION

This method can give us more than we have shown. For example, in the calculation we can obtain the list of galaxies in each group. If we plot the distribution on an equi-area projection, then we can see how the clustering changes as the scale of connection becomes larger and larger. With a minor improvement, we can also give

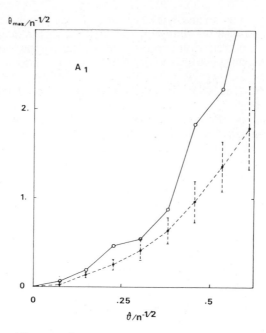

Fig.1 - The $\Theta_{max}/n^{-1/2}$ - $\theta/n^{-1/2}$ curve for sample $A_1$. When $\theta/n^{-1/2}$ becomes larger than about 0.38, the distribution shows string like structure.

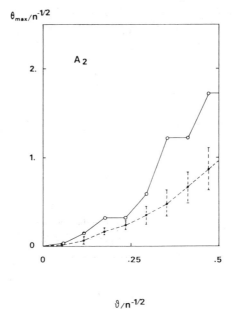

Fig.2 - The $\Theta_{max}/n^{-1/2}$ – $\theta/n^{-1/2}$ curve for sample $A_2$. When $\theta/n^{-1/2}$ becomes larger than about 0.3, the distribution shows string like structure, with more than one small chain.

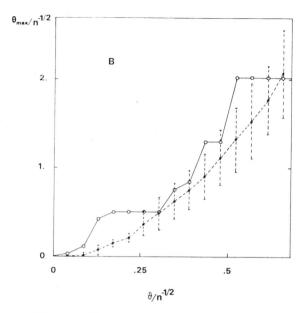

Fig. 3 - The $\Theta_{max}/n^{-1/2}$ - $\theta/n^{-1/2}$ curve for sample B. In this distribution no string like structure has been discovered.

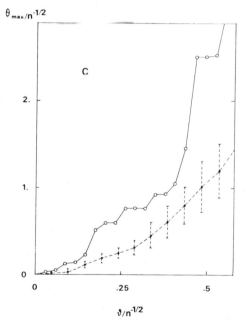

Fig.4 - The $\Theta_{max}/n^{-1/2}$ - $\theta/n^{-1/2}$ curve for sample C. The figure seems to show that there are·many small groups, in larger scale they are connected into a string like structure.

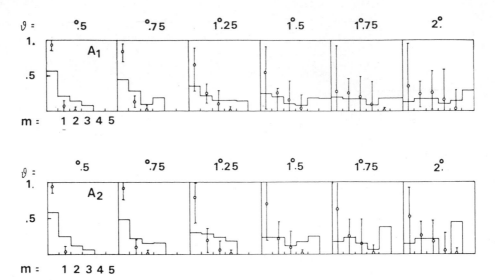

Fig.5 - The abundance spectra for samples $A_1$ and $A_2$ at given values of $\theta$. They show that at small $\theta$, both samples have significant galaxy pairs and small groups. For larger values of $\theta$, both distributions show significant clustering.

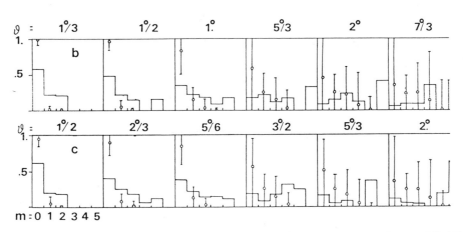

Fig. 6 - The abundance spectra for samples B and C for given values of $\theta$. The conclusions are the same as for Fig.5.

the quantitative morphological classification of these clusters for a given connected scale. The method can easily be generalized to do auto-classification for other objects and other interesting properties. For example, we can use the method with a minor modification to do the auto-classification on a color-color diagram.

In the region $A_1 + A_2$ we have used the method to confirm the linear alignment of galaxies that was called the Lynx-Ursa Major supercluster by Giovanelli and Haynes (1982). This filament was also noted by Focardi, Morano and Vettolani (1986). In region C we have discovered a new, large scale, connecetd filament of galaxies. We have also made preliminary reconnaissance of a filament in the regiopn $4^h30^m <$ R.A. $< 13^h30^m$, Dec. $> 30°$ which stretches more than $40°$ across the northern sky.

## ACKNOWLEDGEMENT

We would like to thank Rudolph Albrecht for discussion of the general problem in computational advice and assistance.

## REFERENCES

Arp, H., 1986, Astron. Astrophys., 156:207.
Focardi, P., Marano, B., and Vettolani, G., ESO preprint No. 420.
Giovanelli, R., and Haynes M.P., 1982, A. J., 87:1355.
Geller and Huchra, J.P., 1983, Ap. J. Suppl., 52:61.
Zeldovich, Ya.B., Einasto, J., and Shandarin, S., 1982, Nature, 300:407.
Zhou, Y.Y., Fang,D.P., Deng,Z.G., and He, X.T., 1986, Ap. J., 311:578.

# OBJECT DETECTION AND ANALYSIS IN DIGITISED IMAGES: ALGORITHMIC COMPONENTS AND SOFTWARE STRUCTURE

F. Murtagh*

Space Telescope — European Coordinating Facility
European Southern Observatory
Karl–Schwarzschild–Straße 2
D–8046 Garching bei München, FRG

## ABSTRACT

The structures of major 2–dimensional image analysis software packages in optical astronomy are overviewed. Reference is made to the varied image analysis and pattern recognition algorithms in use.

## INTRODUCTION

Packages for object recognition and processing are used in a standalone mode or as part of some larger image processing system. Most of the packages looked at in this paper are incorporated into a larger system. However most also started out as self–contained systems. Consequently, they often preserve a user interface different from that of the image processing system, — a different command syntax, the provision of a menu interface, etc. Major astronomical image processing systems in which the packages described here are incorporated include MIDAS (*Munich Image Data Analysis System*, produced at the European Southern Observatory; see MIDAS, 1988); and IRAF (*Image Reduction and Analysis Facility* from Kitt Peak National Observatory, Arizona; see Tody, 1986).

While a very large number of astronomers will by preference or by (resources–constrained) necessity use their own personal software, this article surveys some of the most well known and widely used packages in this area.

Packages INVENTORY, COSMOS and FOCAS — to be looked at below — have star–galaxy separation as their central objective and thus allow extended objects to be analysed. Packages DAOPHOT and ROMAFOT, on the other hand, are

---

* Affiliated to Astrophysics Division, Space Science Department, European Space Agency.

stellar photometry systems. The former group of systems determine parameters or attributes for the objects found and then allow the user to further study these objects in parameter space. Systems of the stellar photometry type attempt to find the Point Spread Function (PSF) as accurately as possible and to fit it to the objects detected. In the latter case, this does not preclude the possibility of a varying PSF (e.g. across the image field).

The problems to be solved in detecting and analysing objects in digitised images include the following: variable, noisy backgrounds, complicated by blended and superimposed objects; nonlinearities and saturation — photographic plates have a nonlinear intensity response and give rise to saturation effects; and plate faults and CCD defects. A comprehensive review of image processing techniques for such images is available in Grosbøl (1987). In this article we will assume that the images have been reasonably well cleaned, so that the pattern recognition problem rather than the image processing one will be under discussion.

## SUMMARY OF PACKAGE CAPABILITIES

In each section, we give the most recent or most comprehensive references to the general capabilities of the package, and to applications involving its use; the developer and maintainer; what sort of images it was designed to process; and then outline the package's structure. Where relevant, some extra comments may be added, including language used.

## INVENTORY

INVENTORY (Kruszewski, 1985; 1987; application in West and Kruszewski, 1981) has been available for some years in the MIDAS system. Developed by A. Kruszewski (Warsaw University Observatory), it was designed for the fast detection of objects on large plates. It has been used for Schmidt, CCD, and electronographic plates. For blended objects, it produces a fast but not optimal solution. It is a modular system, and the commands are as follows:

(1) NORMALIZE: Divide the frame by the average background.

(2) SEARCH: Prepare list of objects. A local sky background is defined, and an object detected if the average intensity over 9 central pixels is greater than the background by a user–defined threshold. A user–defined separation threshold defines distinct objects.

(3) ANALYSE/NOVERIFY: Prepare 20 parameters for the objects which have been listed (a full desciption of these is given in Kruszewski, n.d.).

(4) ANALYSE/VERIFY: As before, but first improve the positions of the centres.

(5) CLASSIFY: Classify into stars, galaxies, defects, and unclassified (using a non–parametric, seed–initiated, iterative algorithm).

112

INVENTORY does not provide the extraction precision of other packages, but is very suitable for the quick scanning of large images. It was originally designed for detecting faint distant clusters of galaxies, and it may be noted that many complementary routines are in use (in the MIDAS environment) for the Southern Sky survey of A. Lauberts, ESO. INVENTORY is written in VAX–11 Fortran, it uses MIDAS command language, and a port and upgrade to C is planned.

## COSMOS

COSMOS ("Coordinates, Size, Magnitude, Orientation and Shape") (Stobie, 1986; Stobie, 1980; COSMOS, n.d.; applications in MacGillivray and Stobie, 1984; MacGillivray et al., 1976) is an integrated system for the scanning and analysis of photographic plates, incorporating a high speed scanning microdensitometer. It was developed and is maintained at the Royal Observatory Edinburgh (ROE). It has been used for large photographic plates, including non–astronomical applications such as aerial photographs and radar pictures. The steps carried out in the analysis of digitised images are as follows.

• The sky background is determined by sampling over the entire picture. Hence a reliable estimate will be thwarted by a large object in the picture. The background is taken as the median of intensities.

• Binary thresholding is carried out: all pixels greater than a user–specified constant times the local sky background are determined. The connectivity of such pixels is then determined, and if objects are too small (i.e. if there are too few contiguous pixels), the object is rejected.

• Parameters relating to position, brightness, shape, orientation, and classification are obtained.

• For the latter stage of classification into stars, galaxies, and other classes, pre-defined discriminators (from among the parameters which are determined) are used.

• Currently under development is a crowded field algorithm which looks for saddle points in blended objects, in order to assign pixels to some one of the latter.

COSMOS is particularly useful for quick–scan processing of large plates — it can be used in a real–time mode as well as a batch mode. It has been extensively described in the literature, and much experience has been built up over more than a decade of use. COSMOS is written in Fortran.

## IRAF Packages: FOCAS and DIGIPHOT

IRAF (see Tody, 1986) will have two packages in the area of digital image photometry: DIGIPHOT is under development, and the implementation of FOCAS —

already well described in the literature — is planned. DIGIPHOT will automatically generate lists of objects, perform synthetic aperture photometry, and allow PSF fitting.

FOCAS ("Faint Object Classification and Analysis System") (Jarvis and Tyson, 1981; applications in Tyson and Jarvis, 1979, and Tyson, 1984) was developed at Kitt Peak National Observatory, and runs now in a UNIX environment. It was designed to provide a faint object handling capability for photographic and CCD images.

Algorithmically, FOCAS is based on a pixel, and thence object, thresholding procedure. Various shape and other parameters are determined. A nonparametric clustering technique is then used to differentiate between objects. Valdes (1982) describes an alternative Bayesian maximum likelihood approach, and these methods are further compared in Tyson (1984).

### Other Discrete Object Systems

Many research workers have, at one time or another, put together their own package. A few are mentioned in the following.

• Many algorithms are available in the highly modular system produced and maintained by A. Bijaoui, Centre de Dépouillement des Clichés Astronomiques (CDCA), Nice.

• A system for faint object detection produced in Rome by G. Pittela and D. Trevese has the name ROMA (but is distinct from the ROMAFOT package described below).

• Herzog and Illingworth (1977) describe a COSMOS–like system for photographic plates, where blended objects are sought for as "bumps" in smooth intensity gradients defining objects.

• Kron (1980) uses a similar approach to the foregoing, stating however that blended objects are not of great importance in his study.

• Malagnini et al. (1985) describe FODS ("Faint Object Detection and Classification System") which contains a range of options for object detection, and a star/galaxy classifier.

### DAOPHOT and ALLSTAR

DAOPHOT (Stetson, 1984; 1987; King, 1986; application in Crotts, 1986) is a stand–alone system. Conversion routines are available for its (ASCII) table data structures and those of MIDAS; and as currently implemented it uses MIDAS .BDF image files (rather than its native mode .DST image format). A port to IRAF is currently being carried out. DAOPHOT was developed and is maintained by P. Stetson (Dominion Astrophysical Observatory, B.C.), and can handle all types of stellar frames, including crowded fields. It is highly modular, and commands available are as follows:

(1) SKY: Estimate the average sky brightness as the mode of sampled intensity values.

(2) FIND: Determine the approximate centroids of objects, using an assumed Gaussian shape and a relative brightness threshold vis à vis the local sky background. This amounts to convolution with a truncated Gaussian. Additionally tested for are hot/cold pixels, and charge–overflow rows and columns.

(3) PHOT: Carry out synthetic aperture photometry — magnitudes for radial distances from the centroids of the objects.

(4) PSF: Define a PSF from sampled bright stars in the frame, where the PSF is taken as a Gaussian plus a table of residuals (hence, allowing a completely non–analytical PSF).

(5) Following the previous steps, crowded field photometry is carried out by the use of GROUP and NSTAR. The former determines self–contained, mutually independent groups of touching or overlapping objects. NSTAR carries out a simultaneous, iterative, linearized least–squares PSF fit to members ($\leq 60$) of groups . The algorithm used for the latter is Eichhorn and Clary (1974).

(6) SUBSTAR: Subtract scaled, reconstructed (via the PSF) images from objects, thus allowing blended stars, stellar–like galaxies, stars with faulty pixels, or the general assessment of the FIND routine, to be seen from the residuals. Commands can be implemented also on the residual image, and the entire procedure iterated.

(7) PEAK: Determine positions and magnitudes by the least–squares fitting of the PSF to the objects.

(8) Various other utilities are also available. These include: SELECT (groups of specified memberships, for processing by NSTAR); SORT (file on specified attribute); LIST (file); OFFSET (add constant values to each coordinate in a list of coordinates); APPEND (files); DUMP (view raw values of part of a frame); ADDSTAR (add synthetic stars at random); and HELP.

ALLSTAR is a new system, independent of DAOPHOT. It allows multiple simultaneous PSF fitting for any number of objects. For crowded fields, it requires the degree of object overlap that is user–acceptable. A critical radius is then determined automatically, — this is the radius at which the object disappears into the background noise.

King (1986) describes DAOPHOT as the "world standard for stellar photometry of digitised images". For galaxy photometry DAOPHOT can only be used as a first stage to remove all stellar objects from the image. Additional software is in use at the Dominion Astrophysical Observatory which takes into account systematic (linear) variations in the PSF. DAOPHOT is written in VAX–11 Fortran.

## ROMAFOT

ROMAFOT (Buonanno et al., 1979; 1988; applications in Buonanno, 1983a; 1983b) is a newly implemented package in MIDAS specially catering for globular photometry. As a system it allows extensive interactive use and is ideally suited for the analysis of crowded fields.

The current form assumed for the PSF function is analytic: it presupposes either Gaussian or Moffat functions (Moffat, 1969). The latter is Gaussian–like but with larger wings:

$$a(1 + \{(\mathbf{x} - \bar{x})^2/s^2\}^{-\beta})$$

where $\beta$ is typically around 4. Commands available include the following.

(1) SELECT: Requests the user to specify parts of the image for analysis. Using cursor input, the user then specifies the objects to be analysed. About 10–15 objects over the whole frame are recommended. Subframes are produced which correspond to the object selected. The user can also enter values for name and V, B and U magnitudes.

(2) FIT determines the PSF. The PSF is taken as a Gaussian or Moffat function. In the case of the latter, the user specifies one or both of the two parameters of this function. The user also indicates whether the background is flat or tilted (in the latter case, implying asymmetric PSF wings).

(3) ANALYSE assesses the fit graphically. There are many subcommands to this interactive command. The original subfield, and the subfield with fitted stars subtracted, may be displayed. Additionally, profiles before and after fitting in the horizontal and vertical directions may also be displayed.

(4) Interactive use of ROMAFOT requires extensive iterative use of the commands FIT and ANALYSE. An automated analysis option uses commands SKY (which determines the sky value in local regions), SEARCH and SORT (which groups together overlapping objects using a user–specified "action radius": the latter indicates the extent of the object) to derive objects from which the PSF is to be determined.

(5) EXAMINE allows interactive and quantitative assessment of the fitting.

(6) REGISTER produces an output file containing a small number of parameters (user–specified identifier and magnitudes in V, B and U bands; magnitude approximation, which is within a multiplicative constant of the real magnitude in the case of a linear detecting instrument; coordinates of the different objects; central intensity; and fit assessment parameters).

116

In its current (January 1988) implementation in MIDAS, ROMAFOT tabular data structures are ASCII files; and its image data structure is the .IMA structure for which a command is available to carry out the conversion to a MIDAS (.BDF) image frame.

## Other Crowded Field Systems

Some other packages include the following.

• Penny and Dickens (1986) briefly describe a system similar to ROMAFOT. A PSF is used which is a mixture of a Lorentzian and a Gaussian. The system is interactive, with user specification of a subimage to be analysed, and with visual assessment.

• Lupton and Gunn (1986) describe another iterative star–subtraction procedure, using an empirically defined PSF derived from user–specified stars.

• Irwin (1985) describes a package for the analysis of crowded fields in CCD data. A maximum likelihood approach is used for PSF fitting. The approach aims at being entirely automated.

## Galaxy Packages

Galaxy photometry is surveyed in Capaccioli (1985). The determination of isophotal magnitudes and equivalent profiles, the fitting of $r^{1/4}$ or exponential laws, or the carrying out of synthetic aperture photometry on galaxy images is handled by a suite of routines in Oracle (1981) and Lauberts (1988). The former is in need of renewal in its structure (originating in the punched card era), and the latter (although using many routines available in INVENTORY) is a set of stand–alone Fortran programs.

## COMPARATIVE STUDIES

Areas of difficulty, which have been studied or are currently under investigation, include the following.

(1) Problems relating to the form of the PSF expected from Hubble Space Telescope data — in particular the expected undersampling of objects — are described in Bendinelli et al. (1985), Rosa and Baade (1986), and STECF (1987). For such data it might be preferable to work with a well–defined empirical PSF, rather than an analytical one.

(2) Crowded fields, with blended and perhaps saturated objects, are often inherently intractable. Other related problems include stars superimposed on galaxies (the latter cannot be subtracted using a PSF), and very non–uniform backgrounds.

Among studies carried out to date, we have Ortolani (1986, 1987) who compared DAOPHOT and INVENTORY. The former was found to be better for more crowded parts of the image, but has fitting problems with objects at faint magnitudes. More accurate aperture magnitudes are also given by the latter. It also handles non–stellar objects. The extra star adding and subtracting routines in DAOPHOT however allow greater precision. In Ortolani (1987) computational efficiency was also in INVENTORY's favour by a factor of five. It is suggested that both packages offer distinct advantages, and should be used in a complementary way.

## CONCLUSION

It rapidly becomes clear that the problems of object recognition and the consequent extraction of information are strongly associated with many mainstream computing and statistical problems. Some of the more important algorithms which have been used to date in this field are overviewed in Murtagh (1988).

### Acknowledgements

Discussions with R. Warmels, S. Ortolani, R. Buonanno and others contributed to the material presented here.

### REFERENCES

Bendinelli, O., Di Iorio, A., Parmeggiani, G., and Zavatti, F., 1985, "Some clues to Hubble Space Telescope image analysis", *Astronomy and Astrophysics*, **153**, 265:268.

Buonanno, R., Corsi, C.E., De Biase, G.A. and Ferraro, I., 1979, "A method for stellar photometry in crowded fields", in G. Sedmak, M. Capaccioli and R.J. Allen (eds.), *International Workshop on Image Processing in Astronomy*, Proc. 5th Coll. Astrophysics, Trieste, June 1979, Osservatorio Astronomico di Trieste, pp. 354:359.

Buonanno, R., Buscema, G., Corsi, C.E. and Iannicola, G., 1983a, "Positions, magnitudes, and colours for stars in the globular cluster M15", *Astrono and Astrophysics Supplement Series*, **51**, 83:92.

Buonanno, R., Buscema, G., Corsi, C.E., Ferraro, I. and Iannicola, G., 1983b, "Automated photographic photometry of stars in globular clusters", *Astronomy and Astrophysics*, **126**, 278:282.

Buonanno, R. *et al.*, 1988, "Crowded field photometry", *MIDAS Manual*, European Southern Observatory, Munich, Chapter 5.

Capaccioli, M., 1985, "2–D photometry", in V. Di Gesù, L. Scarsi, P. Crane, J.H. Friedman and S. Levialdi (eds.), *Data Analysis in Astronomy*, Plenum Press, New York, 363:378.

COSMOS, n.d., *COSMOS User Manual*, Royal Observatory Edinburgh.

Crotts, A.P.S., 1986, "The halo and disk populations of M31", *The Astronomical Journal*, **92**, 292:301.

Eichhorn, H. and Clary, W.G., 1974, "Least squares adjustment with relatively large observation errors, inaccurate initial approximations, or both", *Monthly Notices of the Royal Astronomical Society*, **166**, 425:432.

Fraser, R.D.B. and Suzuki, E., 1966, "Resolution of overlapping absorption bands by least squares procedures", *Analytical Chemistry*, **38**, 1770:1773.

Grosbøl, P., 1987, "Computational methods", in V. Di Gesù, L. Scarsi and P. Crane, *Selected Topics on Data Analysis in Astronomy*, World Scientific Publ. Co., Singapore, 57:85.

Herzog, A.D. and Illingworth, G., 1977, "The structure of globular clusters. I. Direct plate automated reduction techniques", *The Astrophysical Journal Supplement Series*, **33**, 55:67.

Irwin, M.J., 1985, "Automatic analysis of crowded fields", *Monthly Notices of the Royal Astronomical Society*, **214**, 575:604.

Jarvis, J.F. and Tyson, J.A., 1981, "FOCAS: faint object classification and analysis system", *The Astronomical Journal*, 86, 476:49

King, I.R., 1986, "Cluster photometry: present state of the art and future developments", in V. Di Gesù, L. Scarsi, P. Crane, J.H. Friedman and S. Levialdi (eds.), *Data Analysis in Astronomy II*, Plenum Press, New Yor 17:30.

Kron, R.G., 1980, "Photometry of a complete sample of faint galaxies", *The Astrophysical Journal Supplement Series*, **43**, 305:325.

Kruszewski, A., n.d., "Description of searching and classifying INVENTORY software", Report (20 pp.).

Kruszewski, A., 1985, No title, Report (12 pp.).

Kruszewski, A., 1987, "Object search and classification", *MIDAS Manual*, European Southern Observatory, Munich. Chapter 4.

Lauberts, A., 1988, description of suite of galaxy reduction programs (in preparation).

Lupton, R.H. and Gunn, J.E., 1986, "M13: main–sequence photometry and the mass function", *The Astronomical Journal*, **91**, 317:

MacGillivray, H.T., Martin, R., Pratt, N.M., Reddish, V.C., Seddon, H., Alexander, L.W.G., Walker, G.S. and Williams, P.R., 1976, "A method for the automatic separation of the images of galaxies and stars from measurements made with the COSMOS machines", *Monthly Notices of the Royal Astronomic Society*, **176**, 265:274.

MacGillivray, H.T. and Stobie, R.S., 1984, "New results with the COSMOS machine", *Vistas in Astronomy*, **27**, 433:475.

Malagnini, M.L., Pasian, F., Pucillo, M. and Santin, P., 1985, "FODS: a system for faint object detection and classification in astronomy", *Astronomy and Astrophysics*, **144**, 49:56.

MIDAS, 1988, *MIDAS Manual*, European Southern Observatory, Munich.

Moffat, A.F.J., 1969, "A theoretical investigation of focal stellar images in the photographic emulsion and application to photographic photometry", *Astronomy and Astrophysics*, **3**, 455:461.

Murtagh, F., 1988, "Image analysis problems in astronomy", in V. Di Gesù (ed.), *Image Analysis and Processing*, Proceedings, Cefalù, Sept. 1987, in press.

Oracle, 1981, "Groningen Reduction of Digitized Plates", P.C. van der Kruit and B.M.H.R. Wevers, manual.

Ortolani, S., 1986, "Performance tests of DAOPHOT/INVENTORY photometry programs in dense stellar fields", *The Messenger*, No. 43, 23:25.

Ortolani, S., 1987, "Stellar photometry with CCDs", in J.-P. Baluteau and S. D'Odorico (eds.), *The Optimization of the Use of CCD Detectors in Astronomy*, European Southern Observatory, Garching, 183:188.

Penny, A.J. and Dickens, R.J., 1986, "CCD photometry of the globular cluster NGC 6752", *Monthly Notices of the Royal Astronomical Society*, **220**, 845:867.

Rosa, M. and Baade, D., 1986, "Modelling Space Telescope observations", *The Messenger*, No. 45, 22:27.

STECF, 1987, "ST MODEL User's Guide", Release 87APR30, ST-ECF, Garching.

Stetson, P.B., 1984, *DAOPHOT User's Guide*, Dominion Astrophysical Observatory, British Columbia.

Stetson, P.B., 1987, "DAOPHOT: a computer program for crowded-field stellar photometry", *Publications of the Astronomical Society of the Pacific*, **99**, 191:222.

Stobie, R.S., 1980, "Application of moments to the analysis of panoramic astronomical photographs", *Applications of Digital Image Processing to Astronomy*, SPIE Vol. 264, The International Society for Optical Engineering, pp. 208:212.

Stobie, R.S., 1986, "The COSMOS image analyser", *Pattern Recognition Letters*, **4**, 317:324.

Tody, D., 1986, "The IRAF data reduction and analysis system", IRAF User Handbook, National Optical Astronomy Observatories (NOAO), Tucson.

Tyson, J.A., 1984, "Galaxy counts", in M. Capaccioli (ed.), *Astronomy with Schmidt Type Telescopes*, D. Reidel, Dordrecht, 489:498

Tyson, J.A. and Jarvis, J.F., 1979, "Evolution of galaxies: automated faint object counts to 24th magnitude", *The Astrophysical Journal*, **230**, L153:L156.

Valdes, F., 1982, "Resolution classifier", *Instrumentation in Astronomy IV*, SPIE Vol. 331, The International Society for Optical Engineering, 465:471.

West, R.M. and Kruszewski, A., 1981, "Distant clusters of galaxies in the southern hemisphere", *Irish Astronomical Journal*, **15**, 25:35.

# AUTOMATED CLASSIFICATION OF RESOLVED GALAXIES

Peter G. Ossorio [1] and Michael J. Kurtz [2]

(1) University of Colorado and Linguistic Research Institute
P.O. Box 1294, Boulder, CO 80301, U.S.A.
(2) Harvard-Smithsonian Center for Astrophysics
60 Garden Street, Cambridge, MA 02138, U.S.A.

## I. INTRODUCTION

The knowledge that galaxies, then undifferentiated from galactic nebulae, had different forms preceded the era of astronomical photography. Now morphological types establish a fundamental descriptive vocabulary for the study of galaxies and galactic systems (Hubble 1936, de Vaucouleurs 1959, Sandage 1975).

Galaxy types provide a unique, if not et fully understood, view of the local physical conditions, such as angular momentum, at the location of the galaxy. Types also provide insight into more global conditions, through the strong correlation between galaxy type and local galaxy density (Dressler 1980), the cause of this effect remains controversial (for a discussion see Geller 1988).

Galaxy types provide necessary information for the determination of galaxy distances independent of the Hubble constant, and thus for the determination of the non-cosmological velocity field, via the Faber-Jackson relation (Faber and Jackson 1976, Dressler et al. 1987) for ellipticals, and the Tully-Fisher relation for edge on spirals (Tully and Fisher 1977, Aaronson et al. 1986).

Perhaps the most important role for galaxy types will be as a third (along with colour and magnitude) independent parameter extracted from the resolved, faint imaging of future space missions,such as the WideField Planetary Camera on the Hubble Space Telescope. It may well be expected that the changing relation of morphology to colour and magnitude will be of crucial importance to the understanding and deconvolution of the effects of cosmology and evolution;

121

particularly in the absence of redshifts, which for these very faint objects will not be abundant for quite a long time.

Currently all galaxy types are obtained by visual inspection, generally by comparison with an array of standard galaxies observed for that purpose, or with an atlas, such as Sandage (1961). Although there have been some small experiments correlating automatically derived types with visual types (Takase et al. 1984, Thonnat 1985, Murtagh and Lauberts 1986) no automatically derived primary types have ever been published.

Many of the most important uses of galaxy types involve their statistical properties in aggregate. The spatial distribution of galaxies as a function of type has already proved very interesting (Dressler 1980), the differences in colour-magnitude diagrams as a function of type will certainly be of great interest. These kinds of measurements require the existence of large numbers of types, which can only be achieved if morphological types become a routine feature of standard data reduction, as position, magnitude, area, ellipticity, etc. are now standard features of automated galaxy photometry programs (for reviews see Kurtz 1983, 1987).

A number of promising techniques and parameters have been proposed, e.g. descriptive grammars (Balestreri et al. 1979), the luminosity concentration vs. mean surface brightness diagram (Takase et al. 1984), fits of the surface brightness profile to various analytic forms, such as Hubble or de Vaucouleur laws (Watanabe et al. 1982), use of Fourier shape descriptors (Pasian et al. 1986), Principal Component Analysis (Okamura 1985), shape matching (Murtagh and Lauberts 1986) etc. Thonnat (1985) has incorporated a number of these techniques in a rule based classification system; Thonnat's prototype system is perhaps the only working galaxy morphology engine even created.

Nothing approaching a working production system for automated galaxy classification has ever been attempted. A production system will need to be able to recognize and deconvolve overlapping images, deal with noise, and most importantly, discover new types of objects, as well as classify isolated galaxies. In what follows we outline the basics of a methodology which, given substantial effort, we believe can support a true production software system.

## II. A COMBINED APPROACH

The condition of this problem of automatic classification point toward some requirements for a practical solution.

(1.) The final classifications we wish to arrive at are, in general, available through human judgment; this is how it is currently done. They are not, in general, available through other means. This suggests that the procedure for automatic classification must incorporate human judgment. Further the human judgment component must itself be fully automatic; it cannot be a case of human intervention in an otherwise automatic process.

(2.) Presumably, a software package for automatic galaxy classification will operate on pixel level data. Human judgment cannot be expected to operate effectively on pixel level data. This limitation suggests that the procedure for automatic galaxy classification must incorporate one or more forms of analysis which do not involve human judgment.

(3.) The latter analyses must at some point produce results upon which human judgment can be effectively exercised.

Accordingly, in devising an implementation which satisfies these requirements, we have drawn upon several different methodologies. These include (a) syntactic pattern recognition, (b) judgment space methodology, and (c) redescription methodology.

A preview of our proposed solution is as follows.

(1.) The form of the solution is that of a redescription structure. The automatic classification process begins with the lowest level of data, which, for our purposes, is pixel level data. The facts represented by that data are redescribed and the result is a second level of descriptions and descriptors. These in turn generate third level descriptions, and so on. At the top level is a set of descriptions which are the desired object classification or imply them. Our current estimate is that there will be four or five levels.

(2.) Syntactic pattern recognition methods will be used to generate second level descriptions.

(3.) Human judgment methodology will generate redescriptions from the third level to the final classification.

The three component methodologies are described briefly below.

## III. THE THREE COMPONENTS

### A. Syntactic Pattern Recognition

At the base of any image understanding problem is a statistical analysis of groups of adjacent pixels. For isolated normal elliptical galaxies a set of statistical shape parameters contain all the information in the image. For galaxies with structure, or in crowded fields, sets of descriptors and relationships need to be established describing the lower level features, and relating them to each other, and to the global parameters, such as mean surface brightness, or the fit to a deVaucouleur law.

Balestreri, et al. (1979) have suggested a basic vocabulary of low level descriptors, Thonnat (1985) has developed a rule based grammar relating global statistical parameters. Our approach will combine these two basis technique.

We will develop a vocabulary of low level shapes, and sets of statistical measures for their description; we will use normal statistics and syntactic pattern recognition methods (e.g. Fu, 1982) to identify the subpatterns. We will develop a set of measures to relate the subpatterns with each other, such as separation, and to relate the subpatterns to the global object, via radial distance, azimuth, etc.

123

Above the level of basic shape description, where substantial development has occurred over the past 25 years, we expect to use the judgment space methodology, described below, to evaluate the relationships between the multiple measures. From the point of view of the pattern recognition, scene analysis, or image understanding fields the judgment space establishes a combinatorial grammar which implements a series of stochastic and fuzzy rules. From the point of view of a redescription structure these relationships provide mid-level redescriptions.

## B. Redescription

Paradigmatically, a redescription structure has the following background conditions:

(**a.**) There is a phenomenon which we encounter under a certain kind of description. Call this kind a "first level" description.

(**b.**) There is a second kind of description of the phenomenon which is the one we are primarily interested in. Call this kind of description an "end level" description.

(**c.**) If we start with a first level description, the appropriate end level description is not obvious, and we may not be able to generate it or recognize it at all. The problem is to bridge the gap between the first level and end level descriptions.

Under these conditions it is often possible to introduce a sequence of one or more mid level descriptions which have the following characteristics:

(**a.**) Given the first level description, the appropriate mid level description is easier to recognize or generate than is the end level description.

(**b.**) It is easier to recognize or generate the next level description (which may be the end level) if we start with the mid level description than if we start with the first level description.

In short, an intermediate description is easier to generate and makes the next description easier and more obvious. One of the most common applications of redescription structures is in addressing the problem of moving from "hard data" (e.g., measurements or observations) to meaningful or significant "interpretation" of the data. The problem of the automatic classification of resolved galaxies is recognizable as being of this kind. For a commonplace instance of a redescription structure, consider the heuristic example:

## Dinner at 8:30

Imagine that you are having a conversation with a friend of yours in the course of which he says, "You know, I got through at the office last night at 6:00 and got home at 6:30. We had dinner at 8:30, and it was steak, well done." At this point, you mentally shrug your shoulders and say, "So what else is new ? Half the people in this town could have said pretty much the same thing." However, he continues, "Yesterday morning I had a huge argument with my wife and we never got it settled. I usually do get home at 6:30, but we usually have dinner at 7:30, not 8:30. I like steak, but I like it rare and I hate it well done." At this point you smile. Now you can have a very different picture of what was going on at 8:30 last night.

When the "Dinner at 8:30" heuristic is presented to a class of college students, it is common for half the class to begin smiling as soon as they hear, "but we usually have dinner at 7:30, not 8:30." By the time we reach the end, eighty to ninety percent are smiling. When questioned, they unanimously describe the wife's behaviour as hostile.

Not that there is no inference here, since there is no deductive or statistical basis for one, nor is there even any appearance of an inferential process. Indeed, it is not necessarily true that wife's behaviour was hostile, but it is obvious. To be sure, it isn't obvious to everyone. For the person for whom it is not obvious, a redescription is in order.

For example, if you made the obvious comment, your friend might say, "What do you mean, hostile ? She didn't look or act angry." And then you might reply, "Look, you know she had reason to be angry at you, and then she made you wait an hour and then gave you something she knew you hated. If that isn't hostility, I don't know what is." Note that the reply has the characteristics of a mid level description. "She made you wait an hour and then gave you something she knew you hated" is easier to see and it is an easier step from that to "She was really acting out of anger." Because of this, the mid level description makes it possible for someone who did not initially see it to recognize that she was being hostile.

In general, redescriptions are produced by expanding the context and/or by introducing a more general or more meaningful conceptual framework within which to describe the given phenomenon.

## C. Judgment Space Methodology

Judgment space methodology was initially designed as a way of measuring connotative meaning (Osgood, 1957). It was soon revised and extended to the study of denotative meaning (Ossorio, 1961) and to judgments of relationships other than meaning (Ossorio, 1966). In the present context its importance is that it provides a way to incorporate human judgmental abilities into the routine functioning of a computer system.

The earliest application of this sort was the construction of a computer system for automatically indexing free text. This system serves as a prototype for other applications. Briefly, it involves the following. Given either a static body of documents to be indexed, the following procedures are involved:

(a.) Identify the range of subject matter represented by the total set of documents.

(b.) Represent that range as a set of subject matter field.

(c.) For each field construct a set of technical terms which are more or less strongly associated with that field.

(d.) Combine all of these terms into one set designated as the term list or system vocabulary.

(**e.**) For each field, each term on the term list is rated against that field by one or more persons who are knowledgeable in that field. The judgment made by the rater is "What is the degree of subject matter relevance of this term to this field".

(**f.**) The result of the rating procedure is a two dimensional matrix of terms rated against fields.

(**g.**) A standard option, designed to take account of redundancy among the fields, is to intercorrelate the fields and factor analyze the correlation matrix. The result is a factor space in which the fields are represented as vectors. This space is generally a judgment space and specifically a classification space.

(**h.**) By virtue of the ratings of terms against fields, each term can be represented by a set of coordinates which define a point in the N-dimensional judgment space. These coordinates may also be interpreted as a subject matter profile of the term on the reference axes of the space.

(**i.**) Since the terms in the system vocabulary are already indexed in the judgment space, a document can be indexed automatically by (a) scanning the document to determine which system vocabulary terms appear in the document and (b) giving the document a single location in the space as a function of the locations of the system vocabulary terms which appear in it.

(**j.**) In this system, standard retrieval is accomplished by a User specifying what he wants in ordinary language. (Systems of this sort are currently operating in English, French, and Turkish.) The request is indexed in the same way that other documents are indexed. Documents which are located close to the request in the classification space are those which have similar subject matter profiles and are therefore "about" the same thing as the request is. Documents are therefore retrieved in the order of their distance from the request. Pragmatically, this order corresponds to their degree of relevance to the request.

Obviously this general approach is applicable when relationships other than subject matter relevance are involved. For example, one can get judgment ratings on "the degree to which A is characterized by B" and the result will be an Attribute space. Or one can get judgment ratings on "the degree to which A is suitable as a means to B" and the result is a Means-Ends space. For galaxy classification, at least two types of judgment would seem to be relevant. The first is "the extent to which A indicates B." The second is "the degree to which A is compatible with B." A possible third is "the degree to which A is necessary for B." Note that indexing something in a judgment space is, in effect, redescribing it in terms of the dimensions of the space.

For example, in a Classification Space we begin by describing something as being a particular term (or document); when we index it in the C-space we are redescribing the same thing as being something which has a certain subject matter profile. For our present application the "terms" may consist of second level descriptors produced by syntactic pattern recognition methods. The "fields" would correspond to third level descriptors.

Indexing a galaxy in such a judgment space would amount to moving from a second level description to a third level description. Just as the location of a document in a subject matter space is a function of the locations of terms which appear in it, the location of a given galaxy image in a third level judgment space would be a function of the locations of the second level descriptions of that image in that space. In this way, we can proceed from one redescription level to another until we arrive at the level of desired classification.

## IV. RESEARCH PROGRAM

Given this understanding of the component methodologies, the following summarizes our proposed solution.

(1.) The overall logic of the solution is that of a redescription structure. In this structure the first level descriptions are the pixel level data and the end level descriptions are the desired morphological classifications.

(2.) Syntactic pattern recognition methods generate redescriptions starting from the first level descriptions.

(3.) Judgment space methods begin with descriptions generated by the syntactic pattern recognition operations and generate redescriptions at the next level. Those descriptions then generate further redescriptions, and so on.

Of course, it is easier said than done. Implementation of the solution described above will not be simple or easy. In particular, specifying the relevant descriptors at each level is partly intuitive, partly 'a priori', and partly an empirical task. Our strategy is to approach the task both from the first level and from the end level.

(1.) In approaching it from the first level, it is simply not clear in advance how far up the redescription ladder syntactic composition operations can take us. We propose to push this effort as far as it can go.

(2.) In approaching it from the end level, we begin with the final classifications and ask, what prior descriptions would enable a knowledgeable person to make a morphological classification. If one person looking at a photographic image generates the prior descriptions and a second person, using only these descriptions, can correctly classify the morphology at a satisfactory level of success, then we have empirical evidence that those lower level descriptors are an effective set. Then we take each of those and ask, what prior descriptions would enable a knowledgeable person to generate that description,... and so on.

These constructive efforts may lead to new types of analyses in order to generate the various descriptions at the various levels. As a result of these efforts we will have a set of redescriptions relationships implemented by means of judgment spaces and composition algorithms. This will constitute an operational classification system. Calibration of this system will be accomplished by using the measurements in a given existing catalog (e.g., the Lauberts catalog) and attempting to reproduce the morphological classification given in the catalog.

# REFERENCES

Aaronson, M., Bothun, G., Mould, J., Huchra, J., Schommer, R.A., and Cornell, M.E., 1986, Ap. J., 302:536.

Balestreri, M. Della Ventura, A., Fresta, G., and Mussio, P., 1979, in: "Image Processing in Astronomy", G. Sedmak et al. (Eds.), Trieste, Osservatorio Astronomico di Trieste, p. 268.

Dressler, A., 1980, Ap. J., 236:351.

Dressler, A., Lynden-Bell, D., Burstein, D., Davis, R.L., Faber, S.M., Terlevich, R.J., and Wegner, G., 1987, Ap. J., 313:42.

Fu, K.S., 1982, "Syntactic Pattern Recognition and Applications", Englewood Cliffs, Prentice Hall.

Faber, S.M., and Jackson, R.E., 1976, Ap. J., 204:668.

Geller, M.J., 1988, Saas Fee Lectures, to appear.

Hubble, E.P., 1936, "In the Realm of the Nebulae", Oxford University Press.

Kurtz, M.J., 1983, in: "Statistical Methods in Astronomy", E. Rolfe (Ed.), European Space Agency Special Publication, SP-201:47.

Kurtz, M.J., 1987, in: "Large Data Bases in Astronomy, Scientific Objectives and Methodological Approaches", F. Murtagh and A. Heck (Eds.), Garching bei Munchen, European Southern Observatory.

Lauberts, A., 1982, "The ESO-Uppsala Catalog of Galaxies", Garching, European Southern Observatory.

Murtagh, F., and Lauberts, A., 1986, Pattern Recognition Lett., 4:465.

Okamura, S., 1985, in: "ESO Workshop on the Virgo Cluster of Galaxies", O.G. Richter and B. Binggali (Eds.), ESO Conference Proc. 20:201.

Osgood, 1957, "Experimental Psychology".

Ossorio, P.G., 1961, Thesis, University of California at Los Angeles.

Ossorio, P.G., 1966, in "Multivariate Behavioral Research", I.

Pasian, F., Pucillo, M., Santin, P., 1986, Mem. Soc. Astron. Italia, 57:251.

Sandage, A., 1961, "The Hubble Atlas of the Galaxies", Washington, The Carnegie Institution of Washington.

Sandage, A., 1975, in: "Stars and Stellar Systems IX, Galaxies and the Universe", A. Sandage, M. Sandage, and J. Kristian (Eds.), Chicago, University of Chicago Press, 1.

Takase, T., Kodaira, K., and Okamura, S., 1984, "An Atlas of Selected Galaxies", Tokyo, University of Tokyo Press.

Thonnat, M., 1985, INRIA Rapport de Recherche, 387.

Tully, R.B., and Fisher, J.R., 1977, Astron. Astrop., 54:661.

de Vaucouleurs, G., 1959, in: "Handbuch der Physik", 53:275.

Watanabe, M., Kodaira, K., and Okamura, S., 1982, Ap. J. Supp., 50:1.

# STSDAS: THE SPACE TELESCOPE SCIENCE DATA ANALYSIS SYSTEM

Robert J. Hanisch

Space Telescope Science Institute[1]
Baltimore, Maryland 21218 USA

## INTRODUCTION

One of the many components of the software systems used to operate the Hubble Space Telescope (HST) are the applications packages for calibration and scientific data analysis. Although of great importance to the observer, data analysis is at the end of the line in this extensive system. Calibration is a more integral part of telescope operations, in that the STScI will provide all HST observers with nominally calibrated data. However, it is recognized that many observers will wish to verify or modify the calibration themselves, so that the calibration software is also an integral part of the data analysis system. STSDAS – the Space Telescope Science Data Analysis System – is the software tool kit that allows HST observers to calibrate, reduce, and analyze HST observations.

Modern astronomical research is rarely done using data from just one telescope or just one detector. New phenomena, new theories, and astronomical discoveries in general often require the comparison of data from different sources and in different wavelength regimes. Thus, a successful data analysis system must be able to accommodate alternate sources of data, and must have analysis routines both specific enough for use with the primary instrument and general enough for use with other astronomical data. Through its integration with the Image Reduction and Analysis Facility (IRAF), developed at the National Optical Astronomy Observatories in Tucson, and through its full support for data transport in the FITS format (Wells, Greisen, and Harten, 1981), STSDAS strives to meet this goal. Since data analysis algorithms

[1]The Space Telescope Science Institute is operated by the Association of Universities for Research in Astronomy, Inc., for the National Aeronautics and Space Administration

and techniques are constantly improving and expanding, the capability to change and extend the system must be inherent in the design.

Given the large number of observers expected to use the HST, it is not reasonable to provide complete data analysis support for everyone on site at the STScI. Clearly the data analysis system must be *portable*, *i.e.*, available for easy export to other sites and to a variety of computer systems (including systems not yet on the market!). Software portability can be achieved at two levels: individual applications can be written to port from one system to another, or an entire data analysis system can be written to port from one machine to another. It is the latter concept of portability that we have adopted for the STSDAS system. The former is adequate only for applications of rather limited scope, since at some level one has to depend upon such a wide variety of interfaces (tapes, graphics, image displays, *etc.*), that the application becomes bound up within a data analysis environment anyway. In layering STSDAS applications and subsystems on the IRAF Virtual Operating System, the goal of system-level portability is achieved.

## HISTORICAL PERSPECTIVE

The Science Data Analysis Software (SDAS) System was originally planned as a component of the Science Operations Ground System (SOGS), the software system used to plan and schedule HST observations, to support real-time operations, and to apply standard calibration procedures in order to remove the instrumental signature. When development of SDAS began in 1982, the plan was for data analysis applications programs to run under control of the SOGS command language with approximately one-half of a VAX 780 as the computing resource. The analysis routines were to be layered on the I/O subroutine library used by SOGS. SOGS was, and still is, a VAX VMS dependent software system. (The SOGS system was written by TRW, Inc. in California.)

At the time, of course, a VAX 780 was still the state-of-the-art in high end mini-computers, and our concerns about portable software were not fully appreciated by the project management. Early on in the project, STScI recognized that the planned architecture for SOGS was not going to be adequate as a data analysis environment. The design of SOGS constrained users to VAXes running VMS, and the command language was not very well suited for a scientific data analysis system, where flexibility and experimentation is to be encouraged.

### Initial Design and Implementation

As SDAS development got going, two major steps were taken to attempt to isolate it from the rest of the SOGS system and obtain portability, at least at the algorithmic level. First, independent emulations of the SOGS I/O interface routines were implemented, and SDAS applications were layered on these routines via a set of higher

130

level I/O interface routines. This, in fact, *had* to be done because the SOGS routines were not yet available; an interface control document (ICD) had been written describing them, but the actual delivery of the SOGS software and the supporting I/O subroutines did not occur until 1985, some three years after SDAS programming got underway. Second, the search for an alternate command language was initiated. While command languages were being studied (including the possibility of writing a new one), an interim command language based on DCL command procedures was used.

The SOGS-defined I/O subroutine library used in SDAS was a VAX VMS dependent library, making use of the VMS mapped section structures for image I/O. We knew that this introduced an unwanted vendor dependency in our code, but owing to the constraints of an imminent launch of HST (launch imminence has been a continuing presence!) and the continued requirement that SDAS applications be operable within the SOGS system, there was little choice in the matter. In order to minimize operating system dependencies in the data analysis programs, I/O functions were limited to the highest levels of the code. This simplified matters several years later when we ultimately came to convert our code into a more portable form.

## The Command Language

Our initial requirements for a data analysis command language led to early augmentations of the SOGS command language, later called COMET. However, a powerful command language was an essential component in our concept of the HST data analysis system, and we felt that it was important to review a number of existing and planned systems in order to find the command language best suited to our needs. The search for a command language, completed in early 1984, involved a detailed study of a number of systems, including AIPS, COMET, DOMAIN, FIPS, GIPSY, IDL, IRAF, MIDAS, STARLINK, TAE, and TVSYS.

It is extremely difficult, perhaps impossible, to perform a totally objective study of a large computer software system. What one person finds to be convenient another finds to be clumsy, and what one person finds to be simple another finds to be obscure and esoteric. Many "requirements" for the system may not be uniformly appreciated. For us, extensibility and potential longevity were very important considerations. IRAF emerged as the system of choice. Included in the reasoning leading to this decision was the fact that the IRAF system was based on the concept of a virtual operating system, upon which all utility functions, applications programs, and the command language itself were layered. The decision to use IRAF guaranteed that at least the SDAS command language would be portable.

Later in 1984 STScI undertook the first port of the IRAF Virtual Operating System (VOS) outside of the UNIX operating system and implemented the host dependent underpinnings of the VOS for the VAX VMS operating system. Although this first full port took approximately 1.5 man-years to complete, much of this effort was part

131

of the learning experience and due to the very different natures of UNIX and VMS. In doing the port a number of bugs in the VOS were flushed out and repaired, making IRAF a more robust system. Subsequent ports to other operating systems have taken as little as one week.

## Integration with the CL

The next major phase of SDAS development involved the modification of the existing applications programs to operate in the IRAF environment, with the IRAF command language (CL) used for task initiation and parameter setting. This was a fairly major effort, involving most of the programming staff for a period of several months. The result was a set of VMS dependent applications packages running under the control of a host independent command language. Again, the constraints of an imminent launch made it impractical to embark on a more radical revision of the software.

## THE PORTABILITY/PERFORMANCE PROBLEM

A number of problems with system performance and additional delays in the HST launch date, the last of which was caused by the Challenger accident, eventually led us to undertake the major systems revisions that we probably should have done much earlier in the project. We found two major causes of poor performance.

First, the system architecture was excessively modular. Modularity is desirable at the subroutine level, but when this modularity becomes visible at the user interface level it can make the system very clumsy to use. Owing to the various boundary conditions and the software development methodology that had been employed on the SDAS project, algorithm design and user interface design had never been properly decoupled. Thus, users saw a big tool kit with many atomic functions. The only way to make effective use of the system was through extensive use of the CL to construct data processing scripts. While scripts can be useful as exploratory tools, the large amount of file I/O required (each task would read data in from a disk file, do some processing, and write the result to a new file) made the overall performance poor.

Second, we found that VMS mapped sections were not a very efficient way to do image I/O under the typical system loading conditions we encountered. On unloaded machines the mapped section I/O performed about the same as the row-oriented I/O used in the IRAF system image I/O routines. But as system loading increased, mapped section I/O performance degraded much more rapidly than the row-oriented I/O. Since the vast majority of our applications process image data on a row-by-row basis, the advantages of mapped sections for pseudo-random access to image pixel values were lost, and the extra overhead required for routine use of mapped sections led to erratic and unacceptable throughput. Thus, not only did we have a VMS dependent set of applications programs, but the primary VMS dependency was contributing to the performance problems of the system.

With the time afforded us by the shuttle launch delays after the Challenger accident, we undertook a major revision of SDAS. The objectives were to remove our dependencies on VMS, improve performance, and make our applications software more fully integrated with the IRAF environment.

## STSDAS AND IRAF: AN INTEGRATED SYSTEM

### IRAF

An extensive review of the IRAF system architecture is beyond the scope of this paper (see Tody, 1986), but a few aspects of IRAF need to be explained in order for the scope of STSDAS/IRAF integration to be clear. IRAF is basically a three layer system. At the highest layer is the applications software, the majority of which is written in a preprocessed language (SPP). Beneath the applications code is the IRAF Virtual Operating System, which provides all basic I/O functions via a set of interface routines isolated from the actual host dependent system services that are called to implement them. The VOS includes interface libraries for file I/O, image I/O, graphics I/O, tape I/O, and vector operators. At the bottom level is the Host System Interface. All VOS routines are layered on the HSI, and the HSI is implemented for each target machine and operating system. Of a total of some 800,000 lines of code (including documentation) in the IRAF system, only about 3000 lines in the HSI are host system dependent. The rest of the code is totally independent of the host. To date IRAF has been ported to VAX UNIX (Ultrix), VAX VMS, various BSD 4.3 UNIX systems (Sun, Alliant, Convex), System V (Hewlett-Packard), Apollo, and Data General AOS. Ports of IRAF from one host to another with similar operating systems generally take about a week, with ports to very dissimilar operating systems requiring more like a month or two. But in these ports, *not one line of applications source code needs to be changed.*

### The Preprocessor and a Fortran Interface to the VOS

The use of a preprocessor for applications software has been controversial, and many people have argued for and against preprocessors for some years now. The rationale for the preprocessor in IRAF is straightforward: the most obvious programming language for scientific applications software is Fortran, but Fortran lacks a number of features of more modern languages that make it awkward for building large systems. On the other hand, Fortran compilers were and continue to be the most advanced compilers available on most computers. Thus, the SPP preprocessor provides a modern, structured programming language for software development, and produces standard Fortran (Fortran 66, as a matter of fact) for compilation on the host Fortran compiler. The mapping from the preprocessor language to the host compiler is not fixed, however, and could be modified to map onto Fortran 77, C, or some other language. The preprocessor allows for direct fall through of statements to the host compiler. However, code written totally in SPP is, by definition, portable with the IRAF system, since the preprocessor is itself part of the portable system.

133

After much deliberation, we decided at STScI to fully adopt the IRAF programming environment including, for the most part, use of SPP as a programming language. This provided the software development staff with full access to the functions of the IRAF VOS in the design and implementation of their applications. However, we recognized that many end users of the system would not have the time or wish to expend the effort required to learn the SPP language. Also, we had a significant amount of code, especially for HST calibration, that was written in Fortran (with VMS extensions) and which needed to be made portable. For these reasons we developed a Fortran interface to the IRAF VOS that provides most of the capabilities of the VOS. This allows for the full integration of Fortran programs into the IRAF system, both by professional programmers and by end users who wish to build custom applications packages. Programs developed using the Fortran/VOS interface (Fortran 77 is the standard language) are fully portable with IRAF as long as care is taken so that no vendor specific compiler extensions are used in production code. The Fortran/VOS interface is not intended for use in stand alone, host dependent programs; rather, it is designed as an alternative programming interface for use within the IRAF system.

Those familiar with IRAF may question why we wrote our own Fortran interface when the IMFORT programming interface already existed. IMFORT is designed to allow *host* Fortran programs to have access to IRAF image data structures. The ability to truly integrate such a program into the IRAF system is limited. There are no capabilities for doing graphics in IMFORT, and no access to SDAS format data files. In general we felt that IMFORT was too restrictive for what we wanted to accomplish.

## STSDAS

The birth of STSDAS was marked by two events: 1) the decision to convert SDAS applications code fully into the IRAF environment, and 2) the integration of both analysis and calibration software being developed at STScI into a common system. For historical reasons the development of analysis software and calibration software were independent activities. Yet, from the observers' point of view, both calibration and analysis software need to be part of the same system, with a consistent user interface and a natural flow of data from the calibration phase to the analysis phase. The name STSDAS was chosen because 'SDAS' already had some name recognition in the astronomical community, yet we wanted to indicate that all Space Telescope related data reduction could be done with this system.

## SOFTWARE DEVELOPMENT METHODOLOGIES

### The Traditional Approach

The original SDAS development strategy followed standard industry procedures, and included these basic steps: *(1) Requirements Analysis* – the software design team and the end users (in the case of SDAS, the members of the Investigation Definition Teams served in this role) meet to define the scope and content of the system. After some iteration, a final Requirements Specification Document is produced. *(2) Detailed Design* – the software design team turns the set of formal requirements into a detailed system design, with individual applications algorithms defined in a pseudo-programming language. No code is written at this time. The end-users meet with the design team periodically to review the formal design. *(3) Code and Unit Test* – programmers start work on turning the detailed design into functioning code. The programmers themselves test their code, and a system of internal peer review is used to verify that the code works as advertised and has been written in accordance with project programming standards. *(4) Independent Testing and System Integration* – an independent test team exercises the code and verifies that it functions in concert with the rest of the system. When problems are identified, the code is returned to the developer for revision, and an iteration loop starts between the Code/Unit Test step and Independent Testing. *(5) Acceptance Testing* – the end users are called back to verify that the code that has been written satisfies the original requirements and functions as they expect. The approved code is 'baselined' – moved to controlled directories where the developer is not allowed to modify it. *(6) Maintenance and Enhancements* – bugs discovered in routine use of the system, and additions identified as a result of changing requirements, are folded into the system in an incremental fashion. A Configuration Control Board must approve any changes to the baseline copy of the system.

While much of this methodology sounds reasonable, and is in fact reasonable for certain kinds of software development projects, it has serious failings when applied to scientific software development. First and foremost, scientists are rarely willing or able to define the requirements for a data analysis program completely enough for the subsequent design, coding, and testing steps to produce a product that really does what the scientist desires. Indeed, in some respects scientific data analysis software *cannot* meet a strict set of requirements – data analysis is dynamic. Observers are continuously defining new ways to look at their data, and require flexibility in the data analysis environment, even to the point of direct access to the code so that custom changes can be made. The methodology used in the initial SDAS development may well be appropriate for certain production line code or turn-key systems, but is inappropriate for data analysis software. The result of applying this methodology to scientific software development is the creation of a system which may have many functions, but which does not really meet the scientist's expectations. There is too little ongoing interaction with the scientists who will ultimately use the software, and as a result the system diverges from what the scientists had in mind when they wrote down their specifications. There is too long a lead time between requirements definition and delivery of the software, and too little attention paid to the design of a consistent user interface.

## Prototyping

A better approach for scientific software development makes use of rapid prototyping in the context of a well defined data analysis environment, with ongoing close collaboration between the programming staff and the scientific staff. In a sense this method is much more like a scientist developing his own software, except that by including a professional scientific programmer in the process one may have some confidence that the resulting code is robust, well documented, and well integrated with the rest of the data analysis environment. Furthermore, ongoing feedback from system users with timely software updates is essential for the development of a truly useful data analysis environment. Having completed a formal Acceptance Test for the basic SDAS system with NASA, we are now employing a prototyping software development methodology for most of our continuing scientific applications development. Even with prototyping, we still employ peer reviews and code monitoring to insure that software is developed to system wide standards, and releases of the system are kept under configuration control. Even though the procedures are less rigid, the quality of the resulting code is generally equal to or better than that produced by the traditional approach and, more importantly, it comes much closer to doing what the scientists want it to do.

## STSDAS APPLICATIONS PACKAGES

STSDAS, like IRAF, is structured as a set of applications software *packages*, each containing some number of specific, related *tasks*. In addition, STSDAS provides certain system wide capabilities, such as a table I/O system, that can be used by any applications program in the IRAF environment. The STSDAS calibration and analysis packages are listed below.

Calibration and instrument-specific packages:

> *cdbsutil:* Calibration Data Base access utilities. Queries to and updates of the Calibration Data Base. This package is one of two nonportable packages in STSDAS, as it links directly with the libraries needed to support data base I/O on the Britton-Lee data base machines utilized in the SOGS calibration pipeline processing.

> *fgs:* calibration and analysis tasks for the Fine Guidance Sensors.

> *focgeneral:* general calibration and analysis tasks for the Faint Object Camera.

> *focgeom:* geometric correction tasks for the Faint Object Camera.

> *focphot:* photometric reduction tasks for the Faint Object Camera.

> *fos:* calibration and analysis tasks specific to the Faint Object Spectrograph.

> *hrs:* calibration and analysis tasks for the High Resolution Spectrograph.

> *hsp:* calibration and analysis tasks for the High Speed Photometer.

> *rsdputil:* Routine Science Data Processing (RSDP) data base access utili-

ties. Along with *cdbsutil*, a nonportable package because of its direct links to the data base machine.

*wfpc:* calibration and analysis tasks for the Wide Field / Planetary Camera.

Analysis packages:

*astrometry:* analysis software for astrometry (plate solutions, precession, *etc.*)

*fourier:* general Fourier analysis (forward and reverse transforms, cross- and auto-correlation, convolution, editing).

*simulators:* HST instrument throughput simulators.

*spectra:* spectral analysis tasks, curve-of-growth analysis.

*statistics:* statistical analysis software, survival analysis (statistical tests on data sets with lower and upper limits).

*stlocal:* STScI local applications, experimental tasks.

*timeseries:* time series analysis tasks, period determination.

*tools:* generic data handling and utility tools.

*ttools:* table manipulation tools.

### Table I/O System

One of the major components of STSDAS is the table I/O system. Tables are data structures organized, as one might expect, in columns and rows. Columns may be of any IRAF data type (short and long integers, real, double precision, text strings, *etc.*), and rows may have any number of columns. Many packages in STSDAS use tables as a mechanism for transferring data from one task to another. All program access to the content of tables is via column names. This means that no specific information about the content of the table is necessary in order to transfer data throughout the system.

With their column/row structure, tables form a simple relational data base. In order to exploit this fact, a number of data base operators have been provided (merge, join, project, select, *etc.*). Operations between columns may be defined, with results stored in new columns in the table. An interactive table editor is available that limits text changes to the column fields.

### Graphics and Image Display

The IRAF graphics and image display facilities are used directly to support the STSDAS applications packages. In some areas, however, STSDAS has provided extensions to IRAF. For example, the standard vector plotting task has been extended to plot columns from STSDAS tables, and a Mongo-like graphics language that is fully layered on the IRAF graphics subroutine libraries has been implemented.

IRAF and STSDAS image display packages will be layered on the Image Display Interface (IDI) library (Terrett, Shames, and Hanisch, 1988). The IDI provides a device independent interface for a wide range of image display devices, from simple display buffers to high end image processing systems, and includes capabilities for computer workstations.

## PLANS AND PROSPECTS

### Toward a Truly Distributed System

Having largely achieved the desired level of portability in STSDAS, there is still work to do in realizing the full potential of a truly *distributed* software system. A distributed system utilizes high speed network connections to provide resources from a variety of sites, both local and remote, to all users.

High speed networks provide new potential for the use of remote data archives, file servers, compute servers, and image display, graphics, and hardcopy devices. One of the fundamental low level interfaces in the IRAF system is the network I/O interface. This is already in use for accessing files from other hosts on a network and for using remote image displays, printers, plotters, and tape drives. For example, it is now fairly common for IRAF users to use a Sun workstation for local image display and moderately heavy computations, while running more CPU intensive tasks on a large VAX connected via TCP/IP. The network interface also supports remote command execution, with the command output returned to the local host.

In STSDAS we intend to utilize networks to support both on-site and off-site access to the HST data archives and the Guide Star Catalog and Plate Scans. For example, a package in STSDAS will allow users to retrieve subsets of the Guide Star Catalog by posting a data base request on the network, having a server at STScI process the request, and having the server return the catalog subset to the remote user. Similarly, users will be able to retrieve subimages from the actual Guide Star Plate Scans by choosing the field center and image subset size. The Plate Scans are currently being written onto optical disks, so that retrieval of an arbitrary part of the sky should take no more than a few minutes. We hope to make access to the HST data archive and observation data base available via similar mechanisms.

Of course, the practicality of network data retrieval is governed by the availability of high speed, broad bandwidth network services to all data analysis centers. The STScI has been working with NASA to implement such a network to support the astronomical community. In the United States NASA and the NSF are working together to provide 56 kbit network connections to all major astronomy research centers, with 9.6 kbit feeder lines to smaller sites. For image data transfers to be feasible, however, these bandwidths need to increase by roughly an order of magnitude. Although this may not happen for several years, in the interim we can still provide useful remote access to large databases and image archives.

The compute server architecture can also be exploited more fully in IRAF/ STS-DAS. We envision a configuration of workstation clusters in which each cluster would contain of order 6-10 workstations connected with a local strand of Ethernet to a common file server, with some number of the workstations having advanced image display capabilities and local disk storage. These clusters would be tied to a backbone network, giving each workstation access to a central machine and a supercomputer or mini-supercomputer for the execution of CPU intensive data processing tasks. Workstation users would generally be unaware that certain of the tasks they initiate would actually run on the remote supercomputer.

## STScI/NOAO Collaboration, and Other Software for the IRAF Environment

An important aspect of IRAF/STSDAS system and applications development is the ongoing level of cooperation and collaboration between the STScI and NOAO. The software development groups at each site are in constant contact with each other via electronic mail and remote logins. There is continual feedback between the sites on system-level problems and possible enhancements, and we make every attempt to coordinate applications software development in order to avoid redundant efforts. Projects are generally assigned to the group with the most experience and local expertise. The net result is a system which benefits from a broader perspective, and which can take advantage of the increased level of resources made available by coordinating our efforts rather than competing with each other. We try to avoid the introduction of site dependent subsystems in order to maintain the integrity and interoperability of all code, whether written at STScI or NOAO.

A number of other astronomical software development projects are either considering adopting the IRAF environment or have already done so. For example, the Dominion Astrophysical Observatory in Victoria, British Columbia, is rewriting their DAOPHOT package to be integrated into IRAF. DAOPHOT will make use of the IRAF image display system and the STSDAS table I/O system, thus providing for ease of use and immediate availability of its results to the generic table manipulation routines available in the system. Similarly, the ROSAT PROS software under development at the Center for Astrophysics is being designed for the IRAF system. A new data structure tailored for X-ray detectors, a 'photon-oriented event' (POE) file, is being implemented within IRAF to support the PROS package. The POE data format will also nicely support radio visibility data sets.

The STSDAS and IRAF programming groups each have long lists of planned applications development projects. We can expect the IRAF/STSDAS system to grow in richness and breadth and, owing to the high level of extensibility and applications and system portability, we can expect it to be a viable data analysis environment for many years to come.

## ACKNOWLEDGEMENTS

The author would like to acknowledge the encouragement and assistance of the previous SDAS Project Scientists, Rudi Albrecht and Walter Jaffe, and the continuing support of Ethan Schreier, the Chief Data and Operations Scientist at STScI. In developing many of the ideas expressed in this paper the author has benefited from numerous conversations with Doug Tody, Peter Shames, and Don Wells.

## REFERENCES

Terrett, D. L., Shames, P. M. B., and Hanisch, R. J., 1988. "An Image Display Interface for Astronomical Image Processing." *Astron. Astrophys. Suppl.*, in press.

Tody, D., 1986. "The IRAF Data Reduction and Analysis System." In *Instrumentation in Astronomy* **VI**, *SPIE* **627**, Part 2.

Wells, D. C., Greisen, E. W., and Harten, R. H., 1981. "FITS: A Flexible Image Transport System." *Astron. Astrophys. Suppl.*, **44**, 363.

# IRAS CALIBRATION

D.J.M. Kester, TJ.R. Bontekoe, A.R.W. de Jonge, and

P.R. Wesselius

Space Research Department and
Kapteyn Astronomical Institute
P.O. Box 800, 9700 AV Groningen, The Netherlands

## 1 Introduction

The Groningen Exportable Infrared Survey High-resolution Analysis (GEISHA) system consists of a collection of software tools which allow an astronomer to create infrared sky maps from raw IRAS data. The Infra Red Astronomical Satellite (IRAS) observed about 95% of the sky by a semi-overlapping scan technique. There are 62 detectors in the focal plane in 4 wavelength bands at 12, 25, 60 and 100 µm. (See IRAS Explanatory Supplement, 1985, IRAS-ES hereafter).

The database of IRAS has been left unaltered apart from a re-sort. The underlying idea is that in the process of interpreting a certain astronomical object new questions may arise, which require a different processing of the data. Starting from the raw data has already proved to be a fruitful strategy.

One of the central items in the project is the calibration. The procedure for the IRAS raw data calibration consists of flatfielding. For each detector scan a set of calibration constants is extracted by fitting the lower envelopes of the detectors of all scans to each other.

## 2 IRAS Data Access

The basic material for the GEISHA project is the raw IRAS data, containing all preprocessed telemetry data from the survey instrument detectors, and the pointing information. The database of IRAS contains about 6000 scans running from one ecliptic pole to the other at a constant angle to the sun, the solar aspect angle. They were taken at an angular speed of 3.75'/sec and at a rate of 16, 16, 8 and 4 times per second, for the band at 12, 25, 60 and 100 µm, respectively. The scan strategy was organized in such a way that all sources were seen at least 8 times by a detector, yielding a data volume of more than 1000 tapes. To optimise the GEISHA

141

system for detailed investigation of (not too large) sky regions both the detector data for the scans and all auxiliary data necessary for pointing and flux calibration were reorganised into a system of 363 plates, which cover the whole sky. A plate is here defined as a coherent region of the sky, large enough to fill one magnetic tape in FITS format (Wells et al., 1981), which then forms a totally self-contained dataset.

Currently, we have the detector data distributed over the plate tapes, but not yet the auxiliary data. Auxiliary data and software are available for a flux calibration method based upon the zodiacal light, positional calibration and imaging by co-addition. The positional calibration was developed at the IRAS Processing and Analysis Center (IPAC) of Caltech, Pasadena, USA. (See IRAS-ES)

## 3   Calibration

An unavoidable part of the brightness which falls on the detectors is the zodiacal emission. It originates from dust inside the solar system and provides a broad background (or better foreground) feature with a FWHM of $\approx 50°$. It is dominant at 12, 25, and 60 $\mu$m, and somewhat less at 100 $\mu$m. Although it is a nuisance for most astronomers, it can be used as a standard candle for calibration purposes. In the same process a zodiacal emission model (ZEM) can be established.

The GEISHA system will supply the astronomer with a choice of calibration routines and zodiacal emission models (ZEMs) to subtract if desired. All calibrations have a common structure: it is assumed that the raw detector data consist of a detector bias, plus the 'true' infrared brightness multiplied by a gain factor $g$. Both the bias and the gain might vary slowly due to photon and particle induced memory effects, instrumental degradation etc. A whole range of possible ZEMs can be developed, from a simple lower-envelope cubic spline fit to a full-fledged physical model with densities, emissivities, and temperatures as a function of position in the zodiacal dust cloud. Here again the choice is to the astronomer.

The total IRAS database is clearly too large to handle in one program. The first step in the reduction of the database was a cubic spline fit, which brought the sample rate down to one datapoint per second (or 3.75 arcmin). The splines are part of the GEISHA system; they can be used for the study of large scale structures. In total it is still 1 Gbyte.

The second step which was only taken for calibration purposes, involved a lower envelope fit over one degree to the splines. The lower envelope fit entailed that all single point sources dropped out; we are mainly interested in the large scale structure of the zodiacal emission. This last database contains 70 Mbyte or 18 Mbyte in each of the 4 bands.

## 4   Linear Model

In the first GEISHA calibration a linearly changing bias plus a constant gain was assumed for each detector scan.

$$O = a + bt + g(\text{ZEM} + \text{SKY}), \tag{1}$$

where $O$ is the detector output, $a + bt$ is the linearly changing bias (the baseline), $g$ is the gain factor, ZEM is the zodiacal emission and SKY is the astronomical sky. As the astronomical sky is quite empty in the infrared, the SKY term will be ignored in the linear model. The zodiacal emission is a function of the position of the earth, and the pointing direction of the satellite, given by the solar aspect angle $\theta$, and the angle along the scan direction, $\psi$. The angle $\theta$ is constant during a scan. In a first approximation the annual motion of the earth through the somewhat tilted dustcloud (Hauser et al., 1984) can be corrected with a small shift in $\psi$.

For all positions on a scan the brightness of the zodiacal emission increases with decreasing solar aspect angle: closer to the sun the zodiacal emission becomes brighter. It is assumed that the ZEM can be written as the product of two functions:

$$\text{ZEM} = f(\theta)\,\text{ZEM}_{90}(\psi), \tag{2}$$

where $\text{ZEM}_{90}(\psi)$ is the zodiacal emission model at a solar aspect angle $\theta = 90°$, and $f(\theta)$ is a scaling function. Both functions are as yet unknown and will be estimated from the data by an iterative procedure.

Knowing the n-th iterate of the function $\text{ZEM}_{90}(\psi)_n$ we observe that for each individual detector scan the output can be written as a linear equation in the 3 parameters which have to be estimated: $a_{n+1}$, $b_{n+1}$ and the combination of $(g\,f(\theta))_{n+1}$.

$$O = a_{n+1} + b_{n+1}\,t + (g\,f(\theta))_{n+1}\,\text{ZEM}_{90}(\psi)_n. \tag{3}$$

All thus calibrated detector scans are averaged into the next iterate of $\text{ZEM}_{90}(\psi)_{n+1}$.

$$\text{ZEM}_{90}(\psi)_{n+1} = \frac{O - a_{n+1} - b_{n+1}\,t}{(g\,f(\theta))_{n+1}}. \tag{4}$$

In both fitting procedures we are only interested in the common lower envelope. All datapoints which stray from this lower envelope we consider as spurious. They are eliminated with an object function which is more robust than the normal least squared method. We took a one-sided biweight function of Tukey (Goodal, 1983).

When it is sufficiently converged, typically after about 10 iterations, the factor $g\,f(\theta)$ is separated for each detector into a part $f(\theta)$ which comprises the common functional behaviour with $\theta$, and a remaining part. The former is multiplied with $\text{ZEM}_{90}$ and represents the ZEM; the latter is the gain for each detector scan.

This calibration is operational since begin 1987.

## 5   Detector Behaviour

IR photon detectors of the kind IRAS was supplied with are known to show particle and photon induced memory effects (Kester, 1986). A full physical model of these effects still eludes all efforts (Westervelt and Teitsworth, 1985). The model here presented only concerns photon induced memory effects on the gain factor. The particle induced effects due to the passage through the South Atlantic Anomaly seem to influence the bias mainly (See IRAS-ES). We already allow for a linearly changing bias and hope that that will suffice.

The non-linear behaviour of the gain will be corrected according to a two energy level model for IR photon detectors. The model can be described as two coupled first order differential equations.

$$\sigma \frac{dh}{dt} + h = A F + h_{min}, \tag{5}$$

and

$$\tau \frac{dg}{dt} + g = B F + g_{min}. \tag{6}$$

The equations are coupled by

$$\tau = \frac{\tau_0}{h}. \tag{7}$$

Here, $g(t)$ is a time dependent gain which couples the detector output $O$ to the total incident radiation $F$ according to equation (1). Further, $h(t)$ is a function which is inversely proportional to the decay time $\tau(t)$ of the gain. It acts as a saturation on the gain at large input fluxes. The decay time of the saturation function, $h(t)$, itself is $\sigma$. The constants $\sigma$, $\tau_0$, $A$, $B$, $h_{min}$ and $g_{min}$ are parameters to be estimated: one of them can be chosen freely. We take $h_{min} = 1$. The fact that $h$ is always greater than or equal to 1 entails that $g$ changes on time scales equal or smaller than $\tau_0$.

To estimate the parameters of the gain model the two differential equations must be rewritten as difference equations. (Ljung, 1987).

$$h_k = (1 - p)(AF_{k-1} + 1) + p\,h_{k-1}, \tag{8}$$

$$g_k = (1 - q)(BF_{k-1} + g_{min}) + q\,g_{k-1}, \tag{9}$$

where

$$p = \exp(-T/\sigma), \tag{10}$$

$$q = \exp(-Th_{k-1}/\tau_0). \tag{11}$$

$T$ is the sample time interval. From these equations an estimator of the momentary gain, $g_k$, is found in terms of the previous $h_{k-1}$ and $g_{k-1}$, and in terms of the previous input fluxes $F_{k-1}$ and $F_{k-2}$. Via equation (1) an estimator of $O_k$ is found which can be compared with the actual detector scan values. The difference is to be minimized under a robust objective function.

# 6    Non-linear Model

For most of the scans the linear scheme proved to be reasonably successful. But there are broken scans, running only part of a semicircle, for which the baseline and the gain can not be separated any more. This sometimes results in negative gains or other kinds of bad fits, particularly when the short scan covers some remaining galactic emission. Moreover for scans taken at extreme solar aspect angles ($\theta$ less than 75° or more than 105°), the assumption of separability of ZEM, as stated in equation (2), does not hold.

Still a zodiacal emission model was found of which the $ZEM_{90}(\psi)$ part could be very precisely approximated by an ellipse. This led to the conjecture that the zodiacal emission as a whole could be approximated by the emission of a tilted oblate ellipsoidal dust cloud of uniform density and temperature. The ZEM is a function of solar aspect angle $\theta$, the longitude of the sun-referenced coordinate system $\psi$, and the position of the earth represented by the solar elongation $\lambda_\odot$. This idea is supported by Reach and Heiles (1987).

An ellipsoidal model can never fit the so-called sidebands of the zodiacal emission (Hauser et al., 1984). The sidebands have been identified with 3 asteroid families (Dermott et al., 1984). From considerations of celestial mechanics (Hirayama, 1918 and Brouwer, 1951) it seems reasonable to model them with a number of inclined tori, fanned over all nodal angles. A side band model (SBM) incorporating these ideas is still under study.

Apart from the zodiacal emission there is always a contribution of astronomical background (SKY); most of it is galactic emission. The astronomical background is so granulated that only a map of the whole sky can function as a model. A fairly coarse map in an equal area projection with pixels of 1° by 1° will suffice. Each pixel is the mean of all overlapping detector values. As a standard candle the sum of ZEM, SBM and SKY can be taken, while $g(t)$ follows the detector model expounded in the section 5.

The complete model,

$$O = a + b\,t + g(t)(ZEM(\psi, \theta, \lambda_\odot) + SBM(\psi, \theta, \lambda_\odot) + SKY(\lambda, \beta)), \qquad (12)$$

contains as free parameters: 5 for ZEM, 6 for SBM, 40000 pixels for SKY, 5 for each scan ( 2 baseline and 3 starting conditions ), and 5 for each detector ( 2 decaytimes, 2 increments and a minimum gain ). In total about 70000 parameters have to be estimated, still a tiny fraction of the complete database of 13 Gbyte.

The non-linear model asks for a more complicated iterative scheme. Each iteration requires three passes over the data: one to determine SKY, one for ZEM + SBM and one for the baseline and the gain. In each pass the items which are not fitted, attain the values derived in the previous iteration. SKY is found by coadding the detector scans, properly scaled with baseline and gain and stripped of the contribution of ZEM + SBM, into an equal area map of the whole sky. ZEM and SBM are non-linear functions of solar elongation, sun-referenced longitude and latitude. An elegant way to fit non-linear functions is the Levenberg-Marquardt method (e.g. Press et al. 1986).

The derivation of the calibration parameters falls into two parts. For each detector the values of the gain model have to be estimated in the way that was outlined in the section 5. And for each detector scan the baseline has to be estimated plus the fictitious values $h_0$, $g_0$ and $F_0$, which represent the state of the system at the beginning of the detector scan.

Most of the steps outlined above need iterations in themselves. E.g. a non-linear regression model can only be solved in an iterative way. Nontheless it is hoped that one grand iteration loop, in which each set of parameters is estimated only once during an iteration loop will eventually converge. The choice of starting values for the parameters may be quite important.

This calibration and the pertaining ZEM will be available by middle 1988.

# 7  Further Tools

For routine inspection of the data a set of programs is available, applying position and/or flux calibration to the raw detector outputs, with some variation in algorithm used. The end products of these programs are

- a file ("sample file") containing the sampled output of all IRAS detectors covering the region of interest during survey scans, each sample labeled with time, position and orientation of the detector, and errors in these quantities;

- moderate resolution images, formed by averaging of the overlapping samples on an almost arbitrary pixel grid;

- high resolution images obtained by a linear least squares reconstruction technique exploiting the overlap between the samples, and the redundancy in the coverage of the sky.

All intermediate and end products of these programs are in FITS format to fulfill our export requirements. The subroutines used in these standard programs are available as a software library for special research projects.

# 8  References

Brouwer, D. (1951), *Astron. J.* **56**, p 9.

Dermott, S.F., Nicholson, P.D., Burns, J.A., Houck, J.R. (1984), *Nature* **312**, p. 505.

Goodal, C. (1983), in *Understanding Robust and Exploratory Data Analysis*, Hoaglin, D.C., Mosteller, F., Tukey, J.W. Eds. p. 339. John Wiley  Sons, New York.

Hauser, M.G., Gillet, F.C., Low, F.J., Gautier, T.N., Beichman, C.A., Neugebauer, G., Aumann, H.H., Baud, B., Boggess, N., Emerson, J.P., Houck, J.R., Soifer, B.T., Walker, R.G. (1984), *Astrophys. J. Lett.* **278**, p. L15.

Hirayama, K. (1918), *Astron. J.* **31**, p. 185.

IRAS Catalogs and Atlases, Explanatory Supplement 1985, Beichman, C.A., Neugebauer, N., Habing, H.J., Clegg, P.E., Chester, T.J. Eds. U.S. Government Printing Office, Washington D.C.

Kester, D.J.M. (1986), *Geisha Note: 86/05/07*, Space Research, Groningen.

Ljung, L. (1987), *System Identification, Theory for the User*, p. 90. Prentice-Hall, Englewood Cliffs.

Press, W.H., Flannery, B.P., Teukolsky, S.A., Vetterling, W.T. (1986), *Numerical Recipes*, p. 523, Cambridge University Press, Cambridge.

Reach, W. and Heiles, C. (1987), *Proceedings of the 3rd IRAS Conference "Comets to Cosmology"*, (to be published).

Wells, D.C., Greisen, E.W., Harten, R.H. (1981), *Astron. Astrophys. Suppl. Ser.* **44**, p. 363.

Westervelt, R.M. and Teitsworth, S.W. (1985), *J. Appl. Phys.* **57**, p. 5457.

# SIMULATION OF LARGE-SCALE ASTRONOMICAL CCD SURVEYS

Emilio E. Falco and Michael J. Kurtz

Harvard-Smithsonian Center for Astrophysics
Cambridge, MA 02138, U.S.A.

## ABSTRACT

We present the first stage of a project to develop a computer model of a CCD camera system dedicated to galaxy surveys. We plan to improve old and develop new algorithms for reduction and analysis of digital astronomical images. We intend to apply our model to testing theories and observations of the large-scale distribution of galaxies, and data reduction methods. We present here the results of the initial, testing phase of our project, which consists of the development of a benchmark for the main software packages that are commonly used to reduce and analyze astronomical observations. Our initial tests will measure the efficiency and accuracy of such software.

## I. INTRODUCTION

The techniques for reducing observations in a scale as large as is now becoming possible, both from the ground and from space, must be redefined and updated. The classical methods of reduction are too inefficient, and must be automated to perform detection and classification of very faint and blended objects in crowded fields. Our ability to interpret future data that will augment surveys such as the Geller et al. survey will depend more and more on the efficiency and accuracy with which we can reduce, literally, astronomical amounts of data. We intend to attempt to fill what we consider to be a gap between current, available, capabilities, and future, indispensable capabilities for reduction of observations of galaxies.

We have begun by testing algorithms for computer-automated reduction and analysis of large-scale, digital (e.g. CCD) sky surveys. We plan to analyze well-known software packages, such as FOCAS (Jarvis and Tyson, 1979; Valdes, 1982), MEMSYS II (Gull and Skilling, 1987) and INVENTORY (Kruszewski, 1987).

149

We report here on the testing phase of software we are developing to simulate deep CCD observations. We have developed two main computer programs, GENCCD, which generates a distribution of galaxies that would be visible within a prescribed patch of sky, and SIMCCD, which calculates a simulated CCD frame that contains the galaxies generated by GENCCD. We started with a naive cosmological model (Euclidean), with naive galaxy profiles (De Vaucouleurs, 1942) and with a simplified luminosity function (Soneira and Peebles, 1977; hereafter SP). The effect of the telescope optics is modeled by convolution with an appropriate point-spread function (PSF), and the effect of the CCD camera is modeled by the addition of stochastic noise (Gaussian relative to a variable bias, and Poisson shot-noise). Many effects have been neglected for the sake of initial simplicity, but they will be added as we continue to build our model. The development of our software is being carried out on a MicroVAX II; the code is all Fortran 77, vectorized as far as possible, to run on a Convex C2 computer. We describe the cosmological model we have chosen in § II. We present the galaxy profiles we have incorporated in our model in § III. In § IV, we describe the method we use to take into account the telescope optics. We present in § V the noise properties of images generated by our model. We conclude in § VI, with a description of a typical frame of simulated data obtained with our model, and with a discussion of the future stages of our project.

## II. COSMOLOGY

Galaxies are sprinkled in the ball-and-stick fashion of the SP model. The distribution of galaxies is parametrized by the limiting apparent magnitude M of visible galaxies, by the diameter D of a typical cluster of galaxies, and by the number $N_G$ of galaxies desired in the model universe. The latter two numbers are selected to represent the average observed properties of the distribution of galaxies.

The algorithm calls for distributing uniformly at random $N_S$ spheres within a spherical Euclidean volume with radius equal to the maximum distance $R_M$ at which a galaxy with apparent magnitude M would remain visible, given the luminosity function of SP. Within each of the $N_S$ spheres, a direction is chosen at random. A line, or stick, is drawn parallel to that direction, through the center of the sphere. At each end of the stick, new sticks are drawn in random directions, with their midpoints at the end of the previous stick. The lengths of successive sticks are prescribed, monotonically decreasing fractions of D (SP). One galaxy is placed at each end of each new stick; the absolute magnitude of each galaxy is selected at random from the luminosity function of SP. The process continues through five levels of sticks. The construction method builds in the desired properties of the two- and three-point correlation function (SP).

The program GENCCD is given a window of size $W_{RA}$ in right ascension and $W_{DEC}$ in declination, and a limiting magnitude M.Its output is a list of galaxies brighter than the limiting magnitude, within the input window. The output list contains the right ascension, the declination, the distance in Mpc, the apparent magnitude and the absolute magnitude of a collection of galaxies that GENCCD finds within the window.

The cosmological model is Euclidean, and therefore unrealistic. We plan to add non-Euclidean cosmologies in future versions of our software. Thus, we expect to be able to simulate more realistic observations than we have so far.

150

## III. GALAXY PROFILES

The galaxies in our model are simply single-band, monotonically-decreasing surface brightness distributions with De Vaucouleurs (1955), $r^{-1/4}$ profiles, and with elliptical symmetry. The axial ratio and position angle of each profile are selected uniformly at random within respective bounds of {-90° ... +90°} and {0.05 ... 1}. The effective radius (or half-mass major radius) of each profile is selected at random, from a Gaussian distribution with mean equal to 5 kpc, and with one-sigma equal to 1 kpc.

The galaxy profiles are far from realistic, but simple enough for the first tests of our model. In fact, a rather simple shape is more suitable for testing than a real galaxy image. In future versions, the model will include an improved luminosity function, with a morphological dependence, and we expect to test luminosity functions. Galaxy profiles of different morphological type, with realistic spectral properties, and derived from observations, will be used to provide a better approximation to the observed distribution of galaxies.

## IV. TELESCOPE OPTICS

The telescope optics are taken into account by a convolution of the galaxy profiles with a Gaussian PSF,

$$p(x,y) = \frac{e^{\frac{x^2}{2\sigma_x^2}} \; e^{\frac{y^2}{2\sigma_y^2}}}{\sqrt{2\pi}\,\sigma_x \; \sqrt{2\pi}\,\sigma_y} \tag{1}$$

where (x,y) are Cartesian coordinates on the plane of the sky, within the field of view, and the widths $\sigma_x$ and $\sigma_y$ of the PSF are part of the user input. The convolution is implemented by Fourier-transforming the image and the PSF with FFT's, multiplying element by element the matrices of Fourier transforms, and transforming the result back to spatial coordinates. Such an algorithm is ideal for operation with a vector processor.

The PSF model is also limited, because it is far too idealized. Future versions of the model will allow the possibility of using a PSF determined from real CCD observations. We will also allow for other functional forms of the PSF, and for a wavelength dependence of the PSF shape. Other effects of telescope optics on CCD data, such as misalignment of the optics, aberrations, flat-fielding, and guidance problems, will also be included in future versions of our model.

## V. STOCHASTIC NOISE

Artificial noise is added to each frame generated with our model. Two types of stochastic noise that appear in CCD detectors are included: Gaussian and Poisson noise.

Gaussian noise is added to each frame after convolution with the PSF. The mean G and full-width half-maximum (FWHM) $G_{FWHM}$ of this type of noise are part of the user input. G represents the average noise level in a dark CCD frame, and $G_{FWHM}$ the fluctuations away from G. The noise units are electrons. Thus, a conversion factor $E_C$ in electrons per count is necessarily included with the user input. The Gaussian noise for the i-th pixel, $G_i$, is generated with a "portable" generator of Gaussian deviates (Press et al., 1986), and added to the number of signal counts $S_i$ due to the presence of a galaxy that overlaps that pixel.

Poisson noise is added sequentially after the Gaussian noise. The mean of the Poisson noise for the i-th pixel when $S_i > G_i$ is

$$P_i = \sqrt{(S_i - G_i) E_C} \qquad (2)$$

The Poisson noise is also generated with a "portable" generator of deviates with mean $P_i$ (Press et al., 1986), and added to $S_i$ and $G_i$.

The noise model neglects all non-stochastic sources of noise, such as the possible effects of flat-fielding, and physical defects on individual chips that produce dark and bright trenches, and dark and bright columns. These problems vary from chip to chip, and depend on the quality of the manufacturing process, which is beyond the astronomer's control. At this stage of our project, we have chosen to postpone their analysis until our model has reached higher sophistication. Cosmic rays and stars are always present in CCD frames, but were also left out. All of these noise contributions will be incorporated in future versions of our model.

## VI. CONCLUSIONS

Kurtz and Falco (1988) discuss in detail the results of applying the data reduction package FOCAS to the first frames we have generated with our model. We now describe the parameters of these first frames, and discuss their visual appearance.

The accompanying Table shows the list of input parameters for our initial frames. We selected a large, 1°x1° window size, and a limiting apparent magnitude equal to 23 (compared to M≈19 in SP). These parameters resulted in marginally acceptable short run times (of the order of two CPU days each); we have made no significant effort yet to optimize the vectorization of our code for production runs. The flux in each pixel of each frame was calculated from the following input values: the exposure time $T_{exp}$ in seconds, and the (arbitrary for now) number of counts per second $C_{10}$ for a star with absolute magnitude equal to ten. The PSF we adopted was a Gaussian as in Eq. 1, with $\sigma_x = \sigma_y = 2.5"$.

We have generated two frames of simulated data from two adjacent fields centered, respectively, on ($8^h32^m0^s,50°30'0"$) and ($8^h36^m0^s,51°30'0"$), with parameters as in Table 1. We ran the same fields through the model after decreasing G by a factor of two, to begin analyzing the properties of our simulations as a function of signal-to-noise ratio. The generated frames are visually satisfying, and differ from real CCD data mainly due to the absence of non-stochastic noise in our simulations. A quantitative analysis is presented in Kurtz and Falco (1988).

Once we have tested various reduction packages with our model, we plan to publish a comparison chart for them. We will then proceed to build upon the first version of the software, to include the modifications mentioned above. The different parts of the model will be modified in the order in which they were given here, starting with the cosmological model.

The possible useful applications of our model include testing of luminosity functions, testing cosmological theories, testing new designs of CCDs and of reduction software, and predicting statistical results for observational programs. We hope to be able to report results from such tests in the near future.

We thank J. Geary and S. Kent for useful discussions of CCD detectors and observations.

## Table 1

### Input Parameters for SIMCCD

| Parameter | Value | Unit |
| --- | --- | --- |
| $D$ | 11.35 | Mpc |
| $N_G$ | 2400000 | galaxies |
| $M$ | 23 | magnitudes |
| $N_x$ | 512 | pixels |
| $N_y$ | 512 | pixels |
| $W_{RA}$ | 3600 | arcsec |
| $W_{DEC}$ | 3600 | arcsec |
| $C_{10}$ | 12000 | counts |
| $T_{exp}$ | 7200 | seconds |
| $G$ | 5600 | $e^-$ |
| $G_{FWHM}$ | 2800 | $e^-$ |
| $\sigma_x$ | 2.5 | arcsec |
| $\sigma_y$ | 2.5 | arcsec |
| $E_c$ | 28 | $e^-$/count |

## REFERENCES

De Vaucouleurs, G., 1942, Annales d'Astrophysique, 11:247.
Geller, M.J., et al., 1988, (in preparation).
Gull, S., and Skilling, J., 1987, "MEMSYS II: Maximum Entropy Data Consultants", Cambridge, U.K.
Kruszewski, A., 1987, INVENTORY, in: "MIDAS software package".
Kurtz, M.J., and Falco, E.E., 1988, in: these proceedings.
Press, W., et al., 1986, in: "Numerical Recipes", Cambridge University Press.
Soneira, R., and Peebles, J., 1977, The Astrophysical Journal, 211:1.
Tyson, A., and Jarvis, J., 1979, The Astrophysical Journal Letters, 230:L153.
Valdes, F., 1982, Proc. S.P.I.E., 331:435.

# TESTS OF REDUCTION AND ANALYSIS ALGORITHMS

# FOR ASTRONOMICAL IMAGES

Michael J. Kurtz  and  Emilio E. Falco

Harvard-Smithsonian Center for Astrophysics
60 Garden Street, Cambridge, MA 02138, U.S.A.

## INTRODUCTION

The reduction and analysis of the data from large scale CCD surveys will require a degree of automation greater than any yet attempted in optical astronomy. The very large number of objects, plus the demanding uses to which the data will be put combine to present a challenging problem.

We are beginning a long term program to create tools and methodologies for the automated reduction of data from large CCD surveys. It is important at this stage of the work to quantify the capabilities and limitations of the current state of the art procedures for dealing with data of this sort.

## BACKGROUND SUBTRACTION

The proper determination of the background is of paramount importance for measuring magnitudes. This paper reports preliminary results for a series of experiments designed to determine the limits on our ability to measure the background.

We have taken a list of objects obtained by reducing a section of digitized Schmidt plate with FOCAS (Tyson and Jarvis, 1979, Valdes, 1982) and created an image containing only bright ($m_B \lesssim 20$) de Vaucouleur law elliptical galaxies having exactly the same magnitudes, and approximately the same effective radii as were measured in the data. To this was added a uniform background of 22.5 mag/pixel (=23.1 mag/arcsec); all noise is strictly Poisson.

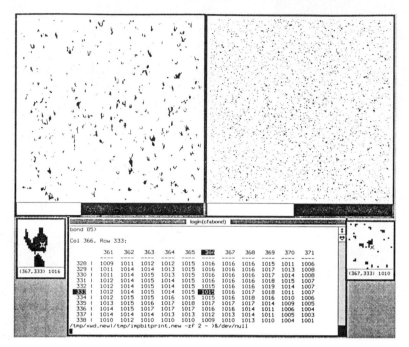

Fig. 1

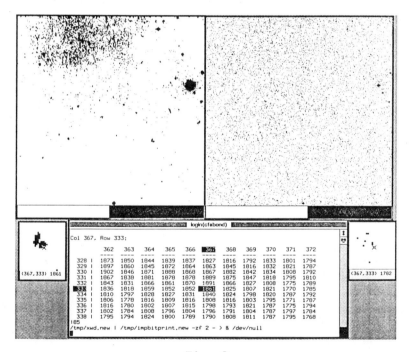

Fig. 2

156

FOCAS was run using the same settings as were used in the original data; these were determined, after some experimentation, to be the best at removing large inhomogeneities, such as the halos around bright stars and the reflection of stars off the telescope supports, from the data.

Figure 1 shows the result. On the right is the model image, and on the left is the sky derived from it (the number matrix is the sky values in the zoom box). The black represents a level above the true background sufficient to cause a 5% error in isophotal photometry. These regions are clearly correlated with the bright galaxies in the model data. In detail the enhanced background is just that part of the extended envelope of the galaxies which is $\lesssim$ 50% of the Poisson fluctuation of the true background (which is 1000).

## CONCLUSIONS

Given ideal model data, with no faint objects, no very bright objects, no gradients across the image, and no plate flaws or internal reflections one can see effects which must be present in the real data, but which cannot be seen, as they lie below the local sky fluctuations, and far below the other systematics in the data. Figure 2 shows the original data, note the sky is dominated by large scale gradients, and the local fluctuations are many times greater than in the model data.

The most obvious systematics in the real data are clearly correlated at spatial scales which are in some way related to the characteristics of the telescope. The unseen systematics discussed here are also spatially correlated, as they are correlated with the positions of the brighter galaxies.

5% systematics in isophotal magnitudes are important for statistical measures of the galaxy distribution. Geller, et al. (1984) have shown that at the magnitude limit of the Shane-Wirtanen counts systematic errors must be kept below 5% to measure the behaviour of the galaxy-galaxy correlation function in the interesting region between 10 and 20 Mpc.

## AKNOWLEDGEMENTS

We thank Bill Wyatt for helping to make the figures. MJK thanks NAGW-201 for partial financial support.

## REFERENCES

Tyson, A., and Jarvis, J., 1979, Ap. J. (Letters), 230:L153.
Geller, M.J., de Lapparent, V., and Kurtz, M.J., 1979, Ap. J. (Letters), 287:L55.
Valdes, F., 1982, Proc. S.P.I.E., 331:435.

# A MODEL FOR THE INFRARED SPACE OBSERVATORY PHOTOMETER

A. Etemadi

Max-Planck-Institut für Kernphysik
Postfach 10 39 80
D-6900 Heidelberg 1, FRG

## Abstract

The Infrared Space Observatoy (ISO) consists of a cryogenically cooled telescope and instrumentation mounted on a 3-axis stablised spacecraft. The launch date for the spacecraft is currently the 1st of April 1993. The expected lifetime of the mission is one and a half years. The Photometer subexperiment (ISOPHOT) was designed to carry out photopolarimetry in the wavelength range 2-200$\mu$m with a photometric accuracy of one percent. A software model of ISOPHOT has been recommended as a means of generating test data for the ISOPHOT scientific data analysis software. Discussions are currently in progress regarding the production of this model, in which case we will attempt to make this model as general as possible so that it may be used in modelling other similar experiments. In this paper we will discuss the problems associated with this work, the overall design concept, and the implementation.

## 1 Introduction

Infrared observations have proved an invaluable tool in improving our knowledge of a large number of astronomical phenomena. Ground based infrared observations are in the first instance severely limited by atmospheric effects and thermal emissions from uncooled telescope optics and the atmosphere. Cold solar system objects, star formation regions, and astronomical sources obscured by dust at visible wavelengths, are some of the few phenomena which may be more fully studied in this wavelength range.

In order to reduce these limitations aircraft and balloon-borne, optical systems have been used which have partially alleviated the problems. Optimal observations, with presently available technology, may only be made by cooling the telescope to cryogenic temperatures and placing it beyond the Earth's atmosphere. The Infrared Astonomical Satellite (IRAS)[1], launched on the 26th of January 1983, was the first satellite designed to overcome these limitations and carry out an unbiased all-sky survey at the wavelengths 12,25,60 and 100$\mu$m. The wealth of data produced by IRAS, and the continued development of infrared technology in Europe, at the German Infrared Laboratory (GIRL), have provided a timely opportunity to follow up these measurements.

The Infrared Space Observatory[2] (ISO) satellite, due for launch in 1993, has been designed to study a whole range of astrophysical objects at unprecedented sensitivity with an observatory range of instruments, including high resolution spectrometers. The 3...200$\mu$m region of the spectrum covered by ISO is of great scientific interest since it is rich in a variety of atomic, ionic, molecular, and solid state lines. Measurements at these wavelengths permit the determination of many physical parameters such as energy balance, temperatures, abundances, densities and velocities.

ISO consists of a 0.6m diameter telescope in a 3-axis stabilised spacecraft. The telescope and scientific instruments are cooled to less than 3K by a dual cryogen system involving liquid helium and liquid hydrogen. The expected lifetime of the mission is approximately one and a half years. By way of introduction we will discuss the ISO scientific instrumentation, with special emphasis on the photometer (ISOPHOT), in section 2. The following section is dedicated to the possible observing modes with ISOPHOT. Details of the software model for ISOPHOT are contained in section 4. In the final secion we will discuss the software modelling work in the context of the ISO mission.

## 2  ISO Scientific Instrumentation

Four independant scientific instruments are included in ISO. The ISO infrared Camera[3] (ISOCAM ) contains two optical channels, one for observations in the wavelength range 3 to 5$\mu$m, and the other in the wavelength range 5 to 17$\mu$m. Interference filters and circular variable filters provide spectral resolutions between 2 and 50. The optics set pixel field of views of 3,6 or 12 arc seconds.

The Short Wavelength Spectrometer[4] (SWS) consists of a pair of grating spectrometers with resolving powers of 1000 from 3$\mu$m to 45$\mu$m. In combination with Fabry–Perot etalons the resolving power will be as high as 30,000 in the 15 to 30$\mu$m range.

The Long Wavelength Spectrometer[5] (LWS) consists of a grating spectrometer covering the wavelength range 45 to 180$\mu$m with a resolving power of around 230. Resolving power of around 15,000 is achieved by switching either of a pair Fabry–Perot interferometers (optimised for the wavelength ranges 45 to 90 and 90 to 180$\mu$m, respectively) into the parallel part of the instrument beam.

The ISO Photometer[6] (ISOPHOT) consits of four separate subinstruments. ISOPHOT–P designed to perform very sensitive high precision photopolarimetry with single detectors, in the wavelength range 3–120$\mu$m. It is possible to choose from a set of fourteen filters and three polarisers for photometry and polarimetry. A large set of available apertures allow diffraction limited observations at all

160

wavelengths as well as wide-beam-observations. ISOPHOT–C, the far infrared camera, consists of three two-dimensional detector arrays, allowing limited photometric imaging in spectral bands $30 < \lambda < 50$, $70 < \lambda < 100$, and $130 < \lambda < 180\mu$m. Polarimetric observations are also possible in all far infrared bands between 30 and $200\mu$m. ISOPHOT–A, a camera, allows highly sensitive two dimensional, diffraction limited mapping in the wavelength range $15...28\mu$m. ISOPHOT–S, is a dual grating spectrometer of resolution around 90, using two linear 64 element detector arrays for simultaneous measurements of spectra covering $2.5...5\mu$m and $6...12\mu$m, respectively. Additionally a focal plane chopper allows differential measurements on the sky. It is also possible to scan one of two fine calibration sources with the chopper in order to monitor the time evolution of detector responsivities. Figure 1 represents the optical flow diagram in the ISOPHOT focal plane unit. The various ISOPHOT optical components are mounted on three separate wheels. The first of the ISOPHOT change wheels carries the polarisers for PHOT–A, PHOT–C, and PHOT–P as well as three groups of path folding flat mirrors for PHOT–A, PHOT–P, and PHOT–S. The beam path for PHOT–C passes through wheel I without deflection. Change wheel II performs a dual function. The field stops for PHOT–P are arranged in an inner circle on the wheel, the spectral filters for PHOT–C, and the path folding mirrors for the shorter wavelength detectors of this subexperiment, are mounted on the outer circumference. Wheel III is equipped with the spectral filters and Fabry mirrors/Fabry lenses for the three individual detectors of PHOT–P.

# 3   ISOPHOT Observing Modes

ISOPHOT observational modes consist of a combination of instrument modes with spacecraft pointing modes. By design ISOPHOT is a very flexible instrument, therefore the transformation of incident photon flux into digitized output is controlled by a large number of instrument parameters. For simplicity, in the following list of parameters, a universal integrating readout scheme has been assumed.

- Chopper waveform: sawtooth, squarewave etc.

- Chopper frequency: $0.1...10.0$ Hz

- Chopper throw: $5...204$ arc seconds

- Polariser settings: none,0,60 and 120 degrees (for

- Filters: 1 out of 14 and 9 for PHOT-C and PHOT-P, respectively.

- Apetures: 1 out of 12 for PHOT-P

- Detector bias voltage

- Detector temperature

- Readout rate

- Reset interval

- Offset subtraction

- Amplifier gain

- Exposure time

The allowed spacecraft pointing modes consist of fixed or raster pointings. Raster pointing is made up of a series of fixed pointings interspersed with slews to the next pointing position. Data taking during these slews will not initially be scientifically analysed, although they will be recorded for later access. Hence we have the additional modes

- Scan step size

- Slew speed

Only a very small subset of this vast multidimensinal parameter space will actually be used in ISO's one and a half year lifetime. Discussions are therefore in progress to reduce these to a few relevant modes. Due to schedule obligations however, work on data analysis software for ISOPHOT is already in progress. Representative ISOPHOT data from the qualification and flight models of the instrument will not be available during the early development stages of the software. Hence at the Max-Planck-Institute für Kernphysik we have begun a pilot project on producing a software model of ISOPHOT, called PHTMODEL, in order to generate test data for the scientific data analysis software. The full extent of this work is yet to be determined (see section (5) for fuller details).

# 4 PHTMODEL

Development of PHTMODEL is expected to proceed by prototyping. Modelling work is similar, in some respects, to data analysis in reverse. A great deal of modelling work involves performing arithmetic upon, creating and displaying images, spectra and auxillary data, so that the initial step in the design of PHTMODEL was to chose an astronomical data processing enviroment which already included these facilities. We considered such packages as AIPS,IRAF,GIPSY and MIDAS. The results of our study indicated that the Munich Image Data Analysis System (MIDAS) was best suited to our work. MIDAS is the system currently recommended by the European Space Agency, and from the standpoint of PHTMODEL, it has the additional advantage that it is the enviroment chosen by the team developing the Space Telescope software model[7] (STMODEL). It is hoped that STMODEL software can to some extent be used as a template for much of our work.

The ultimate goal of PHTMODEL is to accurately represent the functioning of each of the ISOPHOT subinstruments. Additional requirements on PHTMODEL are

- Simple user interface and documentation, with minimal user interaction at runtime

- Maximal error checking and tracing

- Detailed programmers manual so that additional routines may be simply implemented

Some of the possible instrumental and observational effects which it is hoped might be included in the full model are listed in Table 1. Due to the high relative accuracy required from the model (a few precent) very few of these effects may be ignored and the usual approximations, for example, use of Gaussians as point spread functions (PSF's), may not be appropriate. The next stage of PHTMODEL development would involve a study of these effects and how they should be included.

Figure 2 is a simplified overview of PHTMODEL in the form of a flow diagram. The detailed architectural design of the model would eventually be issued as a separate document. Each box in the flow diagram represents an independant software task. The standardisation of input and output within MIDAS (Keywords, FITS images, and MIDAS tables) means that the interface between each box will be clearly defined.

Initially, images and spectral data would be treated separately, and in the case of PHOT–S only spectral information is expected to be required. Box (1), in the flow diagram, is concerned with the generation of model sky images and spectra, although the use of existing images and spectra is not precluded. The facility to generate model images already exists as part of MIDAS. The generation of model spectra would be implemented by application programs written specifically for PHTMODEL.

The inclusion of effects due to the ISOPHOT optical components is the software task represented in box (2). The images/spectra generated above would now be processed to include these effects. Detailed data such as information on point spread functions (PSF's), and throughputs of the individual ISOPHOT optical components would be kept in an ISOPHOT instrument database to be accessed by these routines. The output of this stage would be "dirty" images/spectra convolved with the appropriate PSF's and multiplied by the corresponding throughputs of the optical components in the beam path.

The task of scanning the dirty images/spectra, appropriately time tagging the data, and mapping the illuminating flux unto the appropriate detectors is considered in box (3). At this stage effects such as pointing errors and distortions due to the chopper might be included. On completion of this stage the main result of the

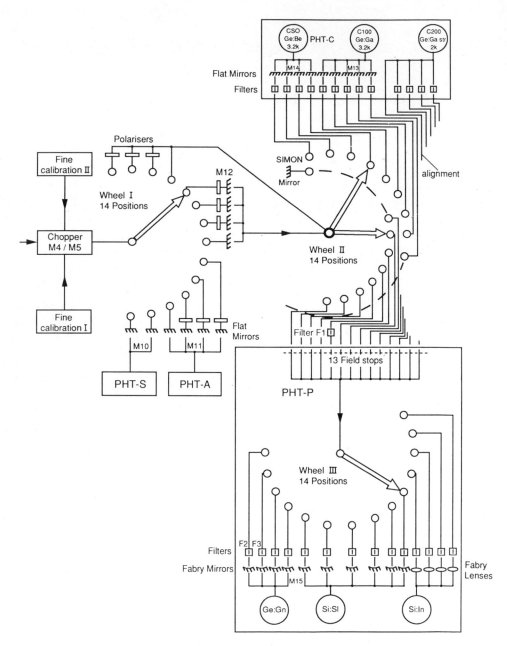

FIGURE 1.

ISOPHOT Optical Flow Diagram

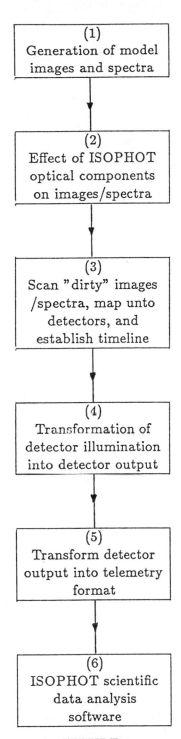

| (1) |
| :---: |
| Generation of model images and spectra |

| (2) |
| :---: |
| Effect of ISOPHOT optical components on images/spectra |

| (3) |
| :---: |
| Scan "dirty" images /spectra, map unto detectors, and establish timeline |

| (4) |
| :---: |
| Transformation of detector illumination into detector output |

| (5) |
| :---: |
| Transform detector output into telemetry format |

| (6) |
| :---: |
| ISOPHOT scientific data analysis software |

FIGURE 2.
PHTMODEL Flow Diagram

TABLE 1.

Instrumental and Observational
Effects

o Noise

- straylight

- thermal

- readout

o Radiation impacts

o Cross-talk

o Sensor malfunction

o History effects

- illumination history

- passage through
radiation belts

o Pointing errors

- gitter

- drift

o Vignetting

o Beam blur

o Zodiacal light

o Cirrus

processing would be in the form of incident flux upon the detectors in appropriate units.

Knowledge of the intensity transfer functions of the individual detectors allows the transformation of illuminating flux into digitised detector output (box (4)). History effects, noise, glitches due to high energy particle impacts, and cross-talk are among the effects which can be included at this stage. The output from this stage would be detector output as a function of time and all appropriate pointing, timing, and subinstrument setting information. Finally this output can be turned into telemetry format (box (5)) and passed on to the data analysis software (box 6)). Initially the telemetry formatting software would be written outside the MIDAS enviroment and later incorporated as part of the model. This step would insure transportibility of this part of the software.

# 5    PHTMODEL in the context of the ISO Mission

In terms of software development, the ISO mission is still in its very early stages. The development of PHTMODEL is critically dependant on available time, manpower and its importance to the mission software development as a whole. The ensuing uncertainties regarding these factors mean that only certain critical parts, and not PHTMODEL in its entirety, might be developed. The software task denoted by box (5) in Figure 2 is the only section of the model which is currently considered critical. At this stage discussions are still in progress regarding further development of the full PHTMODEL.

# References

1.  G. Neugebauer, H.J. Habing, R. van Duinen, H. H. Aumann, B. Baud, C. A. Beichman, D. A. Beintema, N. Boggess, P. E. Clegg, T. de Jong, J. P. Emerson, T. N. Gautier, F. C. Gillett, S. Harris, M. G. Hauser, J. R. Houck, R. E. Jennings, F. J. Low, P. L. Marsden, G. Miley, F. M. Olnon, S. R. Pottasch, E. Raimond, M. Rowan-Robinson, B. T. Soifer, R. G. Walker, P. R. Wesselius, and E. Young, Infrared Astronomical Satellite (IRAS) Mission, The Astronomical Journal, 278:L1–L6 (1984)

2.  W. Aalders, J. Wijnbergen, R. D. Joseph, R. Katterloher, D. Lemke, L. Nordh, G. Olofsson, P. Salinari and F. Sibille, Infrared Space Observatory, European Space Agency publication, SCI(82)6 (1982)

3.  F. Sibille, C. Cesarsky, S. Cazes, D. Cesarsky, A. Chedin, M. Combes, M. Gorisse, T. Hawarden, P. Lena, M. S. Longair, R. Mandolesi, L. Nordh, P. Persi, D. Rouan, A. Sargent, F. Sibille, L. Vigroux and R. Wade, ISOCAM:

An Infrared Camera for ISO, in: "SPIE Vol. 589 Instrumentation for Optical Remote Sensing from Space", 589:170 (1985)

4. Th. de Graauw, D. Beintema, W. Luinge, G. Ploeger, K. Wildeman, J. Wijnbergen, S. Drapatz, L. Haser, F. Melzner, J. Stöcker, K. van der Hucht, Th. Kamperman, C. van Dijkhuizen, H. van Agthoven, H. Visser, C. Smorenburg, The Short Wavelength Spectrometer for ISO, in: "SPIE Vol. 589 Instrumentation for Optical Remote Sensing from Space", 589:174 (1985)

5. R.J. Emery, P. A. R. Ade, I. Furniss, M. Joubert and P. Saraceno, The Long Wavelength Spectrometer (LWS) for ISO, in: "SPIE Vol. 589 Instrumentation for Optical Remote Sensing from Space", 589:194 (1985)

6. ISOPHOT Phase C/D Proposal No. 253-0-87, Prepared for the Deutsche Forshungs- und Versuchsanstalt für Luft- und Raumfahrt by Dornier–System GmbH (1987)

7. M. Rosa and D. Baade, Modelling Space Telescope Observations, in: European Southern Observatory Messenger, 45 (1986)

# ALGORITHMS FOR THE ANALYSIS OF DATA FROM THE

# EXTREME ULTRAVIOLET EXPLORER

Herman L. Marshall

Space Sciences Laboratory
University of California
Berkeley, CA 94720 U.S.A.

## ABSTRACT

An algorithm is described that is intended for use in the analysis of data from the Extreme Ultraviolet Explorer (EUVE) satellite. An improved technique for detecting sources has been developed. This method is shown to be computationally simpler than non-linear methods and more sensitive to point sources than previous linear techniques, because it incorporates more information from the instrument point spread function. It is shown how simulated data help test this algorithm and are useful in the design of the satellite hardware.

## OVERVIEW OF EUVE AND THE DATA ANALYSIS SYSTEM

We have a unique approach to our data analysis system: it is developed rapidly in a succession of revisions that build upon each other. In each release of the software, certain modules are upgraded in response to issues arising during development. For example, the first three releases were devoted to simulating the hardware performance and to developing models of expected astronomical sources and background which provide useful test data. Later releases will concentrate on instrument health and safety issues, archival of satellite data, and subsequent scientific data analysis. We are currently in the process of developing these analysis routines and the algorithms needed for them. An example of preliminary results from our End-to-End System (EES) simulations was described by Marshall et al. (1987). An overview of the EES and the data processing is given by Marshall et al. (1988).

The satellite consists of four telescopes, three for scanning the entire sky during the first six months of the mission, and one for a deep survey during this same time span (see Bowyer et al. 1988). The latter is combined with a spectrometer to be used in a guest observer mode during the remaininder of the mission. The satellite is scheduled for launch on a Delta rocket in August 1991 into low-earth orbit. Photon counting detectors are used, so the electronics assigns times and positions to each event and queued for the downlink data stream. The telemetry bandwidth limits the scientific data stream so that a maximum of about 400 counts per second can be recorded. A history file is generated on the ground that gives a summary of each orbit's aspect, engineering, and data quality.

Because of the extremely large data stream, we do not expect to keep the detailed photon data for the entire sky on-line. The number of photons expected from EUV sources is small compared to the total number of photons that are expected from background. Thus, we plan to create two different data bases for routine analysis and to use the detailed data base in off-line analysis. One of these on-line data bases consists of photons binned in sky coordinates. These data have no time information but may be binned by detector position (e.g., radius) as well. The other data base consists of detailed photon data for individual positions on the sky taken from the preliminary catalog. We call this data base the "pigeonhole" data file. The basic analysis scheme (as outlined by Marshall et al. 1988) relies on use of the pigeonhole data for point source analysis and the binned data for detecting sources. Detailed photon data for new sources are added to the pigeonhole data base using an off-line system and the source positions are added to the preliminary catalog so that later data may be selected for the corresponding pigeonhole file.

## BACKGROUND: THE SLIDING BOX DETECTION METHOD

The detection scheme used in the analysis of data from the *Einstein* Observatory employed a 3 x 3 cell detection window in which the binned sky counts were summed. The sum was tested against a threshold, determined by the background and a preset significance level. The window is shifted one cell at a time until the entire image is covered and the sum is computed for each window position (see Fig. 1). In this method (which has been called the "sliding box" method), the data from any given sky cell is incorporated into nine different sums, so the tests are not all independent. The choice of significance

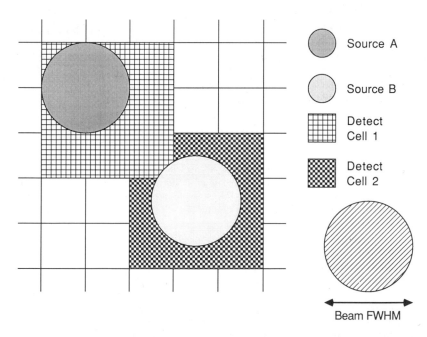

Source A

Source B

Detect Cell 1

Detect Cell 2

Beam FWHM

Fig. 1. Schematic representation of the "sliding box" detection method. The box slides from cell to cell, the sum over nine cells is computed, and this sum is compared with a threshold. At position 1, source A is detectable while B is detectable when the box is at position 2. Note that adjacent boxes may also detect these sources.

level was based on simulations of images and the subsequent application of the algorithm. The thresholds were chosen at many background levels that would produce no more than one false detection (due to background fluctuations) per detector image.

The size of the detection window was chosen empirically by simulations and corresponded to matching the point spread function (PSF) full width at half maximum (FWHM) with the width of two cells (see Fig. 1). The sliding is performed to give ample signal due to a point source even if its position falls on the corners of several cells. Thus, the choice of binning would not play a role in the list of detected sources. This method will now be extended to account for more detail of the shape of the PSF.

## DETECTION WITH A SHAPED APERTURE

A natural extension of the "sliding-box" detection scheme is to apply weights to each pixel before it is tested against a threshold. The test statistic is then $S_k = \Sigma C_{k-i} W_i$, where the $W_i$ are weights applied to the data, $C_k$. (It is implicit that all sums are over the subscript i.) To determine the optimum values for the weights, a figure of merit is formed that can be minimized with respect to the weights. We chose to minimize the flux of a source that could be detected by this scheme. This is calculated by assuming that the telescope has some PSF appropriate to (nearly) all the data to be examined. Using the PSF, one may assign an expected intensity to each pixel by $C_i = I_0 \phi_i + B$, where B is the background per pixel, $I_0$ is the total intensity of a given source to be detected, and $\phi_i$ gives the expected fraction of the source in pixel i. The $\phi_i$ are defined such that they satisfy $\Sigma \phi_i = 1$. The test statistic is to be maximized when the sliding box is centered on the source, so the weights are defined relative to the center of the box. When centered on the source, the detection statistic is then $S = I_0 \Sigma W_i \phi_i + B \Sigma W_i$, so the threshold for detection is $T = K\sigma(B) + B\Sigma W_i$ The quantity $\sigma$ is the standard deviation of the distribution of the weighted sum in the absence of sources; $\sigma^2(B) = B \Sigma W_i^2$ for standard background distributions. The value of the constant K is chosen to give the desired significance level and determines how many detections will be obtained due to background fluctuations. The minimum flux that can be detected (without the benefit of background fluctuations) is then

$$I_{det} = \frac{K \left( B \Sigma W_i^2 \right)^{\frac{1}{2}}}{\Sigma \phi_i W_i} . \tag{1}$$

The optimum weights are determined by minimizing $I_{det}$ over $W_i$, giving $W_i = \phi_i$ and

$$I_{min} = K \left( \frac{B}{\Sigma \phi_i^2} \right)^{\frac{1}{2}} . \tag{2}$$

Note that the weights do not sum to zero so that S is always positive. In the continuous case, $\phi_i$ is replaced by $\phi(\theta)$, $W_i$ is replaced by $W(\theta)$, and the sum is replaced by an integral over $\theta$ (where $\theta$ is the angle from the centroid of a source to the position of a pixel and ranges from 0 to $\theta_{max}$), giving

$$I_{min} = K \left( \frac{B}{2\pi \int_0^{\theta_{max}} \phi^2(\theta) \, \theta d\theta} \right)^{\frac{1}{2}} . \tag{3}$$

171

Using this weighting scheme, it turns out that the size of the box may be extended to the whole sky (as long as the sky background is uniform) because $I_{min}$ increases monotonically with the number of pixels (or $\theta_{max}$). Except for synthesis imaging, the PSF generally drops rapidly as one proceeds away from the source, so the sum can be limited to a "reasonable" region about a test point. The size of this region can be set so that the minimum detectable source is within a certain fraction of the best obtainable. For example, if the PSF has the form

$$\phi(\theta) = \frac{1}{2\pi\theta_0^2} e^{-\theta/\theta_0} \, , \tag{4}$$

then $I_{min}$ is only 5% higher than the theoretical minimum if one integrates to $\theta = 1.15\theta_0$ only.

For comparison, $I_{min}$ can be calculated for a uniform weighting scheme, which is the basic assumption underlying the choice of cell size using the sliding box method. In this case, $\sigma^2(B) = \pi B \theta_{max}^2$ and

$$I_{min} = \frac{K\left(\frac{B}{\pi}\right)^{\frac{1}{2}}}{2\hat{\theta}_{max}\,\phi\left(\hat{\theta}_{max}\right)} \, , \tag{5}$$

where

$$\hat{\theta}_{max}\,\phi\left(\hat{\theta}_{max}\right) = \int_0^{\hat{\theta}_{max}} \phi(\theta)\,\theta d\theta \, . \tag{6}$$

In general, Eq. 6 must be solved numerically. For the form of $\phi(r)$ given by Eq. 4, $\theta_{max} = 1.793\theta_0$. The ratio of $I_{min}$ for the two cases is 0.844, so that the optimum weighting scheme detects sources 15% fainter than the simple uniform weighting scheme. For

$$\phi(\theta) = \frac{1}{\pi\sigma} e^{-\frac{1}{2}\left|\frac{\theta}{\sigma}\right|^2} \, , \tag{7}$$

this ratio is only 0.903 ($\theta_{max} = 1.121\sigma$). This benefit may not seem like much, but if the detection threshold is reduced to account for this sensitivity difference (from $5\sigma$ to $4.2\sigma$ for the exponential PSF), then the probability that a given pixel will show a false detection of a random background fluctuation increases from $2.9 \times 10^{-7}$ to $1.3 \times 10^{-5}$, an increase of a factor of 45 in the number of false detections. Note also that the weighted method has a higher "resolution" than the uniform weighting scheme; i.e., the region which is sampled by the sum has only 64% of the diameter that the uniform method requires.


OPTIMUM LINEAR WEIGHTING FOR SCANNING SURVEYS

This weighting method may be extended for use in a more complex problem. For EUVE and other scanning, imaging telescopes, the PSF depends on the angle of the source from the telescope axis. Thus, in detector coordinates, the PSF depends on distance from the center of the detector (where the boresight falls). One may write the PSF, $\phi_{ij}$, as the fraction of source counts that appear in sky pixel i, relative to the source centroid, and at detector pixel j, relative to the detector center. The test statistic is then

$$S_k = \sum_i \sum_j W_{ij} C_{kij} , \qquad (8)$$

where we find that optimum detection is achieved with $W_{ij} = \phi_{ij}$. The minimum detectable flux would then be

$$I_{min} = K \left( \frac{B}{\sum \phi_{ij}^2} \right)^{\frac{1}{2}} . \qquad (9)$$

The background is units of counts per detector and sky pixel. It can be shown that the flux given by Eq. 9 is always lower than or equal to that given by Eq. 2. Thus, use of *a priori* information will improve the sensitivity of the survey.


COMPARISON WITH NONLINEAR DETECTION USING THE LIKELIHOOD RATIO TEST

Optimum detectability can be achieved using a nonlinear method such as the maximum likelihood method advocated by Cruddace et al. (1987). Simply stated, one forms the likelihood, $\lambda(I_0) = -2\ln(\Pi p_{ij})$, where $p_{ij}$ is the probability that $C_{ij}$ events are found given the expected number of events in sky pixel i and detector pixel j due to both a source (with intensity $I_0$) and background. The value of $\lambda$ is computed for $I_0 = 0$ and then minimized with respect to $I_0$ and the difference is compared to a preset value (see below). The likelihood difference, $\Delta\lambda$, follows a $\chi^2$ distribution, so that the minimum detectable flux is determined by setting $\Delta\lambda = K^2$, where K is the significance level of the detection (say, K = 3 for detection at 99.5% confidence). For a simple, common case where the source is very weak compared to the (uniform) background, B, and assuming Gaussian statistics, then the $p_{ij}$ follow a $\chi^2$ distribution:

$$p_{ij} = \frac{1}{\sqrt{2\pi C_i}} \exp\left[ \frac{\left(C_{ij}-B-I\phi_{ij}\right)^2}{2C_i} \right] . \qquad (10)$$

One then obtains

$$\Delta\lambda \cong \frac{I_m^2}{B} \sum \phi_i^2 . \qquad (11)$$

This results in the same value of $I_{min}$ as given by Eq. 9. Thus, the optimum linear weighting scheme gives nearly the same level of detectability as the maximum likelihood method and there no significant improvement results by using a nonlinear method.


NUMERICAL COMPARISON OF DETECTION METHODS

Simulations of the EUVE type I scanner response to point sources at various off-axis angles

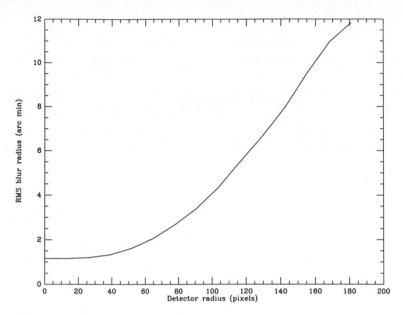

Fig. 2. PSF blur radius (rms) *vs.* detector centroid position (from the center of the detector). The detector is at the nominal focus position.

show that the rms blur radius of the PSF can be fit by a function of the form

$$\sigma = \sigma_0 + \sigma_1 \left(\frac{r}{r_{max}}\right)^2 , \qquad (12)$$

where $r_{max}$ is 180 pixels, $\sigma_0$ = 1', and $\sigma_1$ = 11.3' (see Fig. 2). Applying Eqs. 7, 11, and the appropriate detection method gives $I_{min}/(KB^{1/2})$ = 14.9, 13.2, and 8.8 for the uniform, weighted aperture, and the optimum detection methods, respectively. Thus, using the full information of the PSF results in an improvement in the minimum detectable flux of a factor of 0.66 compared to the uniform weighting sliding box method.

It has been quite useful to settle on a detection scheme far in advance of EUVE launch. There was some question about a hardware modification that would improve source detection. The PSF was generated using the EES for various values of the detector shift along the telescope axis (Fig. 3). The minimum value of the optimum detection statistic from Eq. 9 gave the best shift of z = 0.05 cm (Fig. 4). It turned out that this value differed considerably from the result of a minimization of the PSF standard deviation which suggested that the best shift would be 0.27 cm. The results also show that this shift can be uncertain by up to 0.02 cm before source detection is degraded significantly.

The simulation results also show that the sensitivities estimated from the analytic fit to the PSF shape dependence on detector position are underestimates of the true survey sensitivity. The values obtained for $I_{min}/(KB^{1/2})$ were 8.22, 7.07, and 4.23 for the three linear detection methods previously described. The simulation is the only proper way to determine the *a priori* sensitivities because the EES has software modules to account for obstruction due to the filter frame, which blocks the center of the detector; digitization and other electronic effects; as well as aperture obscurations and other details of the optics. The result is that the optimum detection method that can be applied to scanning telescope data achieves a factor of 2 improvement in sensitivity over a standard linear weighting method. This is equivalent to a factor of 4 increase in telescope effective area and it is achieved entirely by a judicious choice of detection algorithm.

The detectability improvement is quite important when considering a large survey where a large number of independent points on the sky must be tested for new sources. Consider that the effective PSF of the EUVE type I scanning telescopes has a FWHM of about 3', so that there are ~ 6.6 x $10^7$ independent cells in the EUVE sky maps (at Nyquist sampling). Because there are very few known EUV sources (and none known in some bands), it is quite important to reduce the number of false detections to an extremely low level. If we allow, say, 10 false detections over the entire sky, then the implied significance level of detection is 99.999985%, or 5.1σ. To achieve the same level of sensitivity using a uniformly weighted aperture method, one would be forced to reduce the significance level to 2.55σ, giving 3.6 x $10^5$ false detections over the entire sky. Using the lower threshold, the vast majority of the computing time required in the subsequent analysis of the point source fluxes and positions would be spent on weeding out false detections.

FURTHER WORK

Simulations are being carried out to examine thoroughly the distribution of the test statistic for background following the Poisson distribution, which is expected for this mission. Preliminary results indicate that false sources are detected at the expected rates. Further simulations will deal with more practical aspects of the implementation: determining the proper binning size, which may be limited by computation time, and quantifying effects of overlapping test windows. It is expected that exposure effects and background variations will complicate matters but the method may still be applied as long as the distribution of *expected* counts can be generated if given a source intensity and position.

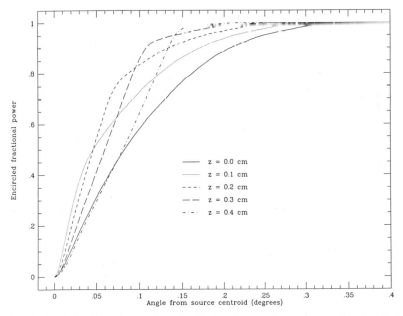

Fig. 3. Cumulative Point Spread Functions for the EUVE type I scanner simulated with the EUVE EES (see text). Various curves are generated for different values of the shift of the detector from the nominal focus position. The source was scanned in a manner nearly identical to that which will be used in the actual all-sky survey.

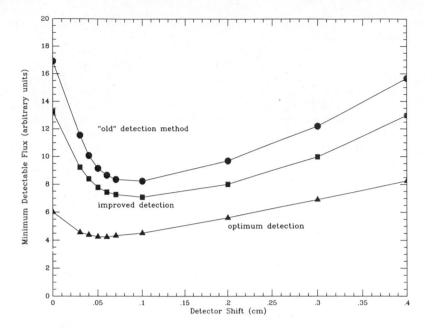

Fig. 4. Minimum detectable flux $I_{min}$, as a function of detector shift. The curves show results for different detection methods. The optimum detection scheme (Eq. 9) gives the highest sensitivity (lowest values of Imin) for all detector shifts.

One interesting approach may be to apply fast Fourier transform (FFT) techniques. This can be done for the simple weighted aperture method. However, it appears impossible for the more general weighting scheme because the PSF variations with sky and detector coordinate are not easily decoupled. This is also difficult when the PSF changes with position on the sky, as will be the case for the EUVE deep survey (see Bowyer et al. 1988). The advantage of a FFT-based technique is that it can be computed quickly for rather large, complicated PSFs. There is a tradeoff involved between the size of the area to be searched for sources and the size of the detection window.

One advantage of this detection scheme is that we may devote more computer time to the analysis of source fluxes and positions because there will be fewer candidates to handle. This method will provide the smallest possible list of sources for further analysis. It is our objective to use the full EES to simulate the observations with high fidelity by incorporating instrument and exposure time effects. We would then explore potential variability and extent in much more detail. Other methods of detecting sources may be used for situations that require more robust results, such as when looking for variable or extended sources.

We would like to thank Mike Lampton, Brenda Hatfield and Pora Park for comments on the manuscript and in its preparation. This work was supported under NASA contracts NAS5-30180 and NAG8-679.

REFERENCES

Bowyer, S., Malina, R. F., and Marshall, H. L. 1988, "The Extreme Ultraviolet Explorer Mission: Instrumentation and Science Goals," in *J.B.I.S.*, **41**, 357.

Cruddace, R. G., Hasinger, G. R., and Schmitt, J. H. M. M. 1988, "The Application of a Maximum Likelihood Analysis to Detection of Sources in the ROSAT Data Base," in proceedings of the ESO conference on *Astronomy from Large Databases*, 177.

Marshall, H. L., Dobson, C.A., Malina, R. F., and Bowyer, S. 1987, "Analysis of Simulated Images from the Extreme Ultraviolet Explorer," in proceedings of the *International Topical Meeting on Image Detection and Quality*, ANRT and SPIE, **702**, 275.

——————. 1988, "Plans for the Extreme Ultraviolet Explorer Data Base," in proceedings of the ESO conference on *Astronomy from Large Databases*, 397.

# ANALYSIS OF ABSORPTION LINES IN QUASAR SPECTRA

V. Müller, J.P. Mücket, H.J. Haubold, and S. Gottlöber

Zentralinstitut fur Astrophysik der AdW der DDR
1591 Potsdam, GDR

## INTRODUCTION

Quasar absorption spectra may offer a clue to the distribution of intergalactic matter clouds at cosmological distances. Essentially three types of absorption lines can be identified:

1 - Broad absorption lines (or "throughs") of highly ionized elements probably connected with the quasar phenomenon itselfes.
2 - Heavy element systems, containing Ly $\alpha$ if in the range of the spectrum, and stemming from gas oh near solar composition. They are characterized by a wide range of ionization states.
3 - Numerous lines of the Ly $\alpha$ "forest" on the short wavelength side of the Ly $\alpha$ emission line and only a small contamination of metal or molecular lines.

The identification of the absorption lines with primordial matter clouds has been suggested long ago [1]. Presently cumulating evidence can be obtained for an intergalactic origin of the last two types of absorbers. It seems natural to connect and to compare the physical properties and the spatial distribution of the matter clouds with the current scenarios of the nonlinear stage of the evolution of large-scale structures. Arising at high redshifts $z \leq 4$ the clouds are very interesting for the epoch of galaxy formation. The finite spectral resolution and smearing effects due to peculiar velocities should not strongly affect the possibility to detect the large supercluster and void structures in the absorption spectra of far quasars [2].

Further the absorption spectra can be used to study the chemical evolution of the intergalactic medium, its physical properties, the intergalactic radiation field, and the mean mass density of the universe. Indeed, the potential applications of absorption line studies in cosmology are just beginning to be realized [3]. Here we want to concentrate on the clustering properties of absorbers and on a model for the nature

179

of the absorbing matter clouds. Modern methods of cluster analysis, special statistical techniques for handling the small data samples and physical models for the absorbing matter have to be used for a deeper understanding of the quasar absorption lines.

## CORRELATION ANALYSIS

For describing the clustering properties of the absorption clouds, the autocorrelation function of the absorbers on the line-of-sight to the quasars seems to be useful means. The spatial distribution of the metal lines and the lines of the Ly $\alpha$ forest and the metal lines show some remarkable physical differences. The metal absorbers are strongly clustered for small velocity splittings up to 150 Km/s [4,5]. This result is mainly due to a fine splitting of absorption lines in the high resolution spectra. One can identify in a quasar spectrum in the average only two or three absorption systems. To obtain a reliable statistics in the correlation analysis, an inhomogeneous sample of metal absorption line data has to be analysed.

On the contrary, the numerous sharp absorption lines of the Ly $\alpha$ forest are only weakly clustered on a wide range of velocity splittings [6]. A detailed analysis reveals characteristic clustering signals in the line distribution of recently gained high resolution spectra [7,8]. We study the distribution of absorbing matter on eight lines_on_sight to the quasars Q 0122-380 (the data stem from [9,10]),Q1101-264 [9,10], Q 0420-388 [11], Q 2000-330 [12], and the projected quasar pairs Q 1236+268,269 [13] and Q 0307-195 A,B [14]. The pairs are especially interesting since they are separated on the plane of the sky by only 173" and 58", respectively.

At first we construct the autocorrelation function of the absorption line distribution. To this end we transform redshift differences $\Delta z$ into comoving distances measured in terms of the Hubble length $H_0/c$ ,

$$\Delta s_0 = \Delta z \, (1 + z)^{-1} \, (1 + \Omega z)^{-1/2} \tag{1}$$

Just because of this transformation the analysis depends essentially on the density parameter $\Omega$, and we choose as characteristic values $\Omega = 0.2$ and $\Omega = 1$. Attempts to perform a correlation analysis without such a transformation[15] are not very promising.

The correlation functions are estimated by two different methods. A continuous density contrast of the absorbers $\delta(\Delta s_0)$ follows from a subdivision of the redshift range of the spectrum in fixed cells, and the correlation function is given by

$$\xi(\Delta s_0) = < \delta(s_0 + \Delta s_0) \, \delta(s_0) > \tag{2}$$

where the average over $s_0$ uses the actual length in which the spectra could be correlated. Alternatively, the correlation function is estimated by the conditional probability of the excess number of absorption line separations $\Delta s_0$ in comparison with the average number

$$\xi_c(\Delta s_0) = n_i \, (\Delta s_0) \, / <n_i> - 1 \qquad (3)$$

For large samples of data, both methods should give identical results. But we use definitions in order to test reliability. In the first method, the distribution of lines on the fixed grid leads to additional noise. On the other hand, the definition has to advantage that the line positions and the regions without lines are dealed with on the same footing, while the second definition, also used in [6], depends only on the separation between the line positions.

In Fig.1 we show the two-point correlation functions of the absorption clouds in Q2000-330, gained with a chosen bin width of $\Delta s_0 = 4.5*10^{-4}$. This quasar is especially interesting, since it represents one of the remotest quasars (the emission lines have a redshift $z_{em} = 3.78$) and the comparatively bright object has a rich Ly $\alpha$ forest: 196 absorption lines with redshifts down to z $\approx$ 2.2. The correlation functions show a clear quasi-periodic structure as indicated by the interpolation curve.One finds positive correlation at $\Delta s_0 \approx (0-3)*10^{-3}$ and at $\Delta s_0 \approx (9-12)*10^{-3}$ with the amplitude I $\xi$ I$\approx$0.2.Smaller bin widths in the correlation functions reveal some remarkable positive signals of the amount $\xi \approx$0.4-0.5 at $\Delta s_0 \approx (1.7, 3.5$ and $19)*10^{-3}$. Much effort has been devoted to obtain senseful estimates of the reliability. Using MonteCarlo simulations we construct model distributions of absorbers with the same spectral resolution as in the real data. The comparison of the mean square root deviations of their correlation functions $\sigma$ are used as error estimate, c.p. the error

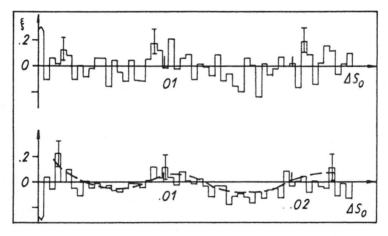

Fig.1. Two-point correlation functions of Ly $\alpha$ absorbers in the spectrum of Q 2000-330 for $\Omega$=1 as function of the comoving distance $\Delta s_0 = \Delta r_0 H_0/c$; above after eq.(2), below after eq. (3).

181

bars in Fig.1. The reliability of our results lies between $2\sigma$ and $3\sigma$. Similarly, $\chi^2$-tests give a reliability of almost 99%. A recent related analysis confirms our positive correlation signal for small line separations of forest lines [16].

It is remarkable, that the collection of all the correlation data in the different spectra leads to better coinciding results if we choose an open cosmological model with $\Omega=0.2$ than for the Einstein-de Sitter model with $\Omega = 1$. Further, it is marginal evident that for smaller z values the amount of the positive correlation at small redshift separations increases. This result is the expected effect of the growth of the clustering during the cosmic evolution.

## CLUSTER ANALYSIS

Special attention has been directed to the double quasars, since the absorbers on the two lines-of-sight are independent and not affected by the finite spectroscopic resolution. The distribution of nearest neighbours can give us information on a connection of the distribution of matter over the transverse separation between the lines-of-sight, which corresponds to a scale $\Delta s_0 = (11\text{-}8){*}10^{-4}$ for Q 1623+268, 269 and $\Delta s_0 = (3\text{-}2){*}10^{-4}$ for Q 0307-195 A,B (the number are given for $\Omega=0.2$ and 1, resp.). In Fig.2 we show the distribution of nearest neighbour separations, where the broken line shows the expectation for a distribution of absorbers with an effective coherence scale $\Delta s_0 = 3{*}10^{-4}$ and $10^{-3}$, resp., for both considered pairs. Collecting the absorption clouds in clusters with such a characteristic scale leads to an orthogonal separation between clusters of the same order of magnitude as the separation between positive correlation regions in the two-point autocorrelation functions. The percentage of clusters with clouds on both lines-of-sight and the transverse separation of the lines provides an estimate of the linear extension of the

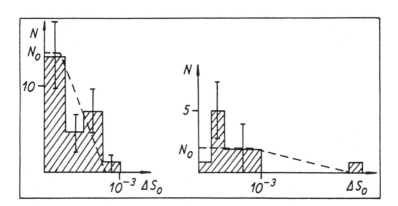

Fig.2. Distribution of the number N of nearest neighbour separations of Ly $\alpha$ absorption clouds in Q 1623+268, 269 (left) and Q 0307-195 A,B (right);

clusters of $\Delta s_0 = (8-6)*10^{-3}$ , much larger than the thickness in accordance with the pancake model of the development of the large-scale structure of the universe. This result underlines the early and farsighted proposal to search in the distribution of absorption lines of quasar spectra for hints of superclustering [2].

The connection of the results with the pancake model of the development of cosmic structure should be stressed [17,18]. To elucidate this connection we discuss specific one-dimensional models [8]. The orthogonal separation of structures in the quasar spectra are in good accord with the spacing of presently observed superclusters of galaxies $\Delta s_0 \approx 10^{-2}$ (see [19], other definition of superclusters give somewhat larger scale [20]).

## METAL ABSORPTION LINES

Sharp peaks in the correlation functions at small separations $\Delta s_0$ may be attributed to characteristic resonance lines, in particular to the C IV doublet with wavelengths 1548 - 1550 Å), the Si III doublet (1190 - 1193 Å), and the separation Si III (1206 Å) - Ly $\alpha$ . The corresponding redshift differences lead to pseudo-separations at $\Delta s_0 = (7,13,17)*10^{-4}$. Direct spectroscopic hints of these metal absorption lines in the unidentified absorption forest shortward of the Ly $\alpha$ emission line has been found recently [21].

For the pancake model, the absorption lines are assumed to result for gas condensations of comparable low temperature in the sock heated hot pancake gas. The thermal instability leads to interface regions between the hot and cold gas, where the hot medium pressure confines the cold clouds. We consider the radiation transport through an extended temperature profile which results from the T-dependent cooling rate $\varepsilon = (c_{ff}T^{-3/2} + c_{fb}T^{-5/2})$ and the heat conductivity $\lambda \propto T^{5/2}$ for isobaric conditions. Then the nonlinear heat conductivity equation has to be solved [22],

$$\frac{5}{2} \frac{dT(x,t)}{dt} + \varepsilon(T) = \frac{\partial}{\partial x} \lambda(T) \frac{\partial T(x,t)}{\partial x} \tag{4}$$

The cooling wave expands the condensation region into the hot gas. The resulting absorption lines width of through-going QSO-radiation depends on the location of the line-of-sight with respect to the cloud centre [23]. For realistic parameters the extension of the cooling wave exceeds the diameter of the clouds leading to the excitation of different ionization levels in a single redshift system. Just this is observed in the different spectra.

To quote an interesting example, an absorption system with a redshift larger than three has been observed in the already discussed quasar Q 2000-330 [24]. We compare the spectroscopically derived clouds and ionization level probabilities as given by the corresponding numerical results [25]:

| ionization level | column density model | log N (cm$^{-2}$) QSO |
|---|---|---|
| H I | 19.30 | 19.78 |
| O I | 14.43 | 14.40 |
| C II | 14.62 | ≤ 13.95 |
| Si II | 13.62 | 13.15 |
| Si III | 13.38 | 13.08 |
| N I | 13.79 | ≤ 13.65 |

(we assumed clouds with 0.1 solar metal abundance).

For encounters further away and for larger clouds in comparison with the cooling region, the ratios of the column densities of metals in comparison with that of hydrogen grows. In this manner we can find realistic models for the metal absorption clouds. Otherwise, the results can explain the misinterpretation of metal lines as Ly $\alpha$ lines found in the correlation analysis. The model of the metal absorption regions leads to improvements of the line identifications in the quasar spectra. On the other hand, it underlines the pancake model with hot or warm dark matter for the development of the large-scale structures in the universe.

## REFERENCES

1. C.R. Lynds, Astrophys. J. (Letters) 164:L73 (1981).
2. J.H. Oort, Astron. Astrophys. 94:359 (1981), 139:211 (1984).
3. W.L.W. Sargent, QSO Absorption Lines and Cosmology, in: "Observational Cosmology", Proc. IAU Symposium No. 124, A. Hewitt, G. Burbidge, and L.Z. Fang, eds., p. 777, Reidel, Dordrecht, Boston, Lancaster (1987).
4. P.J. Young, W.L.W. Sargent, and A. Boksenberg, Astrophys. J. Suppl. 48:445 (1982).
5. A.P. Crotts, Astrophys. J. 298:732 (1985).
6. W.L.W. Sargent, P.J. Young, A. Boksenberg, and D. Tytler, 1980, Astrophys. J. Suppl. 42:41 (1980).
7. V. Müller, J.P. Mücket, and H.J. Haubold, Astron. Nachr. 308:177 (1987).
8. J.P. Mücket, and V. Müller, Astroph. Space Sci. 139:163 (1987).
9. R.F. Carswell, J.A.J. Whelan, M.G. Smith, A. Boksenberg, and D. Tytler, Mon. Not. R. Astr. Soc. 198:91 (1982).
10. R.F. Carswell, D.C. Morton, M.G. Smith, A.N. Stockton, D.A. Turnshek, and R.J. Weymann, Astrophys. J. 278:486 (1984).
11. B. Atwood, J.A. Baldwin, and R.F. Carswell, Astrophys. J. 292:58 (1985).
12. R.W. Hunstead, H.S. Murdoch, B.A. Peterson, J.C. Blades, D.L. Jauncey, A.E. Wright, M. Pettini, and A. Savage, Astrophys. J. 305:496 (1986).

13. W.L.W. Sargent, P.J. Young, and D.P. Scheoder, Astrophys. J. 256:374 (1982).
14. P.A. Shaver, and J.G. Robertson, Astrophys. J. (Letters) 268:L57 (1983).
15. I.E. Waltz, Astron. Zh. 62:19 (1985).
16. J.K. Webb, Clustering of Ly α Absorption Clouds at High Redshifts, in: "Observational Cosmology", Proc. IAU Symposium No. 124, A. Hewitt, G. Burbidge, and L.Z. Fang, eds., p.803, Reidel, Dordrecht, Boston, Lancaster (1987).
17. A.G. Doroshkevich, Astron. Zh. 61:128 (1984).
18. A.G. Doroshkevich, and J.P. Mücket, Pis'ma v Astr. Zh. 11:331 (1985, Sov. Astron. Lett. 11:137).
19. A.G. Doroshkevich, E.V. Kotok, S.F. Shandarin, and Yu. S. Sigov, Mon. Not. R. astr. Soc. 202:537 (1983).
20. N.A. Bahcall, and R.M. Soneira, Astrophys. J. 277:27 (1984).
21. D.M. Meyer, and D.G. York, Astrophys.J. (Letters) 315:L5 (1987).
22. A.G.Doroshkevich, and Ya.B.Zeldovich, Zh. Eksp. Teor. Fiz. 80:801 (1981)
23. V. Müller, J.P. Mücket, S. Gottlöber, and H.J. Haubold, On the Interpretation of QSO Absorption Lines, in: "Evolution of Large-Scale Structures in the Universe" Proc. IAU Symposium No. 130, J. Audouze, and A. Szalay, Reidel, Dordrecht, Boston, Lancaster (1988).
24. R.W. Hunstead, M. Pettini, J.C. Blades, and H.S. Murdoch, Absorption Line Systems at z>3 in the QSO 2000-330, in: "Observational Cosmology", Proc. IAU Symposium No. 124, A. Hewitt, G. Burbidge, and L.Z. Fang, eds., p.799, Reidel, Dordrecht, Boston, Lancaster (1987).
25. P.R. Shapiro, and R.T. Moore, Astrophys. J. 207:460 (1976).

# ATTITUDE DETERMINATION USING STAR TRACKER DATA

M.N. Boyarsky

Space Research Institute
USSR Academy of Sciences
Moscow, USSR

## ABSTRACT

The paper describes the general structure and main algorithms of software system developed in Space Research Institute of the USSR Academy of Sciences (SRI) for processing of star tracker data and spacecraft attitude determination using these data.

## INTRODUCTION

Today SRI performs and prepares several astronomical experiments aboard spacecraft (SC) equipped with star trackers of television type serving for the accurate attitude determination during astronomical observations. In this connection a software system for SC attitude determination using TV star tracker data was designed and has been successfully applied. This report shortly describes the principles of this system and also give an idea of algorithms used which proved their efficiency in processing large amount of real-world data.

## THE GENERAL STRUCTURE OF THE SYSTEM

The main steps of SC attitude determination using star tracker data are shown on the next page diagram.

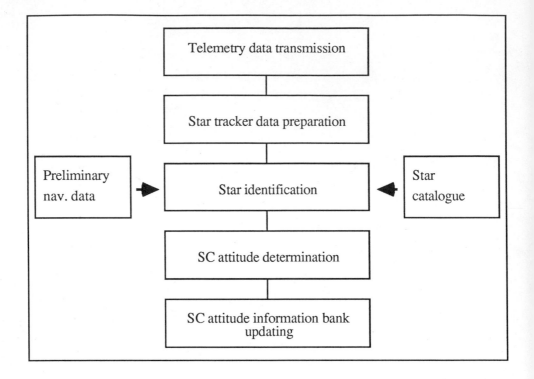

System described here runs under VM/SP operating system and widely employs features of this OS for automation of the data processing. Telemetry information received by one of the ground communication stations is transmitted through wideband TV channel to the Space Research Institute, where this information is written to the mass-storage disk of the mainframe computer dedicated to this experiment; Then this information files undergoes preliminary processing after which a set of disk files is created, each file containing data on a different scientific instrument. Let's designate the virtual machine in which this preliminary processing is done as VMA. During telemetry information input into computer and its preliminary processing virtual machine VMB, which is responsible for attitude determination, remains in a dormant state, waiting for files with start tracker data. After fixed time intervals (now it is every 5 minutes) VMB awakens and compares the list of files created after preliminary processing with the list of files for which attitude determination had already been made. The first file list is created from the corresponding minidisk directory of VMA and it is updated after completion of preliminary processing of each input file. The second file list is stored on one of VMB minidisks and contains for every processed file the name of it and some results showing the quality of attitude determination (such as the number of successful determinations). If this comparison revealed a filename which is present only in first list then this file is assigned as an input file for the attitude determination program and it is respectively processed. The main algorithms for attitude determination are described in next chapter. The nominal data for SC three-axes attitude for this observation session are taken from a bank of preliminary navigational data, in which all information on all SC maneuvers is stored. After completion of the work the list of processed files is updated and file of attitude determination results is created. This file contains the spherical coordinates of star tracker optical center and the angle between star tracker X axis and U axis of the normal coordinate system. This file is then saved in the SC attitude bank which consists of a direct access data set where all attitude results are stored and a sequential access file serving for a quick search of information in the first file.

After completion of all operations for the file VMB machine again falls into waiting status and the above described cycle is repeated. The selection of separate virtual machines for preliminary telemetry data processing and for attitude determination accelerates preliminary data development and excludes possible delays of this process.

## THE MAIN ALGORITHMS OF ATTITUDE DETERMINATION

The input information for attitude determination in case of TV type star tracker is a frame containing the measurements of position and brightness of N images registered by star tracker during a certain time interval. Usually they are the images of N brightest stars (taking into account the spectral sensitivity of device used) stars in the tracker's field of view. Data on each image include its rectangular coordinates on the tracker photosensitive plane and its brightness in certain scale.

The first step of attitude determination is star identification, which means the search of corresponding star for each image. M stars are selected for this process from star catalogue, namely these stars which can be registered by the star tracker giving its sensitivity, field of view size, nominal attitude and attitude execution errors. Usually M is much bigger than N due to the differences in spectral curves of star tracker and star catalogue and because of attitude maneuvers execution errors. for further discussion let's designate star tracker axes as X, Y and Z and let XY plane be the focal plane of the tracker which concurs with its photosensitive plane. As a rule, nominal (approximate) attitude is known for all three axes. We'll call these nominal axes U, V and W respectively. In the first and most frequent case where attitude errors are small (about 0.01 radiant) one may consider the transition from UVW to XYZ as multiplication of independent rotations about U, V and W and account only for linear terms in this transformation. Then rotations about U and V axes lead to equal shifts of all images, while W rotation results in rotation of all images which in case of small field disturbs their positions much less than U and V rotations. So we may consider differences of coordinates of the same name for each star pair to be invariant of this transformation in linear approximation.

With this in mind we designed our first algorithm of star identification which combines high speed with reliability of star recognition. In addition, this algorithm is quite economical in sense of computer memory required for operation of program based on it, thus making this program or its version a possible candidate for onboard computer attitude determination software.

The second algorithm should be used in case of approximately known one-axes attitude, namely, in case of known targeting of W axes and unknown rotation about it. In that case star recognition is made from the comparison of distances between images on XY plane and distances between catalogue stars projections on UV plane. It should be noted that reliability of this process depends strongly on the numbers of stars and images used. Experience shows that when number of images recognized is less than 4 and number of stars used as candidates is greater than 4 then identification made with this algorithm is unreliable for star trackers for which this software was applied, while in case of 5 and more identified stars the probability of correct recognition is close to 100%.

The accurate star tracker attitude is calculated using least-squares method for relation between positions of images and corresponding stars. Since we know how star tracker is situated, it is possible to determine the attitude of the spacecraft itself.

Due to the small size of star tracker field of view the rotation about W axes is poorly defined. The guaranteed approach was applied in order to make accuracy estimation for this rotation determination (for foundations of the approach see Bakhshiyan et al., 1980). This method gave also the possibility to optimize selection of images used for this angle calculation and proved the advantage of employment in one tracker (of Polish design) of combined brightness-distance criterion for onboard selection of images and not the simple selection of N brightest ones.

In addition to first two algorithms of star identification in which some preliminary attitude information is required, our system also includes some procedures to be used when such data is unavailable. That software, quite naturally, is very time consuming and should be used only in special cases.

## PRACTICAL EXPERIENCE

This software system has been used for attitude determination in more than 100 observation sessions and proved to be highly reliable. The average time interval between creation of input file and complete attitude determination turned to be about 20 minutes in time-sharing system with low priority for this task. No operator intervention was usually required, save for some non-standard situations (such as errors in preliminary attitude determination).

## REFERENCES

1.  Bakhshiyan, B.Ts., Nazirov, R.R., and Eliasberg, P.E., 1980, "Opredelenie i korrektsiya dvizheniya", Nauka, Moscow (in russian).

# ARTIFICIAL INTELLIGENCE: WHAT CAN IT DO FOR ASTRONOMY?

R. Albrecht*)

Space Telescope European Coordinating Facility
European Southern Observatory
Karl-Schwarzschild-Str. 2, 8046 Garching/München, FRG

## ABSTRACT

This paper summarizes the development and the applications of artificial intelligence (AI) techniques, and examines their applica-tion to astronomy. Areas of actual and possible use are identified. An attempt is made to extrapolate into future applications, empha-sizing inter-disciplinary aspects and improvements of research methodology.

## INTRODUCTION

Computers were introduced to astronomy in the late sixties and early seventies; enormous progress has been made through this inno-vation, which allowed one to tackle problems which would have been classified as hopeless only a few years earlier. This process is still going on, as computers become more powerful and more afford-able. However, most of today's progress in this area is a progress of degree and not progress of principle.

The time has come to examine again the field of computer science for innovative techniques which might be employed in a useful way in astronomy. One of the most promising new techniques is the area of Artificial Intelligence (AI) which emerged during the last five years.

------------------------------------------------------------------

*) Affiliated to the Astrophysics Division, Space Science Department, European Space Agency.

While the use of computers in astronomy so far has tremendously increased our ability to perform calculations, it has had little or no effect on the way we are doing science: the way we use facts and hypotheses to build models through combination and inference has not changed. AI techniques, in addition to enabling us to carry out research better and quicker, also have the potential of changing the way we are doing science, thus enabling us to gain radically new insights into the universe.

## WHAT IS ARTIFICIAL INTELLIGENCE?

Many definitions have been given which should allow one to make a decision about whether or not a computer performs in an 'intelligent' manner. The most pragmatic one is probably the Turing test: if we carry on a conversation through a terminal and from the responses we get we cannot tell whether we are conversing with a human or with a computer it means that the computer performed intelligently - under the assumption that we attribute intelligence to humans. Along the same lines it is not necessary to simulate human thinking in order to make an intelligent computer, just as it is not necessary to build an airplane like a bird in order to get it to fly. Continuing this pragmatic approach let us call Artificial Intelligence a subfield of Computer Science (as indeed it has been since 1956) and just forget about the mythical notions which still exist; after all, we also got accustomed to horseless carriages.

AI is again subdivided into different fields: Natural Language processing, Machine Vision, Robotics, and Experts Systems (ES) to name the most important ones. Obviously, there are overlaps. For instance, Expert System technology is used for some implementations of machine vision. All these AI applications will become important for science. Natural Language Processing, to give an example, will make access to foreign language publications much easier than it is now. However, the area with the most immediate potential to impact astronomical research is the area of Expert Systems.

## EXPERT SYSTEMS

In a conventional computer program all domain knowledge is generally encoded in the program, the data are taken as mere bit patterns and combined algorithmically. A rule-based program, on the other hand, encapsulates no knowledge of the application domain; this knowledge is provided through a data base containing facts and rules. The program itself only knows what to do with the rules. Incoming new facts are checked against these rules in an "if <condition> then <action>" manner. Actions can range from displaying the result of that comparision, getting more facts, modifying facts, modifying the rule base, or changing the way rules are applied. In

192

an Expert System, the rules are derived from the knowledge of human domain experts through a process called Knowledge Engineering.

Immediately, some problems become obvious. How to best represent knowledge in the form of rules? How to represent facts? How to keep the rule and fact bases consistent (truth maintenance)? How to find one's way through a rule base quickly in order to keep up with the incoming facts? Indeed, much of the expert system development work in recent years has gone into solving exactly these problems.

Concepts like inheritance were developed, which allow to propagate the characteristics of a class into the instances of that class in an automatic manner. Sophisticated search strategies were applied to avoid combinatorial explosion.

It is not the purpose of this paper to provide a detailed description of how Expert Systems work; this can be found elsewhere. Suffice it to say that Expert Systems have reached a considerable stage of maturity and can be purchased like other software products. Most commercial Expert Systems come in the form of Expert System "Shells", which means that they provide the mechanisms to store facts and rules, as well as the mechanism to combine them (the inference engine). Good Expert System Shells also provide their own software development environment, designed to facilitate the configuration of the system for dealing with a particular problem. They also provide tools for constructing interfaces to the end user of the system, usually tools which are based on window-icon-mouse-pointer techniques.

From the above it becomes evident that in order to successfully develop and use an Expert System it is necessary to consider all elements of the hardware/software environment, such as links from the Expert System to other software elements (maybe written in other languages) and to networks. In fact, the typical AI work station consists of a machine which runs LISP and provides an environment which totally integrates the operating system, the language, the editor, the expert system, the debugger, the data base, and the documentation. The user moves through this environment quickly and easily with the help of the mouse, (unsolicited) suggestions provided by the system, prop-up menus, etc.

A word about the languages of AI: although most AI systems today are written in LISP this does not necessarily mean that anything written in LISP is automatically AI. It just so happens that the language LISP offers the kind of data structures and concepts which make it well suited for AI applications. In principle, however, it is possible to write, for instance, an Expert System in any programming language; in fact, in order to increase the operating speed several Expert Systems which originally had been developed in LISP were re-written in C.

## ADVANTAGES AND DISADVANTAGES OF EXPERT SYSTEMS

It is easy to identify some immediate advantages of Expert Systems. The most important one is probably the fact that for the first time the expertise of more than one human domain expert can be combined and made available to the non-expert. This is already being used in a variety of applications, the best known being medical diagnosis. In this area, the level of reliability of diagnoses generated through Expert Systems has reached that of the best medical experts in restricted domains.

In addition to the obvious advantage of being able to combine knowledge bases there is also an immediate application for research: as a byproduct of combining (i.e. processing) a knowledge base the system will flag all instances in which conflicts exist between information provided by different knowledge; on the other hand, machine-based reasoning can be used to infer new knowledge.

A more practical advantage is the fact that because of the above reasons Expert Systems are very attractive to industry and thus can be procured commercially.

The most important disadvantage has to do with the difference between conventional programming and programming an Expert System. While the dominating feature in a conventional program is the loop and the recursion (i.e. program once, do many times), there are generally no loop constructs in an Expert System. This means that if the problem is relatively simple it is usually more efficient to tackle it without the Expert System (unless the operators are totally non-experts, which is generally not the case in astronomy).

For relatively simple problems an Expert System can help only if the number of operations is very high; in this case the loop construct is not in the programming, but rather procedural. Another way of overcoming this disadvantage is to use the Expert System only for very complex problems. In this context, complex means easy to understand but cumbersome to actually do because of multiple, interacting considerations; in contrast to complicated in the sense of hard to understand. An example is scheduling, where the individual constraints are trivial. The gain is derived by freeing the user from having to track down all the information which is needed to solve the problem.

There is another disadvantage which is often overlooked: the information threshold required to implement an ES is considerable and at this point in time very often exceeds the level of computer literacy of most astronomers. However, not only will astronomers become more comfortable with advanced computer tools, the tools themselves will also have very good user interfaces which will facilitate not only the end use of the system, but also the implementation of new capabilities.

194

While it is an advantage for scientists that Expert System development is being driven by industry requirements, there is also a disadvantage associated with this: S/W prices are considerable. However, there is a trend right now to market Expert Systems for PC class computers, which is bound to bring down the price level.

APPLICATIONS OF EXPERT SYSTEMS

From the considerations above follows that the ideal application of an ES is the area of monitoring, diagnosing and operational assistance.

Applications under development in industry include medical diagnosis (already mentioned), fault detection and analysis, process monitoring, and intelligent "assistants" (e.g. the pilot's assistance, a system which frees the pilot of an aircraft from some routine tasks). Expert System based decision support systems are being used extensively in the banking and insurance industry.

There have not been many applications in astronomy. In late 1986 the ST-ECF started a pilot project to obtain experience with AI technology; a number of projects were carried out. An example for gain derived by multiple application to a relatively simple program is the HST Expert (Adorf & di Serego, 1988). Even if only one single HST exposure is planned the user still has to go through the planning process many times in order to try out the alternatives and assess the trade-offs. The Expert System effectively serves as a decision support system.

An example for the second application is the calibration of HST data. While the problem is not too difficult and readily understandable to astronomers it is very cumbersome to actually carry out an optimum calibration of a data set extracted from the HST Archive because of questions like: which calibration files and which calibration procedures are appropriate for the time at which the data were taken? Do the calibration files need to be processed? If so, in which way? Were there instrumental malfunctions at the time? Which procedures have to be applied to correct for them? The prototype of such a system, the Data Analysis Assistant, was also developed within the ST-ECF AI pilot project (Johnston 1987).

The pilot project also contained an application to astronomical research, the classification of IUE spectra using an Expert System (Rampazzo et al., 1988). The problem is of the first category (i.e. moderately complex, but to be carried out consistently and very often), and it reveals another need which comes with the practical use of Expert Systems: the need to have a network link to a data analysis environment. In the case of the spectral classification problem it is necessary to derive the classification parameters from the data. To do this in LISP would be redundant: these capabilities (and other specialized data analysis tools) already exist (in MIDAS, for instance) and only need to be accessed.

Beyond the immediate astronomical application there are several applications which are related to astronomy in the sense that they provide tools which allow to do astronomical research faster and easier. Among these applications are Expert System assisted user interfaces to complex distributed archives, which will allow the astronomer to cross-query archives of scientific data using domain terminology, with the system mapping the requests into the query language of the individual data base management systems, as well as taking care of network connections and protocols.

Another application is somewhat removed from astronomy, but of great practical importance. It is the area of software development, where we will see the advent of Expert System assisted software development environments. An example of such a system is the Computer Aided Software Engineering (CASE) environment (Bibel, 1985; Hindin, 1986; Fischer, 1987). Already, S/W development methodology is taking over many of the principles of AI S/W development. In fact, some people are of the opinion that these new S/W paradigms are the most important result out of the AI development.

CURRENT DEVELOPMENTS OF INTEREST

Even though Expert Systems at this point in time are not being used widely in astronomy, their possible application is understood and it is just a matter of time until Expert Systems will be common-place, both recognizable as such, like in research support tools, and transparent, like in fault diagnosing and network monitoring. There are, however, many other developments going on at this time which will produce technologies or procedures which can be applied to astronomical research. Of course it is beyond the scope of this paper to go through these developments in great detail and assess their importance; on the other hand, there is at least one obvious candidate for promising new technology coming out of AI related developments, and that is the emerging field of Neural Networks.

The concept of Neural Networks has been around for decades, but it was not until recently that the potential of this technology was fully recognized. Only in 1987 did Neural Networks become noticeable in the computer science literature; today, a year or so later, we are seeing an enormous amount of work being done.

Detailed descriptions of Neural Networks and their operating principles can be found elsewhere (Lippmann, 1987). Basically, Neural Networks consist of a relatively large number of simple processing elements (neurons) which are massively connected to each other, for instance in such a way that the output of each neuron is fed into every other neuron (Hopfield & Tank, 1986). The "strength"

196

of these connections determined the performance of the net (i.e. the "program"), serving at the same time as an associative memory. Many layers of such interconnected neurons can by built up as the problem to be solved increases in complexity. The way the net operates is that it turns input stimuli into output signals in a manner which is very similar to the way in which biological neural systems work, thus the name. Similar to biological systems Neural Nets are very good at producing reasonable results with incomplete information in a "best guess" type scenario.

The possible applications of Neural Net technology is in areas where one solution has to be derived in the presence of ambiguous, and conflicting information, like in deciphering handwriting or understanding speech; or when a set of input stimuli allows more than one feasible solution, like scheduling a spacecraft (Johnston 1988). This will be of importance for HST scheduling, out also for the operation of complex ground based facilities such as the ESO VLT. Promising applications in astronomy also include machine assisted object classification (Adorf & Murtagh, 1988).

Even the simulation of Neural Networks on conventional machines demonstrates that the technology is competitive for the above applications. However, because of the relative simplicity of the processing elements, Neural Networks lend themselves to implementation on massively parallel hardware. Not only is this computationally very interesting, it will also lead, in the near future, to developments like the combination of panoramic detectors with Neural Nets into cameras which not only produce an image but also parameterize ("recognize") the objects they are looking at. Military technology is obviously the driver for these developments, but some interesting astronomical applications can be thought of (for instance, to scan large image data bases for certain classes of objects).

More in line with the first section of this paper, there is research going on at this time to use Neural Network techniques in order to improve the performance of Expert Systems. Specific areas are: rule base search, conflict resolution, reasoning with uncertainty, and rule firing strategy.

WHAT CAN BE DONE ?

Even though the above considerations may be interesting, and the initial results encouraging, at some point questions have to be asked such as: what are the most promising near-term applications, and how do we go about implementing them?

The amount of software available to the astronomer has grown enormously, to the extent that some visitors at ESO are not complaining any longer about the lack of available options, but about their multitude and complexity. For these reasons the desirability of "user-friendliness" has been proclaimed. Indeed, many packages offer extensive on-line help as well as nice hard-copy documentation; command languages and user interfaces of various degrees of complexity and functionality abound. However, not only is the sheer volume of some of the manuals frightening, the documentation of the different systems follows different and inconsistent guidelines. In most cases, one has to pretty much know what one wants and how to do it in order to find the required information.

In addition to the problems mentioned above, there are significant areas of astronomical research in which we do not use the computer at all, although the technology required is already available and the capabilities could be implemented easily. One of the most obvious areas is library work, where so far we are using the computer only in order to generate lists of preprints and for some administrative work. Obvious simple improvements, already being offered through commercial packages, are systems which scan the titles and the abstracts of incoming publications for certain keywords and alert the users to potentially interesting papers.

THE CONCEPT OF THE COMPUTER-BASED RESEARCH ASSISTANT

The only long-term solution to the problem of how to use the computer more efficiently is to get the computer to help us.

Trying to define a mental model of the kind of help which would be ideal, one quickly finds that the most ideal help would be a system resembling a Research Assistant. The ideal (human) research assistants are of senior graduate or junior undergraduate level, so they can be talked to in domain terminology (jargon). They know, for instance, that RA and magnitudes run upside down, and why. They can be asked to look something up in the library, and they can be relied on that they also consider the context correctly. We expect them to provide unsolicited input if relevant, and to check our reasoning. Most importantly, they should know how to get a publication quality plot out of the computer, how to run MIDAS, or how to contact our colleagues in California through SPAN. Another quality which the research assistant has to possess is superb communication skill.

Not surprisingly, the "assistant" concept is being used in a number of areas with similar requirements: "Mudman" is a software system for support and evaluation of geology tests; "Pilots Assistant" is a system which aids pilots in the operation of complex aircraft; "Programmers Apprentice" controls the functions of a software development environment.

As already described, we have done the first experiments here: the Data Analysis Assistant can perform the calibration of generic CCD frames and map the operation into the MIDAS or IRAF systems.

POSSIBLE TRIAL IMPLEMENTATION

The area in which the single most important gain for the largest number of people can be achieved within an organisation like ESO is the area of data analysis (i.e. MIDAS) support. This is facilitated by the fact that we fully control the system, in the sense that there is no proprietory software involved, and we have the domain experts (the original authors, or experienced users of the MIDAS application programs).

MIDAS could be covered with an Expert System driven user interface, which has access to all MIDAS documentation and which shields the user from the idiosynchracies of the MIDAS command language. The system should have limited natural language understanding capabilities in order to be able to extract from the user the information which MIDAS needs to successfully perform data analysis operations.

This can be achieved relatively simply by replacing the command input terminals of the MIDAS work stations with PC class machines. Using window-icon-mouse-pointer interface technology, the user would be free to either talk directly to MIDAS (or to other tools, for that matter), or to use varying degrees of expert help through the "Assistant".

As we gain more experience with this approach, we can tie more and more of the tools which are currently residing on the computer into this system: HST archive queries, network access, etc. It goes without saying that alternative data analysis systems (AIPS, IRAF) could be made available through such a user interface in a painless (for the user) manner.

The result of this would be the possibility for the casual user to optimally exploit all available tools: in the end, most users are casual users, in the sense that even an expert on MIDAS will need help trying to use AIPS, or STARCAT. An intelligent interface of the sort described above has additional important advantages: for instance, it would be possible to maintain MIDAS as is, i.e. it can be used on computers which do not have sophisticated user terminals; systems could cooperate user-invisibly beneath the interface layer without the need of making drastic changes in these systems, etc.

WIDENING THE BASE

The activities proposed above are, of course, only the very beginning of an integrated research support system. As already indicated, most of the pieces required for such an initial system are available on the market and can be implemented with little or no development work.

Several considerations indicate that the issue should be regarded on a somewhat larger scale. The total investment in observing facilities, for instance, is in the order of $10^9$ dollars for HST, and $10^8$ dollars for the VLT (or for the combined facilities currently available on La Silla). Compared to these expenditures, the current data analysis investment is miniscule, in the range of one percent and less. While the optimum ratio of investment between data acquisition and data exploitation is probably hard to define, it is equally probable that one percent is a bit on the low side.

While the first phase attempted to make the already existing tools available to the users in a more efficient manner, the second phase should attempt to integrate into the data analysis environment additional elements which so far are considered apart from data analysis. The most important single such element is the scientific library.

While reading incoming journals is relatively enjoyable, it is right now extremely cumbersome to do a thorough library search on a subject. Even having found what one wanted to find it is usually extremely time consuming and inefficient to sort the useless from the useful. To enlist the aid of a computerized system in that process is a long overdue step. For instance, the abstracts of articles published in the major refereed journals could be archived on a data base (abstracts in machine-readable form are already available for some journals; also, text scanners could be used). As a very first step this data base can be scanned with an Expert System which is set up to recognize certain contexts and issues a first alert to those scientists which have an interest in the matter. Beyond this, the data base could not only be used for literature search operations, it could actually be a science tool in itself: it could be processed with context understanders, turned into a fact base, and be examined for consistency and inferences in a fully automatic manner. Also, new findings could be checked against such a fact base.

Carrying the concept a step further, we should begin to build knowledge bases of selected subfields of astronomy. Not only could we, in this manner, combine the expertise of more than one specialist in the field in such a knowledge base, we could also generate new knowledge by processing the combined knowledge. This will be especially productive when knowledge bases are combined across disciplinary boundaries. So far, we have been achieving progress by narrowing down areas of expertise, essentially tayloring them to human capabilities; computer based knowledge processing will for the first time enable us to reverse that process.

In the area of software engineering we should aim for the possibility of generating analysis software in real time, during an analysis session, through the use of pre-fabricated building blocks, which can be combined "like pieces of clay" to form a customized application routine. The rudimentary beginnings exist already, using procedure languages, but right now the casual user is not in a position to use this capability in an efficient way.

We should build up the first prototype of a "research station", which in many ways is the extrapolation of the MIDAS work station. The major difference, however, would be that the research station would not be a set of peripherals of one particular computer, but rather would be an independent entity, connected to a number of other machines and data bases (locally and remote) through networks and comprising local computer power. The capabilities of the research station should be under the control of an expert-system driven user interface, which will enable the researcher to interact efficiently with the dynamic environment. In addition, there should be the possibility for the researchers to load their personal knowledge bases into the knowledge server of the station, thus customizing the station to their individual needs and ways of doing things.

LONG TERM CONSIDERATIONS

This phase will get us to the mid-nineties, seven to eight years from now. To make the following plans sound less like science fiction it is appropriate to remember that much of what we are doing with computers today would have sounded pretty much like science-fiction seven or eight years ago.

The main aim of this phase should be to not only use the computer to support the research process, but indeed to alter this process, generating more, and unexpected insights. Let me illustrate this.

In astronomy, as in all natural sciences, we are trying to build models of reality with the ultimate objective of coincidence between the model and reality, i.e. finding the "truth". This process suffers from a variety of shortcomings: the fact that the reasoning being used is predominantly inductive; the fact that models might entirely miss important aspects of reality, and even if they do not, there cannot be proof that they don't; not to mention the fact that our model building capabilities were evolved on Earth, so their application to the rest of the universe is questionable at best. However, not only has this process allowed us to gain considerable insights into the universe, it is also the only one available to us.

Models are being constructed from observed facts (data) by applying principles of model building. These are tools which, when applied to the data, ultimately allow us to make predictions about unobserved (or unobservable) aspects of the model, spatially, in wavelength, and with time. These tools include principles like: constancy, steady change, periodicity, dependence, analogy. They are usually expressed in mathematical notation, and, when combined, quite often lead to complex expressions and operations.

To turn raw data into facts that can be used for model building, they have to be pre-processed. This is known as calibration, parameterization, taxonomy and classification. A problem in this process is that it usually is not free of underlying assumptions, i.e. a previous model (a hypothesis) exists, which influences the observing strategy, the selection of parameters to determine, and certainly the classification. Conversely, quite often a classification is predetermined by the representation of the data, resulting in an unreasonable model.

There are two possible ways to classify data: the purely morphological one (or even physiological one, if the data representation is the dominant element), and one which tries to capture the physics behind the data. Historically, classifications usually start out as morphological/physiological classifications. When refined, they quite often collapse and get replaced by classifications which are more meaningful in terms of physics. This is usually paralleled by a significant change of the underlying model.

These simple considerations illustrate the importance of proper classification in the interrelation between facts and models. The question is, how to define "proper", and, once defined, how to implement it. At the same time the classification should be independent of data representation and human physiology, and allow an interpretation in terms of physics. In principle, we could establish an open ended classification scheme, with the number of dimensions determined by the number of classification parameters, and the bin size governed by the uncertainty of the data. Again, this can be done with and without regard for the physics behind the data. If the physics is taken into account, and the underlying model is reasonably correct, such a classification will result in a refinement of the model, i.e. it will be possible to predict more accurately the behaviour of the object. If done without regard for the physics of the object, in other words, if no assumption is made about a previous model, such a classification might yield a totally new model, which covers previously not understood aspects of the object. The classification parameters of the second type of classification can be derived from multivariate statistics, or Bayesian classification.

Clearly it is a reasonable first step to try to simulate as closely as possible the classification techniques which have been traditionally used in astronomy, if for no other reason, so as to demonstrate that the system is indeed capable of performing the operation. The problem there is mainly one of knowledge engineering: to capture, fully and consistently, the rules which human experts use to classify data, and to map them into one of the existing expert system shells.

202

In the long run, however, the second approach will yield the better scientific return, in the sense that it will allow us to create models which otherwise would have been inaccessible. Needless to say, the realisation of such a system is a long-term project and will require substantial preparatory work, both in the area of statistical analysis, and in the utilisation of the tools.

While the automatized search for the optimum data classification will be an enormous advantage, we will, in the long run, proceed to automatize the model building process. As already stated, our model building capability was developed as a result of evolution, and has served us well in the effort of surviving in the Earth-based environment. Research into model building mechanisms will allow us to identify additional model building tools, and to apply them in contexts which our minds are not conditioned to recognize. Combined with the automatized optimum classification, this will make it possible to derive multiple models, based on different aspects of the observed objects. The scientist will then be able to concentrate on finding out which model is closer to the truth.

In a similar manner, we should examine other steps in the research process and identify computer based solutions, which, in addition to saving time, might allow us to tackle problems which are inaccessible to the human mind because of their complexity.

In general, the approach to unveiling the universe so far has been to look harder in order to find evidence for existing hypotheses or indications of new ones. What is being suggested is to try to improve the process of hypothesis generation in order to gain new insights.

CONCLUSION

It is very hard at this point in time to suggest concrete actions since the planning of the third phase will have to be based on the results of and experiences with the second phase. However, it is already possible to make a few predictions.

The bottleneck in the interaction between the researcher and the computer in the solution of complex problems will be the man-machine interface - we already see this on a small scale trying to evaluate the output of automatic classification. In the computer science laboratories there are already experiments being made with man-machine interfaces of a radically different kind. One major aspect of phase three would thus be the development of the research station outlined above into a research facility using "artificial reality" interface techniques (Foley, 1987).

As far as computing itself is concerned, the main emphasis will be on the manipulation of concepts and on reasoning. Number crunching will still be important, but only in a role of supporting (or disproving) assertions which came out of the reasoning process.

Another foreseeable necessity is that in parallel to publishing the results of our research on paper (or on CD-ROMs, for that matter), we will also have to "publish" by distributing new findings in the form of updates to knowledge bases across networks.

ACKNOWLEDGEMENTS

Thanks are due to H.-M. Adorf (Space Telescope European Coordinating Facility) and to M. Johnston (Space Telescope Science Institute) for many stimulating discussions.

LITERATURE

Adorf, H.-M., Murtagh, F.: Clustering Based on Neural Network Processing. COMPSTAT 1988 (Preprint)

Bibel, W.: Wissensbasierte Software-Entwicklung. Informatik-Fachberichte 112, p. 17, Brauer & Radig (eds.), Springer-Verlag, 1985.

Fischer, G.: Cognitive View of Re-Use and Re-Design. IEEE Software, p. 60, July 1987

Foley, J., Interfaces for Advanced Computing. Scientific American, pg. 127, October 1987

Hindin, H.: Intelligent Tools Automate High-Level Language Programming. Computer Design, p. 45, May 1986.

Hopfield, J., Tank, D.: Computing with Neural Circuits: A Model. Science 233, p. 625, August 1986

Johnston, M.: An Expert System Approach to Astronomical Data Analysis. In: Proc. Goddard Space Flight Center Conference on Space Applications of AI, 1987.

Johnston, M.: AI Approaches to Spacecraft Scheduling (Preprint 1988)

Lippmann, R.: An Introduction to Computing with Neural Nets. IEEE ASSP Magazine, p. 4, April 1987

Rampazzo, R., Heck, A., Murtagh, F., Albrecht, R.: Rule-Based Classification of IUE Spectra. In: Astronomy from Large Data Base (Murtagh & Heck, eds.), ESO Proceedings No. 28, pp. 227, 1988.

# ARTIFICIAL INTELLIGENCE APPROACHES TO ASTRONOMICAL OBSERVATION SCHEDULING

Mark D. Johnston
Space Telescope Science Institute[1]

Glenn Miller[2]
Astronomy Programs, Computer Sciences Corporation

**Abstract:** Automated scheduling will play an increasing role in future ground- and space-based observatory operations. Due to the complexity of the problem, artificial intelligence technology currently offers the greatest potential for the development of scheduling tools with sufficient power and flexibility to handle realistic scheduling situations. This paper summarizes the main features of the observatory scheduling problem, how AI techniques can be applied, and recent progress on AI scheduling for Hubble Space Telescope.

## 1. Introduction

The purpose of automating observatory scheduling is to increase the effective utilization and, ultimately, scientific return from one or more telescopes. The development of increasingly sophisticated satellite observatories, as well as the planned high level of automation of ground-based telescopes, has led to a demand for flexible scheduling so that astronomers can optimally exploit the capabilities that these facilities have to offer.

The fundamental requirements of optimal telescope scheduling are similar in many ways to those of other scheduling problems, e.g. those encountered in commercial and industrial domains. These problems have been found to be notoriously difficult to solve in practical settings. In this paper we discuss the source of some of these

---

[1]Space Telescope Science Institute, 3700 San Martin Drive, Baltimore, MD 21218 USA, is operated by the Association of Universities for Research in Astronomy for the National Aeronautics and Space Administration

[2]Staff member of the Space Telescope Science Institute

difficulties and how the use of advanced software technology ("artificial intelligence", or AI) can be applied to help overcome them. We describe the progress made at Space Telescope Science Institute (STScI) in developing AI scheduling tools for Hubble Space Telescope (HST), and conclude with a discussion of the prospects for integrating automated scheduling into the overall context of observatory operations.

## 2. Formulation of the Problem

The scheduling problem we are concerned with may be briefly stated as follows:

> Given a collection (pool) of *programs*, each consisting of a collection of desired observations, schedule the execution of these observations over some specified time period so that no strict constraints are violated and that preferences on when observations are scheduled, or on the conditions obtaining at these times, are satisfied to the greatest extent possible.

Hidden in this formulation is a wealth of complexity. *Strict constraints* refer to conditions that must not be violated under any circumstances. They determine the *feasibility* of a schedule. They can range from the obvious "don't observe targets too close to the sun" to more subtle statements such as "don't schedule simultaneous instrument changes on different telescopes which would overload the available operations staff." *Preference constraints*, or simply *preferences*, refer to conditions which are more desirable than others to hold in the final schedule. Examples of this type of constraint include scheduling observations as close to the zenith as possible, or scheduling a followup observation as soon as possible after its predecessor. Both strict and preference constraints can be expressed in terms of absolute time or relative to other observations, either past, already scheduled for the future, or part of the same scheduling pool. They can also refer to resource consumption or loading limitations.

In addition to the variety of constraints that may be relevant, other factors can complicate scheduling. Some observations may be *conditional* upon others or upon external factors: they may or may not be executed depending on whether the necessary conditions are met. Observations may be defined at differing *priority levels* which can change as the schedule evolves (e.g. it may be a high priority goal to obtain the first 80% of a statistical sample, but lower priority to obtain the last 20%.) There can also be shifts in emphasis for science reasons: targets of opportunity can disrupt the most carefully arranged schedule. Some of the most important complications are due to intrinsic uncertainty and therefore unpredictability. On the ground the weather is the most obvious such factor, but there are often many others.

## 3. Classical Approaches to Scheduling

Computer techniques for optimal scheduling have been investigated for many years for a number of applications (see, e.g., [1] for a comprehensive review and bibliography). Much of this classical work has focused on versions of the idealized "job-shop"

206

scheduling problem, i.e. the problem of scheduling $n$ tasks on $m$ machines. This problem and related ones are NP-complete, meaning essentially that there are no efficient algorithms for finding optimal solutions (see, e.g., [2]).

The basic problem with these classical results is that they require key features of the problem to be abstracted away, so that even an "exact" solution to the abstracted problem would be of little relevance to the original "real" problem. Approximate solutions to the abstracted problem suffer from the same limitations. For example, there exist good approximate methods for finding near-optimal solutions to the well-known "travelling salesman" problem, to which the telescope scheduling problem is isomorphic if minimizing target-to-target slew time is the *sole* scheduling criterion. But this is rarely the case: other constraints enter into the problem in an essential way; these generally cannot be formulated within the framework of the abstracted problem.

It is clear that classical approaches can be useful for problems which are sufficiently simple: in practice this often means that schedule optimization is driven by a *single* overriding criterion. For the problem of scheduling complex modern observatories, however, this is not the case: more powerful techniques are required.

## 4. Artificial Intelligence and Scheduling

In recent years a variety of new software methodologies have been developed under the general term of "artificial intelligence" (AI). This refers to a collection of software development techniques and tools that have evolved in the course of computer science research as effective ways to represent and solve certain kinds of problems. These techniques have moved from the laboratory into widespread use in applications as their effectiveness has been demonstrated. For the purposes of automated scheduling, the most important of these are:

- a language (Lisp) that is particularly appropriate for manipulating complex data structures and symbolic data

- object oriented programming with inheritance and message passing; this facilitates the incremental development of complex problem representations

- new ways to represent and manipulate knowledge of various types, e.g. frames, associative networks, rules, demons, etc.

- methods for reasoning when facts and/or inferences are uncertain

- improved algorithms and heuristics for searching large and complex problem spaces

- integrated graphics and windows technology for facilitating user interaction

Several artificial intelligence research efforts have considered scheduling as a domain where AI techniques can be fruitfully applied. Of particular interest is the factory scheduling work of Fox, Smith, and co-workers (e.g. [3,4]) who have developed a rich constraint representation and versatile reasoning process for attacking realistic factory scheduling problems. While factory scheduling shares a number of common features with telescope scheduling (most notably a similar set of precedence and efficiency constraints), there are some important differences. Certain important factory scheduling constraints (e.g. minimizing work-in-progress or inventory) are not relevant for telescope scheduling, while the latter has a significant number of highly predictable constraints (e.g. those based on the motions of celestial objects) that can be exploited to limit the search for alternative schedules.

At Space Telescope Science Institute we initiated a project (SPIKE) in early 1987 for the purpose of developing AI scheduling tools [5,6,7] for Hubble Space Telescope (HST). HST scheduling is an extremely demanding task, requiring the scheduling of some tens of thousands of observations per year subject to a large number of proposer-specified and operational constraints [8]. Our overall approach to HST scheduling was inspired by the work of Smith and co-workers on the factory scheduling problem but has drawn on a number of other lines of research as well: as part of the SPIKE project we have developed a new framework for representing and reasoning with scheduling constraints [9] (based on discrete uncertainty reasoning for rule-based expert systems) and new techniques for searching the space of possible schedules [10] (based on recent developments in artificial neural networks).

In the following section we highlight some aspects of telescope scheduling that contribute most significantly to its complexity. This is followed by a discussion of how AI techniques and methods can be effectively applied to the problem.

## 5. Why is Scheduling a Hard Problem?

There are four notable features of the telescope scheduling problem that make it difficult: interacting constraints, uncertainty, optimization criteria, and search.

### Interacting Constraints

As discussed above in Section 2, realistic scheduling problems will typically involve a large number of different types of constraints, both strict and preference. The first problem to overcome is that of *representation*, i.e. accurately describing the relevant constraints so that they can be interpreted by the scheduling software. A suitable representation must include not only binary (yes/no) decision criteria but must also express varying degrees of preference. The second issue is that of *trade-offs*, i.e. the knowledge of how to judge among competing or conflicting constraints. This is necessary because constraints must be considered simultaneously, not individually. The third issue can be termed *constraint reasoning*: this refers to the process of deducing, at any stage of the scheduling process, the implications of constraints and prior scheduling decisions in order to determine permitted and excluded scheduling times. This must also include possible inferences about the degree of preference of the permitted times.

It is evident that the point of constraint representation and trade-off is to capture the knowledge that human schedulers would use when faced with constructing a schedule by hand. Constraint reasoning deals with how to manipulate this knowledge dynamically as the schedule evolves in order to provide a useful view into the available scheduling choices.

## Uncertainty

Because we cannot predict the future with certainty there are occasions when we cannot be sure that required preconditions for an activity will be satisfied at any given time. This unpredictability can influence scheduling in variety of ways. Unpredictable constraints can exhibit completely chaotic behavior (e.g. unexpected hardware breakdown) or can more-or-less smoothly diverge from some predicted state (e.g. a satellite position prediction). This behavior can be characterized by a *coherence time*, i.e. the timescale over which typical "significant" changes in the constraint are expected. The severity of significant changes can be qualitatively characterized by degree of *impact*, i.e. the extent to which the schedule is sensitive to changes (taking into account the trade-offs of the unpredictable constraint with any others that may be relevant). Clearly the most difficult cases to handle are short-coherence-time high-impact constraints: they may well be ignored altogether in advance scheduling and simply invoke a reaction when they occur. Constraints with coherence times that are a significant fraction of the scheduling horizon must often be incorporated explicitly.

## Optimization Criteria

Since the primary purpose of automating telescope scheduling is to optimize telescope utilization, it is clearly important that a scheduling system adequately represent what is meant by "optimal". This is less straightforward than it might seem at first: scheduling goals vary depending on the circumstances, so that a schedule which is optimal in some sense can be far from optimal in another. For example, at different times the most important optimization criterion could be some combination of overall telescope throughput, picking up a disrupted schedule, diagnosing an instrument problem, and scheduling a best match to changing environmental conditions. It is thus important that a scheduling system be flexible in terms of the high-level criteria by which schedule optimality is judged, and that *multiple* criteria be utilized depending on the circumstances.

## Search

The process of scheduling can naturally be viewed as search, where at each step some decision must be made about (a) which activities to consider and (b) how to restrict their allowed scheduling times. Because of the large number of possible choices for (a) and (b) at each step, the effort required to search the space of possible schedules is typically exponential in the size of the problem. This "combinatorial explosion" is the problem most directly addressed by classical approaches to scheduling (in contrast to interacting constraints and uncertainty which are usually idealized away).

Effective search requires the early identification of both "good" decision paths as well as the early pruning of "deadend" paths, i.e. partial schedules must be judged by their *potential* for being completed beneficially as well as by their current state. This is complicated by the fact that scheduling conflicts may not be detected until many steps into the search, at which point a large amount of effort may have already been expended. It is desirable in this case to identify a minimal number of past decisions to "undo" to resolve the conflict and thus repair the partial schedule, instead of simply backing up and throwing the partial schedule away.

Another aspect of the search problem is that there may be no solution because the problem is overconstrained. Since it is generally infeasible to enumerate all possible deadends to prove that this is the case, there is a need to identify and diagnose overconstrained problems without becoming bogged down in a fruitless search.

## 6. AI Strategies for Optimal Scheduling

In this section we survey some of the AI techniques that can help deal with the problems described in Section 5. Many of these techniques have been implemented in the SPIKE scheduling tools; some are planned for future development.

### Separate constraint reasoning from strategic search

This is a statement about the overall architecture of the scheduler. The intent is to separate those aspects of the system that reason about *constraints* from those that reason about (partial or complete) *schedules*. The reason for this separation is that these reasoning processes take place on very different levels. Constraint reasoning is low-level and determines feasible and preferred scheduling times among which choices can be made; strategic reasoning evaluates one or more schedules and actually makes the choices. There may be more than one source of strategic knowledge available to work on one scheduling problem: all would, however, make common use of the results of constraint reasoning.

### Use uncertainty reasoning methods for reasoning about constraints

There has accumulated a large body of theoretical and practical results on reasoning with uncertainty in the context of discrete rule-based expert systems (e.g. Mycin [11] and Prospector [12]). Based on this work we have developed a continuum version of uncertainty reasoning that can efficiently represent a wide variety of scheduling constraints [9]. Our framework is well-suited to the weighing of evidence for and against different scheduling hypotheses, thus providing essential inputs to trade-off decisions.

We associate with each activity (or group of activities) to be scheduled a *suitability function*, a function of time whose value represents how desirable it is to start an activity at that time (or possibly that it is forbidden to start at that time, i.e. would violate a strict constraint). Suitability functions are derived from constraints, an arbitrary number of which may be associated with each activity depending on the type of activity and any specific factors that can affect when it is scheduled. The suitability function of an activity is the product of the suitability functions derived from its

210

constraints. This not only mirrors an intuitive notion of how to combine different sources of evidence for and against scheduling an activity at a given time, it can also be shown to be logically required by the plausible assumptions that combination of evidence should be associative and monotonic [9].

The suitability function framework provides several important capabilities:

- a uniform way to capture human value judgements that enter into the definition of strict and preference constraints and into the trade-offs among conflicting constraints

- a straightforward mechanism for the propagation of constraints, i.e. the deduction of consequences of constraints and strategic scheduling decisions

- the explicit representation of some classes of intrinsically unpredictable constraints, in terms of maximizing the probability that desirable conditions will be met

- a mechanism to track the probable and/or certain consumption or loading of critical resources

## Provide multiple control mechanisms for strategic scheduling and search

Based on the constraint-reasoning layer it is possible to implement a variety of strategic search mechanisms, any of which may be invoked depending on the nature and state of the problem. To date we have implemented three such mechanisms in the SPIKE scheduling tools:

- procedural search: this includes standard search techniques such as best-first or most-constrained-first algorithms. These tend to be computationally expensive and often encounter deadends which result in grossly sub-optimal schedules.

- rule-based heuristic search: this mechanism includes *search rules* to examine the state of a collection of partial schedules to identify the most "promising", and *commitment rules* to decide how to extend the schedule by making some scheduling decision [7]. The rules communicate with the constraint-reasoning layer through "frames" or "schema" that hold summary information about the partial schedules. This general approach is well suited to the representation of quite complex scheduling heuristics. It also has the advantage that it can be easily extended to handle new situations as they are encountered.

- neural networks: a very different approach makes use of an "artificial neural network" [13] to represent a set of discrete scheduling choices. These networks are conceptually composed of a large number of simple processing elements operating in parallel whose computational power comes from their massive interconnection. These connections can be derived directly from the suitability functions of the activities to schedule [10]. The advantages of this approach are rapid execution, the ability to easily reschedule, and, on hardware now in the development stage, the possibility for a true parallel implementation.

Other mechanisms could also be implemented (and are currently under investigation at STScI). The most important of these are:

- plan repair mechanism: this is a facility to examine an infeasible schedule, analyze it to identify which prior decisions contributed to the conflict, and undo a sufficient number of those decisions to make the schedule feasible again. Various criteria could be employed for which types of decisions are undoable and which activities can be rescheduled. This mechanism would be useful not only for revising an infeasible schedule, but also for "reactive rescheduling" when an ongoing schedule is disrupted.

- alternative perspective focus: this mechanism would focus on certain classes of constraints (strict and preference) to help with optimizing scheduling decisions and identifying "bottlenecks" that can be determined from that class alone. For example, potential resource overloads could be detected early and then avoided by judicious scheduling of activities which use that resource. Constraints grouped into classes of this type can be regarded as corresponding to the *perspectives* of Smith et al. [4].

## Formulate and attack the problem hierarchically

A common and important problem-solving strategy is to formulate and solve a simpler higher-level problem, then attack the resulting lower-level subproblems by constraining them with the higher-level solution. In the scheduling domain there are two obvious ways to accomplish this: by scheduling *groups* of related activities at once, and by limiting the *time granularity* of the schedule. For example, it is possible to cluster appropriately related activities into a single "meta-activity" which can be scheduled initially as a single entity. It is also possible to limit the initial decisions on when to schedule activities to e.g. one-week intervals out of a six-month schedule. Then, once a satisfactory allocation of activities to weeks has been determined, each week can be scheduled individually in detail.

This approach has the major advantage that some constraints which are important at the detailed scheduling level can be treated in an average or statistical sense when activities are allocated only at a sufficiently coarse-grained level. This will generally further simplify the calculation and propagation of constraints. The drawback of hierarchical scheduling is that levels may destructively interact: a deadend in a detailed schedule may require revising the higher-level schedule, which can potentially invalidate other detailed schedules. It is thus important that the higher-level problem reflect as accurately as possible the constraints that will be important in detailed scheduling.

## Provide explicit user visibility and control

The approach we have taken in SPIKE is that automated scheduling is fundamentally a support tool for the people who are responsible for making scheduling decisions. In this approach one of the most important characteristics of the scheduler is how it interacts with the user. The user must have *visibility* into all aspects of the scheduling problem and the evolving schedule. The user must also have *control*, i.e. the ability to override any decisions made by the scheduler, and the ability to create and evaluate alternative schedule fragments. Because of the large volume of information required to specify even modest-sized realistic scheduling problems, it is almost essential to utilize graphical display and interaction capabilities. This makes it necessary to implement scheduling tools of this type on single-user workstations, where high-speed graphics and dedicated processing power can both be exploited.

## 7. Conclusions

It is clear that software technology and approaches to scheduling have reached a sufficient level of development that automated telescope scheduling is a realistic goal. The use of artificial intelligence techniques makes it possible to develop and adapt software, such as the HST SPIKE scheduling tools, for a variety of telescope scheduling problems (see [14] for a discussion of the experimental use of SPIKE on ground-based telescope scheduling). The advantages of using these techniques are primarily a rapid software development cycle, a concise but expressive representation of scheduling data, flexibility in the definition and modification of scheduling constraints, powerful facilities for expressing search strategies, and the ability to incorporate a graphics-oriented user interface to help the user understand and modify the schedule.

Observatory scheduling is not an isolated task: for it to be ultimately successful it must be integrated into the overall operations environment. At the simplest level this integration must include the ability to inform the scheduler of what must be scheduled and what has been executed. A fuller integration should include capabilities for [15]:

- automatic access to data on environmental conditions and predictions that can be used to update scheduling constraints

- user support for proposal preparation, including observation design tools and simulators

- integrated planning capabilities, so that e.g. observations of various types are included in the scheduling pool along with appropriate calibration observations

- feedback from the schedule as executed, so that deviations and discrepancies can be recognized and diagnosed as early as possible

While this represents an ambitious program, it is not beyond the reach of current technology or its modest extrapolation.

Acknowledgements: The authors are grateful to J. Sponsler, S. Vick, K. Lindenmayer, and R. Jackson (STScI), D. Rosenthal (NASA Ames Research Center) and H.-M. Adorf (ST-ECF) for many useful discussions, and for the hospitality and support of the Space Telescope European Coordinating Facility and the European Southern Observatory (Garching) where some of this work was conducted.

# References

[1] King, J.R., and Spachis, A.S. 1980: "Scheduling: Bibliography and Review," Int. Journal of Physical Distribution and Materials Management **10**, p. 105.

[2] Garey, M., and Johnson, D. 1979: *Computers and Intractability*, (W.H. Freeman & Co.: San Francisco).

[3] Fox, M, and Smith, S. 1984: "ISIS: A Knowledge-Based System for Factory Scheduling," Expert Systems **1**, p. 25.

[4] Smith, S., Fox, M., and Ow, P. 1986: "Constructing and Maintaining Detailed Construction Plans," AI Magazine, Fall 1986, p. 45

[5] Miller, G., Rosenthal, D., Cohen, W., and Johnston, M. 1987: "Expert System Tools for Hubble Space Telescope Observation Scheduling," in *Proc. 1987 Goddard Conference on Space Applications of Artificial Intelligence*; reprinted in Telematics and Infomatics **4**, p. 301 (1987).

[6] Johnston, M., 1988: "Automated Telescope Scheduling," in *Proc. Conf. on Coordination of Observational Projects*, Strasbourg, Nov. 1987, in press.

[7] Miller, G., Johnston, M., Vick, S., Sponsler, J., and Lindenmayer, K. 1988: "Knowledge Based Tools for Hubble Space Telescope Planning and Scheduling: Constraints and Strategies", in *Proc. 1988 Goddard Conference on Space Applications of Artificial Intelligence*.

[8] "HST Planning Constraints", 1987, Space Telescope Science Institute, SPIKE Report 87-1.

[9] Johnston, M. 1988: "Reasoning with Scheduling Constraints and Preferences," in preparation.

[10] Johnston, M., and Adorf, H.-M. 1988: "Scheduling with Neural Networks", in preparation.

[11] Shortliffe, E. 1987: *Computer-Based Medical Consultations: MYCIN* (American Elsevier: New York).

[12] Duda, R., Gaschnig, J., and Hart, P. 1980: "Model design in the Prospector consultant system for mineral exploration", in *Expert Systems in the Microelectronic Age*, ed. Michie, D. (Edinburgh University Press).

[13] Hopfield, J., and Tank, D. 1985: "Neural Computation of Decisions in Optimization Problems," Biological Cybernetics **52**, p. 141.

[14] Johnston, M. 1988: "Automated Observation Scheduling for the VLT", in *Proc. ESO Conference on Very Large Telescopes and their Instrumentation*, Garching, March 1988.

[15] Fosbury, R.A.E., Adorf, H.-M., and Johnston, M. 1988: "VLT Operations - the Astronomers' Environment," in *Proc. ESO Conference on Very Large Telescopes and their Instrumentation*, Garching, March 1988.

# OCAPI : AN ARTIFICIAL INTELLIGENCE TOOL

# FOR THE AUTOMATIC SELECTION AND CONTROL

# OF IMAGE PROCESSING PROCEDURES

Monique Thonnat and Véronique Clément

INRIA : 2004 Route des Lucioles
Parc de Sophia Antipolis , F-06560 Valbonne , France

Abstract : We present a general artificial intelligence tool, named OCAPI to manipulate the knowledge on image processing methods and on the use of libraries of procedures. The functionalities and the architecture of OCAPI are described. Then an example of utilization is developed for a pattern recognition problem in astronomy : the processing of an image of a galaxy for its description.

## INTRODUCTION

When we develop a new image processing tool, usually we start from already existing sets of procedures collected in libraries. A new application (i.e. working on new images to extract some information) will need to adapt the use of these procedures and build specialized ones. The time needed to perform the first operation can be notably speed up if an intelligent tool is able to manipulate the knowledge about the library and to perform automatically the tasks of planning (selection of a sequence of programs), adjustment (choice of the input parameters and options) and control of execution (evaluation of the results). When the new programs are built, the knowledge on their context of use, their effects, the way to initialize and adjust their options and input parameters, the syntax and the way to evaluate the results can be added to the knowledge base to extend it.

Such a monitoring tool is now feasible using the expert systems techniques [Hay 84]. Planning has been largely studied in artificial intelligence. Previous work [Fri 79] or current systems [Bro 85] provide some of the features needed for a monitoring tool [Tho 88a]; some studies in progress [End 85] are related to the problem of control of execution. But as there is still no system adapted to both all the tasks of planning, adjustment and evaluation, we have developed OCAPI, an expert system generator, designed for monitoring, which is independent of the application domain and of the libraries of procedures.

## OCAPI

In this section, we present OCAPI : an implementation of a monitoring tool for the automatic control of image processing procedures. The context of its use, the structure and the contents of the knowledge base and the control mechanisms are presented in the next three sections.

## The Context

We propose a tool to build knowledge based systems for the automatic monitoring of libraries of procedures. This tool OCAPI, which is independent of the library, is a kernel to which specialized knowledges are added to make expert systems (see Figure 1). It has three components:
- a specific library of image processing procedures,
- a knowledge module describing the procedures of the library,
- a general kernel for the reasoning in monitoring.

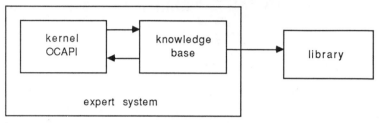

Figure 1. Synoptic of a monitoring system

The Expert System : Usually, the user directly interacts on the images by running libraries of procedures. With an expert system built with OCAPI (see Figure 2), a knowledge-based layer is added between the user and the data and programs. According to a request of the user to process an input image, the expert system uses its knowledge on the procedures of the libraries to define a plan of processing and to execute it.

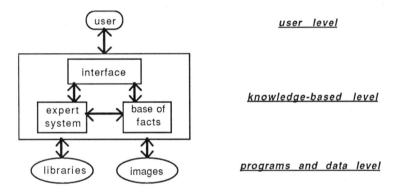

Figure 2. Expert system running

The Libraries : The libraries contain the code of a set of procedures. These procedures can be of various kinds : image processing procedures from a standard library, programs developed by the user, operating system commands, and written in different languages : C, Fortran77, Le-Lisp[1] or operating system command.

The Images : They are the input image, the intermediate ones and the final results. They are stored out of the expert system.

The Base of Facts : This base contains the *description* of the input and computed images, directly stored in the expert system. These facts are those given by the user or deduced by the system.

---

[1] Le-Lisp is a registered trade-mark of INRIA

## The Knowledge Base

In the knowledge base, we have many kinds of knowledge to express : the description of the procedures of the libraries, the image processing sequences, and heuristic criteria to choose between methods, to evaluate results and to adapt the values of the parameters. and we need a set of representation schemes to describe them naturally. Three schemes are specially well adapted:

* the frames (structured objects with numerical, symbolic or complex slots) to represent in a declarative mode the objects of the domain and their components,
* the semantic nets (graphs made of symbolic nodes and labeled links) to represent the various relations between objects,
* the production rules (if condition then conclusion) to represent the heuristic (or operating) knowledge.

The goals : Each goal represents an image processing functionality; it is implemented as a frame with specific slots which can contain symbols, frames or rules. The main slots of the object are: the functionality, the input and output parameters (each one is a frame of type Parameter), the knowledge to choose between methods and to evaluate the results (implemented with rules: choice rules and evaluation rules).

> **Goal** *symbol*
> functionality : *symbol*
> input-parameters : *list of Parameters*
> output-parameters : *list of Parameters*
> choice-rules : *list of rules*
> evaluation : *list of rules*

Figure 3 . Structure of a goal with in italic the type of each slot

The operators : We define by operator the general term for each action that can be performed: at the lowest level, an operator is a program; at a higher level it is a vision algorithm (relaxation algorithm...) or an image processing (contour detection...) or more generally a mean to reach a goal. Each operator is implemented with a frame. The general slots are: the name, the functionality, the input and output parameters, the characteristics which explicit the effects of the operator (in terms of the parameters of the goal), the adjustment which expresses the way to modify the result by changing the values of the input parameters (implemented with rules). These operators are divided into two subclasses: the elementary or terminal operators and the complex ones.

> **Operator** *symbol*
> function : *symbol*
> effects : *list of frames*
> input-parameters : *list of Parameters*
> output-parameters : *list of Parameters*
> adjustment-rules : *list of rules*
> syntax : *frame* (for terminal operator only)
> decomposition : *semantic net* (for complex operator only)

Figure 4 . Structure of an operator with in italic the type of each slot

*Terminal operators* : They correspond to a program of the library. They inherit the general slots of the operators but they have the particular slot *syntax* to describe the programming language and the way to run the procedure (name, options ...).

*Complex operators* : They correspond to a particular decomposition of a goal into steps. Each step is a special request of a subgoal. The decomposition is implemented with a semantic net where the nodes are *requests* (see Figure 7) and the links are *and* or *then* links.

operator A :

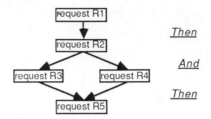

Figure 5. Decomposition of a complex operator

Parameters are described with the following slots : direction (input or output), visualization (the way to show the value, for example command print for ascii value, specific display program for data ...), clear (file management option), the function which enables the connection between the parameters of the goal and those of the corresponding operator.

**Parameter** *symbol*
    type : *symbol*
    direction : *symbol*
    range : *frame*
    default : *symbol or number*
    visualization : *frame*
    clear : *frame*
    function : *symbol*

Figure 6 . Structure of a parameter with in italic the type of each slot

The requests : The knowledge found in the goals and the operators will be used to satisfy requests, coming from the user or the system. Requests are goals to be reached in a certain restricted context. They are implemented with frames. Their main slots are : the goal and the characteristics of the context, which are the restrictions on the input parameters and the specifications on the results (output parameters of the goal).

**Request** *symbol*
    goal : *symbol*
    specifications : *list of frames*
    restrictions : *list of frames*

Figure 7 . Structure of a request with in italic the type of each slot

## The Control Mechanisms

The role of the OCAPI system is to run programs to satisfy a request; i.e. to reach a goal within a certain context. The control mechanisms are built around three components (the Planner, the Parameter Settler, the Pilot), each of them performing a precise task and using for that the appropriate knowledge found in the knowledge base.

The Planner : Given a goal expressed in a request, the role of the planner is to find a tree of operators (the plan).

request R :

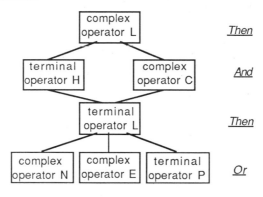

Figure 8. A plan

The Planner acts in two steps: a simple selection phase and a more complex one using inferences. The first phase selects the operators matching with the current goal. The second one restricts these operators, based on the rules of choice attached to the goal, the characteristics of the operators and the characteristics of the request.

The Parameter Settler : The first role of the Parameter Settler is in order to use a selected operator to perform a goal, to connect the parameters of the goal and of this operator, i.e. to find for each input or output parameter of the goal, the corresponding parameters of the operators which may be more numerous.

The second role of the Parameter Settler is to set the values of all the input parameters of the operator. In an initialization phase it means that for each input parameter a value is established using: the default value and the range defined in the operator, and the restrictions coming from the request. In an adjustment phase (decided by the pilot after failure), a new value for each input parameter is determined using: the previous values, the characteristics of the failure, and the rules of adjustment attached to the operator.

The Pilot : The Pilot controls the global execution of the plan; i.e. it tries to execute the plan in such a way that the specifications on the results expressed in the request are satisfied. It selects in the plan an operator to execute, calls the Parameter Settler and controls its execution. The choice of the next operator to execute is done through the hierarchy of the plan. The operators are chosen in sequence from a node *then*, exhaustively from a node *and*, and using the general strategy rules of the associated goal from a node *or*. The control of the execution of an operator is presented on Figure 9.

*To execute a terminal operator*, the knowledge on the implementation language and the syntax is used to run the binary code of the procedure of the library.

*To execute a complex operator*, for each request of its decomposition, the planner is called to give a plan, then each of these subplans is processed by the pilot.

*The evaluation of the results* is achieved using the evaluation rules of the associated goal comparing the output parameters of the goal with the specifications of the initial request, or by interaction with the user who will visualize the results and will appreciate them. If they are found too much different, another execution will be performed after adjustment of the input parameters of the operator, depending on the modification desired.

219

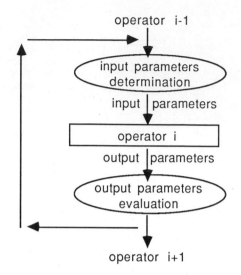

Figure 9. Control of the execution of an operator

## AN EXAMPLE IN ASTRONOMY

We present in this section an application of OCAPI in the field of astronomy.

### The Context

In this application we want to automate the processing of images containing a galaxy.

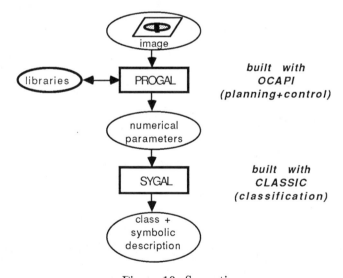

Figure 10. Synoptic

The role of this processing is to detect, isolate and describe the galaxy before its classification by the expert system SYGAL [Tho 88b] built with CLASSIC [Ilo 87], a classification tool generator. For each image the goal to reach is the same : the description of the galaxy. But as there is a great variability in the images, the sequence of procedures and the values of the parameters need to be adapted. In fact, there are low or high resolution images, noisy or good images, CCD or photographic plates, centered or uncentered galaxy on the image, eventual presence of stars; these features can be expressed in the request. So we have built an expert system PRO-GAL, using OCAPI, to pilot the processing of an image containing a galaxy. In this application the set of procedures to pilot is compound by specific procedures (C and Fortran77), standard vision procedures of a library (INRIMAGE) and operating system commands (Unix).

## The Knowledge Base

Requests, Goals and Operators : Figure 11 shows a high level description of the global processing of an image performed by PROGAL.

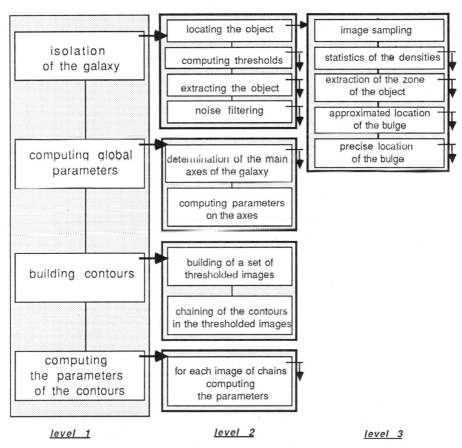

level 1          level 2          level 3

Figure 11. Hierarchical structure of the global processing

This processing may be decomposed into more precise steps. In the knowledge base, we describe in such a hierarchical way the decomposition of a processing into

221

substeps. This description is much more natural than just a list of commands stored in a command file (script), where the goal of the processing is not explicit.

Each step of this decomposition is a request to reach a goal; for instance at level3 (Figure 11) the request "extraction of the zone of the object" corresponds to the goal OZE which has such a functionality; Figure 12a shows the contents of this goal. Many operators can perform this goal; one of them, the operator OZE-1 is shown in Figure 12b. This operator is a complex one, i.e. its decomposition into substeps (requests) is a sequence of a thresholding, an extraction of chains of contours, the research of the location of the largest chain and the extraction of a sub-image containing this chain. The thresholding is a goal (see Figure 13a) for which there is a terminal operator Thr-1 (Figure 13b).

Goal *OZE*
| | | |
|---|---|---|
| function | *Object-zone-extraction* | |
| in-parameters | *in-image :* | type *image ...* |
| out-parameters | *out-image:* | type *image ...* |
| | *Xpos :* | type *integer ...* |
| | *Ypos :* | type *integer ...* |
| | *Xsize :* | type *integer ...* |
| | *Ysize :* | type *integer ...* |
| choice-rules | if *centered galaxy* | |
| | then *use-operator OZE-1* | |
| | else *use-operator OZE-2* | |
| | ... | |
| evaluation | if *zone-size < 200* | |
| | then *object too-small* | |

Operator *OZE-1*
| | |
|---|---|
| function | *Object-zone-extraction* |
| in-parameters | *in-image : ...* |
| out-parameters | *out-image : ...* |
| | *Xpos : ...* |
| | *Ypos : ...* |
| | *Xsize : ...* |
| | *Ysize : ...* |

decomposition
> *thresholding in-image i2*
> then *chains-extract i2 i3*
> then *max-chain-location i3 Xpos Ypos Xsize Ysize*
> then *extract-sub-image in-image Xpos Ypos Xsize Ysize out-image*

adjustment
> if *object too-small*
> then *less-strict thresholding*
> ...

a)                                        b)

Figure 12. Extraction of the zone of the object
a) the goal ; b) an operator

Goal *Thr*
| | | |
|---|---|---|
| function | *Thresholding* | |
| in-parameters | *in-image :* | type *image ...* |
| | *threshold :* | type *float ...* |
| out-parameters | *image-out :* | type *image ...* |
| evaluation | if *image-out.mean ~= image-out.min* | |
| | then *number of selected pixels too small* | |
| | | |
| | if *image-out.mean < request.val* | |
| | then *number of selected pixels too small* | |
| | ... | |

Operator *Thr-1*
| | | |
|---|---|---|
| function | *Thresholding* | |
| in-parameters | *in-image : ...* | |
| | *threshold :* | type *float* |
| | | range *]-inf , +inf [* |
| | | default *0* |
| out-parameters | *image-out : ...* | |
| syntax | language: *shell* | |
| | call: *muls -vs threshold in-image image-out* | |
| ajustment-rules | if *number of selected pixels too-small* | |
| | then *dichotomic-raise threshold* | |
| | ... | |
| effects | *image-out.type = binary  ...* | |

a)                                        b)

Figure 13. Thresholding
a) the goal ; b) an operator

Evaluation : As we have seen in these examples, the knowledge on the control of execution of a processing is given by two means : the evaluation rules and the adjustment rules. For instance there are very general rules of evaluation as these of goal Thr in Figure 13a : if the result of the thresholding is a black image (if the mean of the values in the final image is close to zero (min)) then the number of the selected pixels is too small. For the operator Thr-1 there is an adjustment rule to explicit the way to operate when the number of the selected pixels is too small. The new value of the threshold is computed by dichotomy (Figure 13b).

222

The evaluation of the quality of the results of a processing may be defined at the different levels of the processing: each goal checks that its request is achieved by the selected operator (terminal or complex). For instance the extraction of the zone of the object may be wrong because of a too small number of selected pixels but the detection of the bad result may be done at a highest level by an evaluation rule attached to the goal OZE object zone extraction (Figure 12a). Then according to the operator, adjustment rules will choose the way to adjust the parameters or to restrict the requests.

An illustration is given Figure 14. The input image is presented Figure 14a. A first attempt of thresholding is shown on Figure 14b, the extraction is shown on Figure 14c; then evaluation rules attached to the goal object-zone-extraction detects a bad extraction. After adjustment the new thresholding is displayed Figure 14d and the good result in Figure 14e.

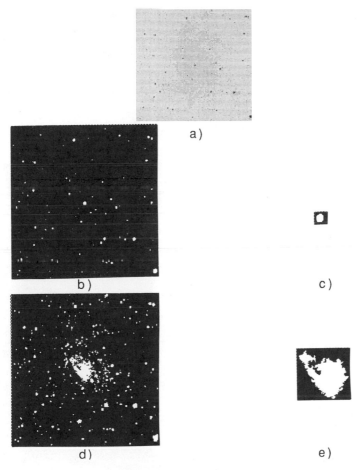

a)

b)

c)

d)

e)

Figure 14. Evaluation process
a) input image ; b) first thresholded image ; c) first extracted zone
d) thresholded image ; e) extracted zone

Because of the large number of stars the evaluation rule of the operator of thresholding has not detected that the selected pixels were not enough numerous. The rule

223

of the higher level testing the size of the object detects the error (the big star in the lower right-hand corner was selected instead of the galaxy), and a new attempt leads to the detection of the galaxy. When the system cannot evaluate a result, the user, visualizing it, will perform the evaluation, the adjustment being done in any case by the adjustment rules. This feature enables the progressive building of a knowledge base in control.

## CONCLUSION

We have developed the architecture of OCAPI to have both the facilities of planning and the automatic adaptation of the plan during its execution. OCAPI is a general tool which enables the building of expert systems. These systems facilitate the use of existing procedures libraries. Moreover, because of their structure, they are easy to modify or to extend with new programs and knowledge. As it has been shown in the example the description of the knowledge is done in a very natural way and consequently it avoids errors or incoherence in the knowledge base and enables the building of large automatic systems. As we aim at developing real intelligent vision systems, we have designed this tool to integrate it in a system handling several expertise domains and various reasonings, as planning, control and classification.

# References

[Bro 85]   L.Brownston, R.Farrel, E.Kant, N.Martin, 'Programming expert systems in OPS5', Addison-Wesley, 1985.

[End 85]   M.Ender, 'Design and implementation of an auto-configuring knowledge-based vision system', 2nd Int. Technical Symposium on Optical and Electro-Optical Applied Science and Engineering, Conference Computer Vision for robots, Cannes, France, December 1985.

[Fri 79]   P.E.Friedland, 'Knowledge-based experiment design in molecular genetics', Rep.79-771, Computer Science Dept., Standford University, 1979.

[Hay 84]   F.Hayes-Roth, D.B.Lenat, D.A.Waterman, 'Building expert systems', Addison-Wesley, 1984.

[Ilo 87]   ILOG, 'CLASSIC Manuel de l'utilisateur', Paris 1987.

[Tho 88a]  M.Thonnat, V.Clément, 'OCAPI : a monitoring tool for the automatic control of image processing procedures', 12th IMACS World Congress on Scientific Computation, Paris, France, July 1988.

[Tho 88b]  M.Thonnat, 'Towards an automated classification of galaxies', in Le monde des galaxies, PHYSICS, Springer-Verlag, New York Inc. Eds, 1988.

# AN EXPERT ASSISTANT SUPPORTING

# HUBBLE SPACE TELESCOPE PROPOSAL PREPARATION

H.M. Adorf [1]  and  S. Di Serego Alighieri [1,2]

[1] Space Telescope - European Coordinating Facility
European Southern Observatory
Karl-Schwarzschild-Str. 2
D-8046 Garching bei Munchen, F.R. Germany

[2] Affiliated to the Astrophysics Division
Space Science Department
European Space Agency

## ABSTRACT

The *IIST Expert Assistant* is an advice giving expert system currently under development within the Artificial Intelligence Pilot Project of the Space Telescope - European Coordinating Facility. The system aims at supporting astronomers who are planning observations with the Hubble Space Telescope. User requirements, knowledge elicitation and design issues, current status and future plans are described and discussed.

## INTRODUCTION

A scientist wishing to use the Hubble Space Telescope (HST) must apply for observing time. This is a two stage process: In the first, he is requested to fill in a proposal form describing the scientific rationale together with sufficient details of the proposed observation to allow the assessment of feasibility. If the proposal is accepted by peer review, more detailed information has to be supplied in order to allow the observations to be entered in the telescope observing schedule which eventually becomes the telemetry command stream to the spacecraft.

The process of writing an observing proposal for HST differs from that for a ground-based telescope, which every astronomer is familiar with, in two respects: Firstly, the information on the observations to be performed must be more detailed and precise for HST, since the observing schedule will be nearly completely defined from it, with particular care on exposure sequence optimization. Secondly, although HST is a complicate instrument and no real experience with it is available yet, HST is described in detail in many documentation volumes and technical brochures. These and some other considerations led us to believe that an expert system (see e.g. Stefik et al. 1982; Buchanan & Duda 1983; Hayes-Roth 1985; Shoval 1985; Thompson & Thompson 1985; Ramsey et al. 1986) might provide an effective way to synthesize large amounts of information and to convey it to the inexperienced user, thereby providing assistance in the definition of the technical means to achieve a specified astronomical goal with HST. We are fully aware that developing an expert system in the astronomical domain is a challenging task, primarily because no precursor exists.

In the following we will report on the current status of the *HST - Expert Assistant* (HST-EA), an expert system in support of astronomers who are planning to use HST for their observations. As observation planning is to the astronomer what experimental design is to the physicist, the HST-EA might be compared with the *EXPER* expert system (Esposito et al. 1986), aiding users in the design of experiments. The HST-EA is one of two expert systems which have been initiated within the Artificial Intelligence Pilot Project (Adorf et al. 1986; Adorf & Johnston 1987) of the ST-ECF.

## THE PROPOSAL PREPARATION TASK

The following two examples, quoted from the HST *Proposal Instructions* (Whitmore 1985), may illustrate the kind of mapping from ends to means, which an expert assistant would have to support. The first couple of lines represent the input (the goal), and the exposure lines represent the output (the means) ready to enter the exposure logsheet. (Here FOC and WFC abbreviate Faint Object Camera and Wide Field Camera, respectively).

**Photometry of a faint A0 star** (FOC: simple exposure)
The FOC is used to observe a 25th magnitude A0 star.
EXPOSURE LINE NUMBER:
> (1) A 14.5 min FOC exposure of a faint A0 star is requested using the FOC/48 configuration and the IMAGE mode. The F430W filter is used with the 512x512 pixel field of view. A S/N of 10 is required.

**Quasar Survey** (WFC:generic target parallel observation)
The WFC is used to make observations of 25 fields with galactic latitude |b|>35°. These fields will be used to identify quasars. The observations are made in parallel with the primary observations of other observers,on a non-interference basis.
EXPOSURE LINE NUMBERS:
> (1) Two 12 min exposures with the WFC and the red grating (G450L) are requested during dark time (when the HST is in the Earth's shadow).
> (2) Two 5 min exposures of the same field are made using the *B* filter (F439W).
> (3) The four exposures specified on lines 1-2 are repeated for 24 other fields. Note that line 3 defines 100 exposures.

The purpose of the HST-EA is to assist the user during the formation stage of ideas for observations. In order to achieve the required mapping, information from the following areas must be available: (i) association of observational goals, described in general astronomical terms, with specific details of HST *instrument modes*, (ii) location and use of the HST *instrument database*, (iii) functionality provided by and use of the HST *instrument models*, and (iv) status of HST *calibration data*.

The purpose of the HST-EA is to collect the relevant information in an extensible, maintainable and consultable knowledge base. The HST-EA rest upon an extensible knowledge base of *static* facts about HST instruments, configurations, modes, filters and other knowledge relevant for proposal preparation. Its dynamic reasoning capabilities let the expert system effectively function as an advice giving *decision support system*.

## KNOWLEDGE ACQUISITION

Knowledge Acquisition for the HST-EA differs somewhat from standard knowledge engineering practice of interviewing domain experts: Firstly, there is no record of past proposal preparation cases, for, in cause of the Space Shuttle disaster, ST-EFC domain experts have hardly had an opportunity to assist outside astronomers in HST proposal preparation tasks, and thus problems which may occur and their solutions have to be imagined. Secondly, as already mentioned, much of the information needed for the construction of the HST-EA already exists in computer files, which happen to contain a large number of tables.

Therefore a keystone of the knowledge acquisition component of the HST-EA is a *table-to-knowledge-base converter*, which allows tabular data from text files to be entered into the expert system fact base (see Fig. 1) with little or no human intervention. The existence of such a converter has a couple of advantages: when the instrument data, on which the expert system rests, changes - and changes have in fact already occurred - an easy way is required and provided by the converter for updating the knowledge base; also interfacing with an image processing system such as MIDAS is facilitated by a simple data structure, which effectively establishes a kind of communication protocol.

## USER INTERFACE AND CONTROL

The construction of a suitable man-machine interface and flexible control mechanism is crucial for the success or failure of any expert system. The user interface problems (for a general discussion of user interface design issues see e.g. Casey & Dasarathy 1982; Durham et al. 1983; Good et al. 1984; Bertino 1985; Draper & Norman 1985; Crawford & Becker 1986) involved may be illustrated by comparing the interaction of an astronomer with a standard data analysis system (DAS) and with a computerized expert system (ES), respectively.

In case of a DAS, the astronomer is supposed to know the system's capabilities, e.g. from reading introductory material. So the astronomer will act as an 'expert commander' and the system re-acts as an executing 'slave'. The DAS will never 'take over'; it is reactive, but passive otherwise.

Interactions with an ES are quite different and the communication situation resembles in a way that of a co-operative human-human dialogue with mixed initiative (see e.g. Wahlster 1985; Wahlster & Kobsa 1986; Webber 1986). At the start of a consulting session the uninitiated user has no conceptual model of the system's capabilities; conversely the system has no knowledge of the user's problem. Both, user and system have to 'introduce each other' at the beginning of the session. Also, the user may have an ill-defined perception of his/her problem, which, ideally, the system would detect and correct. If an error has been made or the user found him/herself in a dead end it is necessary to backstep and allow for a correction. Thus the the ES must use non-monotonic reasoning capabilities.

Control is another issue of concern. For example a user might already know every detail of his/her exposure except the exposure time. Thus he/she would like to enter all necessary information in one go and retrieve the required result as quickly as possible. Another user might only know that be/she wants to observe a target in a particular wavelength range, but being undecided about whether to take a (low-resolution) spectrum or a series of narrow-band photometric exposures. He/she could not decide on the instrument early on, but only after having explored several possibilities.

So neither a system which is always commanded by the user (control is completely with the user) nor a system always querying the user (control is completely with the system) is feasible for an ES to be used for proposal preparation. Instead a more flexible control mechanism would seem preferrable. However, as in real life, mixed-initiative is not easy to achieve, and certainly requires further consideration and thought.

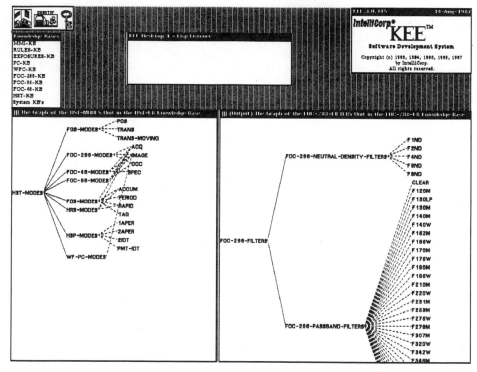

Fig.1. Part of the Knowledge Base of the HST-Expert Assistant. Solid lines indicate class-subclass relations, dashed lines indicate class-instance relations.

## IMPLEMENTATION

Expert system work in these days can rely upon commercial expert system shells (for reviews on such shells see Cross 1986; Gilmore et al. 1986; Richer 1986, Gevarter 1987; Szolovits 1987), the larger of which almost exclusively are written in Lisp. The HST-EA is implemented in Common Lisp and embedded into the Knowledge Engineering Environment (KEE) expert system shell. Development work originally started on a Symbolics 3620 workstation and is now being carried out on a Texas Instruments Explorer dedicated Lisp machine.

The HST-EA mainly consists of an extensible fact base containing facts attributed attached to objects, also called *frames* or *units* (see e.g. Fikes & Kehler 1985). Frames are hierarchically organised into classes and classes of classes and represent a conceptual model of the knowledge domain.

We have decided to define the knowledge base (KB) completely in terms of logical (class-subclass or 'is-a' and class-membership) relations, which are the only native relations built into KEE. A more complete semantic net (see e.g. Ritchie & Hanna 1983), allowing other relations such as 'consists-of' or 'has-part' etc., is constructed from the logical description using a set of *semantic net rules*.

At any time the knowledge base can completely be reconstructed from text files. This facilitates maintenance and allows the use of Lisp macros and functions to help building the KB. Also, a transition of the KB into a system different from KEE, if that were required at some later stage, should be not too difficult.

The HST-EA in its current state comprises the following modules:
- a fact base about the HST instruments in general and detailed information about the Faint Object Camera (FOC) and the Wide Field and Planetary Camera (WF/PC), stored in KEE data structures,
- an exposure time calculator for direct imaging observations of point sources with for FOC and WF/PC instruments,
- the complete Guaranteed Time Observers (GTO) catalogue, stored in the Relational Table Management System (RTMS) of the Explorer and
- an *HST dictionary*, comprising a large part of the Proposal Instructions and almost the complete Call for Proposals and stored in the Explorer's Glossary hypertext system.

The latter two modules are still in the experimental stage and it has not yet been decided whether development on these will continue.

## FUNCTIONALITY PROVIDED BY THE HST-EA

*Selecting an instrument configuration*

A very basic step in proposal preparation consists in finding the most appropriate instrument configuration and mode, a task, which can be described as establishing a mapping between some exposure characteristics such as 'high-resolution time-resolved spectroscopic' observation and instrument configurations/modes capable of carrying out the required observation. In case of HST, as with many ground-based telescopes, for a given observational characteristic there may be many instrument configurations, or exactly one or none satisfying the request. In other words, the relation from exposure characteristic to configurations may be one-to-many, one-to-one or one-to-none.

We have implemented a simple module which is based on a decision table, indicating the suitability of an instrument configuration for a specific task. The user may require an observational mode like *imaging, spectroscopy* etc., and may enter additional exposure characteristics such as *the target is a point source, the observation requires high resolution* etc. by clicking with the mouse on the appropriate menu item in the instrument configuration selection window (Fig. 2). The system almost instantaneously displays a selection of those instrument configurations which suit the specified exposure characteristics. For those interested in the internals: the system uses 'frame inheritance' for its reasoning.

Of course the configuration selection function of the HST-EA is of little use to an instrument scientist, who probably knows the feasible configurations by heart. An inexperienced user, however, might the surprised to learn that the HST Fine Guidance Sensors or the High Speed Photometer may be used for imaging of point sources - a potentially useful choice which might be easily overlooked otherwise.

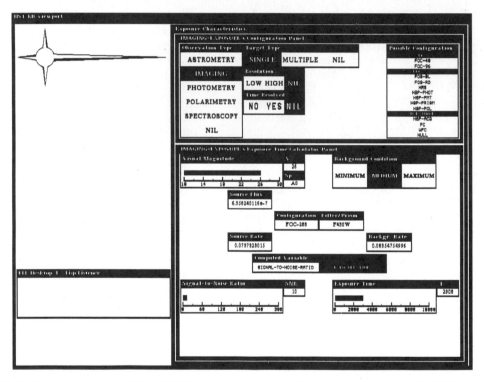

Fig.2. The user interface of the HST-Expert Assistant. Its major components are the *ST-ECF logo* (top-left), the *Configuration Panel* (top-right) and the *Exposure Time Calculator Panel* (bottom-right). The *ST-ECF logo* provides a catch-all help and command facility. The *Configuration Panel* allows the specification of observational type, target type, spatial or spectral resolution and time resolution, from which the HST-EA derives possible instrument configurations. The *Exposure Time Calculator Panel* allows the specification of some target properties such as apparent target brightness and spectral type, which are used in conjunction with other quantities to arrive at an estimated exposure time or a expected signal-to-noise ratio.

## Exposure time calculation

An important part of proposal preparation is to acquire a feeling about what signal-to-noise is achievable in a reasonable exposure time with a particular instrument configuration. This requires as system component which is easy to operate (as always) and which quickly responds to changed user input. The HST-EA has been equipped with an exposure time calculator (ETC) module which closely models the flow of photons on their way from the telescope aperture through filters, prisms, polarizers onto the detector. Currently exposure times for the FOC and WF/PC instruments in imaging mode can be estimated. A near-term goal consists in extending the expert system's capabilities to other instruments and modes.

The ETC needs as input the spectral type and apparent brightness of the target, the background level (low, medium, high), the instrument configuration (FOC in f/48, f/96, f/288 imaging mode, WFC and PC) and the selected filter. From these quantities the ETC estimates the exposure time necessary to achieve a given output signal-to-noise ratio, or, conversely, the signal-to-noise achievable in a given exposure time.

A few words on the different background options: For HST background, being essential for the calculation of signal-to-noise ratios or exposure times, cannot be foreseen exactly, when the calculations are performed. We therefore decided to implement three different levels of background-levels: 'Medium background' is the expected background under standard conditions. No problems are foreseen for the HST scheduler. 'Maximum background' is the worst-case background where observations are just possible. This option is included essentially in order to allow the user to check how critical the background conditions are for the exposure. 'Minimum background' occurs only during darktime and, if especially required, may have an adverse effect on the schedulability of the exposure.

Each input quantity can be specified by typing into a Lisp listener. The much preferable mode of interaction is by clicking on the active image (see Fig. 2) which displays the current value of the quantity to be changed. The effect of a changed input is almost instantaneously propagated to the selected output quantity. Internally, calculations are performed within a *constraint network*, a standard technique in Artificial Intelligence: 'daemons' individually watch input and intermediate quantities and issue an action if they detect a change. In effect, calculations propagate like an avalanche. Other daemons permanently watch fixed constraints, such as detector saturation, and, if violations are found, alert the user, e.g. by flashing a simple value display or by sending a message to a special text window.

## Accessing the GTO database

Checking the GTO catalogue for duplication is a recommended step in proposal preparation. In order to facilitate some initial experimentation with the GTO catalogue, the 1985 "GTO Observing Program" was downloaded from the IDM database machine, where it is regularly kept, into the Lisp machine and stored in the Explorer's RTMS, an object oriented database management system. This allowed us to construct a GTO catalogue access module which made querying almost trivial: The fact that some fields of the GTO-catalogue may contain only one out of few different entries, allowed the construction of a customized menu driver to the GTO catalogue. For most fields the user specifies a retrieval request by clicking on the appropriate menu item. Queries for fields which can hold rather arbitrary entries, e.g. the target name, can be specified by a user supplied text string. An interesting implementational feature of the GTO module is that it translates a user request into a syntactically correct query in the RTMS query language. Another convenient feature of the menu interface is that the fields of retrieved data records are displayed in the order in which they were specified in the query.

With improved and more reliable local area networking capabilities it is foreseen to replace the access to a local copy of the GTO catalogue by a remote access to the original version on the IDM database machine. Also a more sophisticated interface between the KEE system and the database machine may be envisaged and, indeed, is commercially already available (see Adorf et al. 1988).

*An HST dictionary in a hypertext system*

The basic idea of a hypertext system (see e.g.Conklin 1987) is to provide a non-sequential access to textual information, similarly to a printed dictionary with its cross-references. At least in one case a hypertext system has been used as a tool supporting knowledge acquisition (Anjewierden 1987).

We felt that it would be worth the effort to experiment with a dictionary of definitions, explanations and acronyms - and there are a lot of them within the HST project - and to insert it into the glossary hypertext-like system provided on the Explorer. To this end, part of the Proposal Instructions and the Call for Proposal were converted into a form suitable as input to the hypertext system.

An active HST dictionary resulted which can be browsed (Fig.3) for textual explanatory information relevant for proposal preparation. A first keyword may be selected from thumb index and the corresponding text paragraph is displayed in a large window. A keyword occurring in that text portion are also mousable and actually mousing it will pull the corresponding text block onto the screen and so on. Thus the user may walk around in a network of interrelated notions.

Our experience was that a conversion from linear text to hypertext does not pose major problems. In principle the functionality provided by the system tested was sufficient apart from the following two limitations: (i) Mathematical formulae, of which the HST Proposal Instructions contain many, cannot be represented in a satisfactory way. (ii) Non-textual information such as graphs or even images cannot be included into the text. However, a new hypertext system has just been delivered with the new version of the Explorer's operating system and it will be of interest to see how it compares with the glossary system discussed here.

Right now the HST dictionary is an experimental stand-alone facility, i.e., there are no links to the KEE knowledge base of the HST-EA. Such a connection is, however, highly desirable, if, e.g. the HST dictionary is to be used as an explanation component within the expert system. This is possible in principle, for all structures of the hypertext system are normal Lisp data structures.

It should be remarked that, similar to the case of configuration selection above, a dictionary of HST terms is of little use to in-house domain experts. On the other hand, an inexperienced prospective user of the HST might find a browsable dictionary quite helpful and preferrable to extensive printed documentation.

## SUMMARY AND OUTLOOK

Some aspects of the task of planning astronomical observations and preparing proposals have been analyzed from the perspective of a user in need of support and of a 'knowledge engineer' who is supposed to construct a proposal preparation decision support system. We have described and discussed the HST-Expert

Assistant, a prototype expert system for HST proposal preparation. The effort which went into its construction has had a twofold effect: First, a useful tool was created which currently covers imaging of point sources with the Faint Object Camera or the Wide Field and Planetary Camera and which will initially aid ST-ECF staff in their task of supporting astronomers who wish to use HST. Second, experience was gained with a variety of advanced software techniques, an experience which is almost certainly bound to pay off in the future on quite different tasks.

Though designed to be a portable, the HST-EA is currently residing on the development machine and is only internally usable by ST-ECF staff. We are investigating the question of in-house delivery and monitor the market and also the situation within the European astronomical community with respect to the availability of workstations or Personal Computers, which would allow a delivery of the HST-EA on a wider basis.

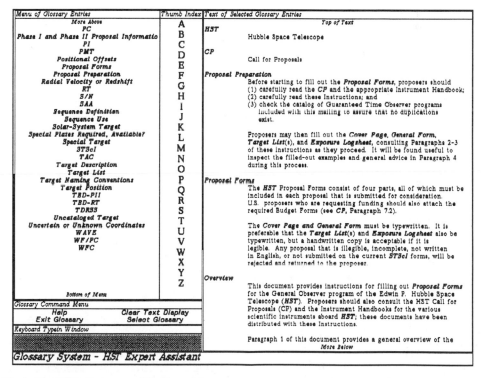

Fig.3. The HST-dictionary. Clicking with the mouse on a letter in the *Thumb Index* (top-middle), the user selects a group of notions, which pop up in the *Menu of Glossary Entries* (top-left). Clicking on a notion of interest will cause the system to display the explanation text in the *Text of Selected Glossary Entries* (top-right). All text portions are automatically cross-referenced by the hypertext system and mousable keywords are displayed in bold typeface.

## ACKNOWLEDGEMENTS

During the work on the HST-EA we have benefited from numerous discussions with our colleagues at the ST-ECF and STScl. We thank especially Bob Fosbury for his continuous encouragement and contributions, Mark Johnston for numerous suggestions and technical hints, and last, but not least, Piero Benvenuti and Rudi Albrecht for promoting the idea of advanced computing at the ST-ECF.

## Appendix: A glossary of abbreviations

AI — Artificial Intelligence
DBMS — Database Management System
ES — Expert System
ETC — Exposure Time Calculator
FOC — Faint Object Camera
GTO — Guaranteed Time Observer
HST — Hubble Space Telescope
HST-EA — Hubble Space Telescope - Expert Assistant
IDM — Britton Lee Intelligent Database Machine
KB — Knowledge Base
KEE — Knowledge Engineering Environment
MIDAS — Munich Image Data Analysis System
RTMS — Relational Table Management System
SQL — Standard Query Language
ST-ECF — Space Telescope - European Coordinating Facility
STScI — Space Telescope Science Institute
WF-PC — Wide Field and Planetary Camera

## REFERENCES

Adorf,H.M., Albrecht, R., Fosbury, R.A.E.: 1986; "Artificial Intelligence Applications within the Space Telescope - European Coordinating Facility", ST-ECF internal document

Adorf, H.M., Johnston, M.D.: 1987, "The Artificial Intelligence Pilot Project of the ST-ECF", ST-ECF Newsl. 8, 4-5

Adorf, H.M., Albrecht, R., Johnston, M.D., Rampazzo, R.: 1988, "Towards Heterogeneous Distributed Very Large Data Bases", in: Proc. ESO Conf. "Astronomy from Large Databases", Garching, Oct. 1987, F. Murtagh & A.Heck (eds.), pp. 137-142

Anjewierden, A.: 1987, "Knowledge Acquisition Tools", Al Communications O, No. 1, 29-38

Bertino, E.: 1985, "Design Issues in Interactive User Interfaces", Interfaces in Computing ???, 37-53

Buchanan, B.G., Duda, R.O.: 1983, "Principles of Rule-Based Expert Systems", "Call for Proposals": 1955, Space Telescope Science Institute, Baltimore, 1-66

Casey, B.E., Dasarathy, B.: 1982, "Modelling and Validating the Man-Machine Interface", Software-Practice and Experience 12, 557-569 Adv. Computers 22, 163-216

Conklin, J.: 1987, "Hypertext: An Introduction and Survey", Computer 20, No. 9, 17-41

Cross, G.R.: 1986, "Tools for constructing knowledge-based systems", Opt. Engineering 25, 436-444

Crawford, R.G., Becker, H.S.: 1986, "A Novice User's Interface to Information Retrieval Systems", Inform. Proc. Managem. 22, 287-298

Draper, S.W., Norman, D.A.: 1985, "Software Engineering for User Interfaces", IEEE Trans. Softw. Engin. SE-11, 252-258

Durham, I., Lamb, D.A., Saxe, J.B.: 1983, "Spelling Correction in User Interfaces", Commun. ACM 26, 764-773

Esposito, F., Capozza, F., Altini, F.: 1986, "EXPER: An Expert System in the Experimental Design", COMPSTAT 86, Short communications and poster, F. De Antoni et al. (eds.), 83-84

Fikes, R., Kehler, T.: 1985, "The Role of Frame-based Representation in Reasoning", Commun. ACM 28, 904-??

Fosbury, R.A.E., Adorf, H.M., Johnston, M.D.: 1988, "VLT Operations - The Astronomers' Environment", in: "Proc.ESO Conf. "Very Large Telescope and their Instrumentation', M.H. Ullrich (ed.), (in press)

Gevarter, W.B.: 1987, "The Nature and Evaluation of Commercial Expert System Building Tools", Computer May, 24-41

Ritchie, G.D., Hanna, F.K.: 1983, "Semantic Networks - a general definition and a survey", Inform. Technol.: Research and Development 2, 187

Shoval, P.: 1985, "Principle, Procedures and Rules in an Expert System for Information Retrieval", Inform. Proc. & Management 21, 475-487

Stefik, M., Aikens, J.S., Balzer, R., Benoit, J., Birnbaum, L., ...: 1982, "The Organization of Expert Systems: A Tutorial", Artif. Intell. 18, 135-173

Szolovits, P.: 1987, "Knowledge-Based Systems: A Survey", in: On Knowledge Base Management Systems, M.L. Brodie, I Mylopoulos (eds.), pp. 339-352

"The GTO Observing Program": 1985, Space Telescope Science Institute, Baltimore, M.D., pp. 1-256

Thompson, B.A., Thompson, W.A.: 1985, "Inside an Expert System", Byte 10, 315-330

Wahlster, W.: 1985, "Cooperative Access System", in: Artificial Intelligence Toward Practical Applications, T. Bernold, G. Albers (eds.), pp. 33-45

Wahlster, W., Kobsa, A.: 1986, "Dialog-Based User Models", Berichte des KI-Labors am Lehrstuhl fur Informatik IV, Saarbrucken 3, 1-34

Whitmore, B.C.: 1985, "Proposal Instructions", Space Telescope Science Institute, Baltimore, MD, pp. 1 99

# DECISION PROBLEMS IN THE SEARCH FOR PERIODICITIES

# IN  GAMMA-RAY ASTRONOMY

M.C. Maccarone,  and  R. Buccheri

Istituto di Fisica Cosmica ed Applicazioni dell'Informatica, C.N.R.
Via Mariano Stabile 172, 90139 Palermo, Italy

## ABSTRACT

The basic steps of the analysis in the search for periodic signals present in a set of single arrival times are here described, with particular attention to those nodes of the procedure for which external and subjective decisions are needed.
Final aim is to study the possibility to connect the various aspects of the analysis in a support system able to help the decision making phase by using Artificial Intelligence tools and methods.

## INTRODUCTION

In most scientific fields there exists a wide choice between different techniques that may be applied to analyse the same data. Often it is not known a priori what is the more suitable one for the problem we are faced with, and a misuse of the used techniques may lead to incorrect results.

As a matter of fact, the choice can be dependent not only on the goal one wants to reach, but also on several other elements that can be interconnected. The final choice can therefore be dictated by:
- the results of previous analyses (personal or from literature) on the same data ,
- similar applications as known from the literature,
- statistical considerations about the data,
- the direct experience of the analyst whose knowledge is not always formally described because most of the times mathematical and heuristic reasoning are mixed,
- and also, at a certain extent, the specific analysis packages available; in fact, sometimes, we just use the packages we have, with their more or less heavy and known limits, rather than to implement new procedures.

An interesting example is present in gamma-ray astronomy when we want to search for periodicities [1]. From the direction of a suspect pulsar, gamma-ray photons are detected (see Fig.1 which shows the main steps of the procedure for periodicity searches in gamma-ray Astronomy). By using the knowledge of the experiment and of the coordinates of the source under observation, the raw data are preprocessed giving us a list of arrival times $t_i$ which include, beside those emitted by the pulsar, those coming from the cosmic background radiation. These arrival times are then folded with chosen values of period $P$ and its time derivative $\dot{P}$ so giving us a distribution of residual phases $\phi_i$ on which an uniformity test is applied. If the residual phases $\phi_i$ distribution is not uniform, a periodicity is claimed to be detected at a given confidence level and the features of the detected pulsar have to be described.

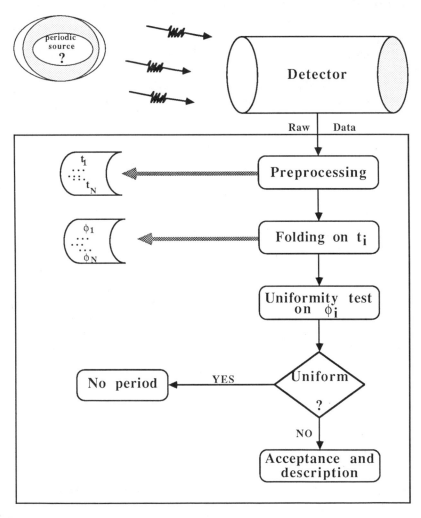

Fig. 1 - Search for periodicities: the general scheme.

# DECISION MAKING IN THE SEARCH OF PERIODICITY

General and specific knowledge is required at each step of the analysis, both extracted from previous experience and coming from the external world; this knowledge is needed to decide about the technique to use and/or the parameter-values to set for it.

Let us consider the main decision problems in the procedure. The **folding** operation allows the derivation of a set of residual phases from the set of arrival times. It requires, as input, a test value for the period P and for its time derivative $\dot{P}$ , and is done by the use of the following formula [1]:

$$\phi_i = \phi_0 + \frac{1}{P}(t_i - t_0) - \frac{\dot{P}}{2P^2}(t_i - t_0)^2$$

In general, we have not a purely confirmatory case, in which to confirm given values of P and its time derivative. Even if it is possible to have some information from other experiments, often our search is mainly exploratory. Then we have to consider many sets of residual phases, derived in correspondence of all the values of the parameters to be tested. In particular, if we have a measure of P and $\dot{P}$ at a given epoch from a radio measurement, the range of their variability increases with the time with a given law and the number of period values to test against the presence of periodicity at a different epoch can be quite large. Therefore, coming from a purely confirmatory case to a purely exploratory one, the sensitivity of the search goes down while the width of scanning intervals, the number of steps and the C.P.U. time increase. The definition of the scanning intervals and of the values of parameters is often subjective and requires an external decision that is of great importance; in fact
- it addresses the search to a particular range of periods (corresponding to a particular physical meaning for a possible periodicity)
- and it affects the acceptance of a positive detection on the basis of the number of trials done.

Let's now consider the **test** step. This is the operation by which every set of residual phases is tested for presence of periodicity. To accomplish it a statistics has to be defined depending on the knowledge about the strength and the harmonic content of the signal. Pearson's, Rayleigh's and $Z^2_k$ tests are statistics normally used in search for periodicity [2]; in the general case where the knowledge about the harmonic content is poor, the definition of the test statistics to use carries a high degree of subjectivity and may result in non-detection of an existing, although weak, signal.
For example, in Fig. 2 the application of the Rayleigh test and of the $Z^2_{10}$ harmonics, on two different situations is shown. Fig.2.a is referred to the presence of a Gaussian broad peak of 0.3 FWHM (Full Width Half Maximum), containing 13.4% of the total counts. The abscissa describes the distance from the true period in dummy units, the ordinate is the probability for chance occurrence, in logarithmic scale.

239

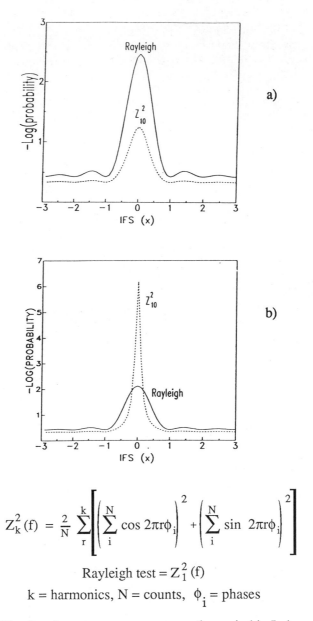

$$Z_k^2(f) = \frac{2}{N} \sum_r^k \left[ \left( \sum_i^N \cos 2\pi r\phi_i \right)^2 + \left( \sum_i^N \sin 2\pi r\phi_i \right)^2 \right]$$

Rayleigh test $= Z_1^2(f)$

k = harmonics, N = counts, $\phi_i$ = phases

Fig. 2 - Probability for chance occurrence versus the period in Independent Fourier Spacing (IFS) units in a search using Rayleigh and Z-square with 10 harmonics to detect a periodic signal. (reproduced from [2] ).

Let's assume that the duty-cycle (here equal to 0.3 FWHM ) is not known a priori. In this case the Rayleigh test, that considers only 1 harmonic, gives better results against the application of the $Z^2$test with 10 harmonics. In fact, it may accept as positive this broad signal at a confidence level of 99.7%.

In Fig.2.b the situation overturns and a much narrower Gaussian peak (only 5% of FWHM) and containing only the 9% of the total counts, is accepted at a confidence level of 99.9%, by applying the $Z^2$ test with 10 harmonics.

The problem to choose the appropriate technique exists also when we want to estimate the physical parameters connected with the structures of the detected periodicity. And the choice is depending on the scale of the structure to analyze. Fig.3, for example, is referred to the light curve of Vela pulsar above 100 MeV as observed by COS-B satellite by three different estimators [3]. The most commonly used **estimation method**, the counts histogram in Fig.3.a, is the easiest to apply but also rather inaccurate for what concerns the derivation of the intensity and location of the structures especially for low statistics. If one wants to study the general topology of the light curve of the detected pulsar, smoothing down all disturbing small scale fluctuations, Kernel Density Estimation [4] may be the correct tool, as shown in Fig.3.c. If, on the other side, it is considered important to look for microstructures, adaptive cluster analysis [5], like in Fig.3.b may be more useful. In any case an external decision is needed.

At the end of this short overview on the search for periodicities, it is important to note that a very crucial node of the entire procedure lies in the evaluation of the independent trials implicit in the process of scanning. This information, subjective in some cases, is used also to accept as positive (at the defined significance level) the result of the application of the chosen test. A too much conservative **definition of the significance level** may conduct to reject the existence of pulsed signals when they are true. On the other hand, inverse errors increase by defining a low significance level.

The significance level definition is often strongly affected by the personal taste of the analyst also if a previous good knowledge on the statistical behaviour of the data in this context is crucial.

## LINK WITH ARTIFICIAL INTELLIGENCE METHODS AND TOOLS

It is clear that, in a search for periodicities, many general and specific knowledges are required, in correspondence with decisions to be made. These are also dictated by the personal taste of the analyst; so we have always to deal with formalizable and unformalizable knowledge. At each step of the procedure, our reasoning follows more or less different paths, and it uses, for each of them, knowledge that may also be shared among various steps. For these reasons, we find useful to deal with the decision problem in the search of periodicities by using some Artificial Intelligence tools [6] like planning, machine learning, frames and Expert Systems techniques to assist the user in the analysis procedures.

**Planning** [7] is the process of finding a sequence of operations that leads from an initial state to a goal state. Its relevance in the choice of a method pertinent to the case under study is evident!

241

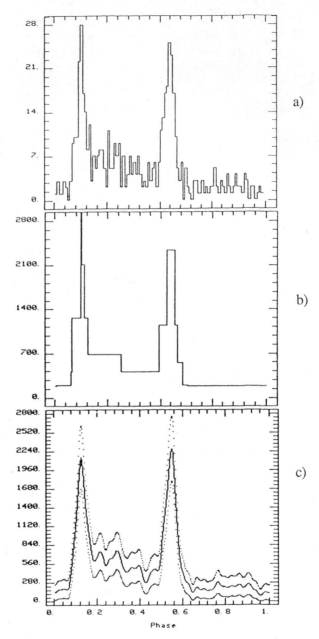

Fig. 3 - Light curve of the Vela pulsar above 100 MeV as observed by COS-B from Sept. 17 to Oct. 7, 1975, by the following estimators: a) histogram with 100 bins; b) cluster analysis (0.05 probability threshold); c) Kernel Density Estimator (with 2 $\sigma$ confidence bands. (reproduced from [3]).

The goal of the **machine learning** [8] is to automatize the acquisition of knowledge, directly in the form required by the system. Among the several sources of knowledge, it is here opportune to remember books or textual files, that can contain literature and other information related to methods of data analysis and/or to the particular problem to deal with.

**Frames** technique [9] to represent objects is based on the concept that human mind, in the analysis of new situations, does not start from zero neither builds new knowledge structures, but rather it uses and adapts predefined schemas according to the expectations. A frame describes a class of objects or situations on the ground of their different aspects. The advantage in using frames is that, attributes of an entity in one frame can be inherited from an entity in another frame. Therefore frames can be very useful in the case of data analysis to take into account data descriptors, relations, references and procedures related to them.

The knowledge representation via frames can also be combined with the technique of production and with Knowledge Based Systems [10]. In the field of data analysis, the role of an **Expert System** should be to specify, to each analysis, a set of methods which can be applied profitably [11,12,13]. In it new methodologies may be added when necessary. Furthermore, the assistance in the choice of a method, and in the interpretation of the results, uses inferential processes and needs, other than the manipulation of a-priori knowledge, also that of the data, as performed in the conventional procedural systems. In this sense an **open system** [14] may be required, that takes into account both data and knowledge manipulation, and in which, different inferential processes can share knowledge.

## FINAL REMARKS

Until today, classification problems have been mainly posed from the astronomers to the Artificial Intelligence scientists. Search for periodicity in gamma-ray astronomy is only a restricted example of data analysis performed by astronomers. Nevertheless it presents many problems also coming from the very-low statistics and it requires to make several decisions, by using different kind of knowledge. Therefore it seems to be suitable for new applications of Artificial Intelligence techniques.
We plan to study each step of this problem in further detail to choose the most adequate types of (expert) decision support systems, if they exist among the available ones, to assist the analyst in this application.

## REFERENCES

[1]   R. Buccheri et al., "Search for Pulsed Gamma-ray Emission from Radio Pulsars in the COS-B Data", Astron. and Astroph., 128:245-251 (1983).

[2]   O.C. De Jager, "The Analysis and Interpretation of VHE Gamma Ray Measurements", PhD thesis, University of Potchefstroom, S.A. (1987).

[3]  R. Buccheri, O.C. De Jager, "Detection and Description of Periodicities in Sparse Data. Suggested Solutions to Some Basic Problems", Proc. of the ASI-NATO "Timing Neutron Stars", IZMIR, Turkey, 1988

[4]  O.C. De Jager, J.W.H. Swanepoel, B.C. Rauben-heimer, "Kernel Density Estimators Applied to Gamma-Ray Light Curves", Astron. and Astroph. , 170: 187-196 (1986).

[5]  R.Buccheri, V.Di Gesu', M.C.Maccarone, B.Sacco, " High Resolution Cluster Method for Topological Studies of Light Curves of Gamma-Ray Pulsars", Astron. and Astroph., 201, 194, 1988

[6]  D.A. Rosenthal, "Applying Artificial Intelligence to Astronomical Databases - A Survey of Applicable Technology", in 'Proc. Astronomy from Large Databases Conference', Garching b. Munchen, F.R.G., Oct. 12-14, (1987).

[7]  F. Hayes-Roth, D. Waterman, D.B. Lenat, "Building Expert Systems", Addison, (1983).

[8]  S. Tanimoto, "Artificial Intelligence - Course Notes", Dept. Computer Science, Univ. of Washington, Seattle, (1986).

[9]  M. Minsky, "A Framework for Representing Knowledge", in: 'The Psychology of Computer Vision', P.H. Winston (ed.), McGraw-Hill, (1975).

[10] D.A. Waterman, "A Guide to Expert Systems", Addison-Wesley Publ. Company, (1986).

[11] J.M. Chassery, "Expert Systems for Data Analysis", in: 'Data Analysis in Astronomy II' V. Di Gesù, L. Scarsi, P. Crane, J.H. Friedman, S. Levialdi (eds.), Plenum Press, (1986).

[12] H. Ralambondrainy, E. Demonchaux, G. Jomier, "Data Analysis, Data Bases and Expert Systems: The Common Interface", in: 'Proc. Astronomy from Large Databases Conference', Garching b. Munchen, F.R.G., October 12-14, (1987).

[13] N. Altman, "Expert Systems and Statistical Expertise. Part I: Statistical Expert Systems", Dept. of Statistics, Stanford University, Tech. Rep. 17, (1985).

[14] V. Di Gesù, M.C. Maccarone, "An Approach to Random Images Analysis", in: 'Image Analysis and Processing II', V. Cantoni, V. Di Gesù, S. Levialdi (Eds.), Plenum Press, pp.111-118, (1988).

244

# RULE-BASED DESCRIPTION AND PLAUSIBLE CLASSIFICATION OF OBJECTS IN DIGITIZED ASTRONOMICAL IMAGES

A. Accomazzi (+) , G. Bordogna (+ +) , P. Mussio (+) , and
A. Rampini (+ +)

(+)  Universita' degli Studi di Milano, Dipartimento di Fisica, Via Viotti 5, 20133 Milano, Italy
(+ +) Istituto di Fisica Cosmica e Tecnologie Relative - C.N.R., Via Ampere 56, 20131 Milano, Italy

## ABSTRACT

A general approach to image interpretation founded on rule-based description and its plausible evaluation is tailored to astronomical image case by the study of the image model adopted by the astrophysicist and of his procedure of visual inspection of astronomical plates.
The formal tools on which the approach is based are Bidimensional L-systems and Multivalued Logical Trees. They are briefly introduced and some examples of application are discussed.

## 1.INTRODUCTION: THE PROPOSED APPROACH

An approach to automatic interpretation of digitized astronomical images is presented and formal tools on which it is based are introduced. The approach is based on the observation that the astrophysicist deals both with imprecise data (King &Raff,77) (Mussio,84) and with problems whose solution is based on plausible reasoning by means of heuristic methods (Godwin et al,83) (Kron,80).
In this paper it is first shown how sets of pixels (in the following called structures) candidate to be source tracks are defined by means of bidimensional L-systems (BLS). BLSs are generative devices which define languages whose strings are bidimensional sets of pixels. The idea of describing such a language goes back to (Dacey,70), was studied in (Merelli et al,85) and here is exploited in the astronomical field. Next, Multivalued Logical Trees (MVLT) formalize heuristic considerations which allow a bidimensional structure to be judged as a meaningful or a noisy object by the evaluation of its properties. These tools have been widely discussed in the image understanding field (Michalski,80); here we follow the approach discussed in (Garribba et al,85).
In other words a two stage interpretation procedure is as follows: in the first stage the total or partial description of the digital image structures is performed, in the second one the structures are given the "meanings" with the reference to a real scene. Description is in

its turn a two step process: first in a segmentation step structures which are candidates to be meaningful are selected. Next, they are all described by computing the attributes values (properties of the structures) of a set of features associated to the structures themselves.

The main characteristic of such general approach (i.e. applying to different types of images) is that image interpretation can be performed by evaluating a total or partial collection of properties which can be imprecisely known. The combination of the collected properties, suitably weighted by Multivalued Logical Trees (MVLT) allows the interpretation, i.e. the decision about the nature of a real object present in the scene from which the image originated.

This general approach is tailored to the interpretation of astronomical images first by studying the image model (section 2), which describes the astronomical plates and their digitized counterpart.

This model suggests to avoid any thresholding on the image intensity for segmentation. Adaptive thresholding and/or previous filtering may lead to introduction of spurious signal, faint object loss and increase of overlap of close object tracks. A unique and general method is therefore required to accomplish astronomical object detection when dealing with crowded fields in which both extended and point-like objects are present, as required for example in (Sedmak,86),(Oegerle at al,86).

In section 3, therefore, a segmentation procedure is introduced which is based on the collection of hints about any significant digital structure. In this stage the image is seen as a 3-D surface and structures are described on the basis of their topological (e.g. connectness) and differential (e.g. variations in colour of neighbour pixels) properties.

In section 4 it is shown how the segmented structures are described and, on the basis of this description, judged and classified into three classes: Real tracks, i.e. plausible tracks of real sources; Uncertain tracks, i.e. those which the data in the image do not allow a precise verdict about, and noise tracks. Real objects are thereafter examined by a new description phase in which structured galaxies are identified and classified following Hubble's taxonomy (Sandage,75). Nearly round objects are classified as star, galaxy or unclassifiable analyzing the fuzzyness of the track profile (Bianchi et al,83) (see Fig.1).

## 2. THE ADOPTED ASTRONOMICAL IMAGE MODEL

Automatic interpretation is based on an hypothesis regarded as acceptable: the examined plate contains tracks of well identified astronomical objects characterized by a high signal to noise ratio. These objects correspond to the highest excesses from the background.

On the other hand, astronomical plates contain tracks of objects separated by background zones corresponding to unstructured regions in the digital images.

Astronomical object tracks have no crisp boundaries and faint objects give rise to weak tracks characterized by some structural regularities; star tracks have a more crisp cross-section than galaxy ones; finally galaxy tracks may be highly structured; the latter are described following Hubble's taxonomy (Sandage,75). In order to detect the faint objects it is necessary to take into account the effects due to the photographic grain which yields structures comparable to the most faint object ones.

The digitization process maps object tracks into sets of pixels (i.e. structures) characterized by similar features in the digital image. For example, most of the extended structures emerging from the background consist of sets of adjacent pixels whose gradient indicates a common growth toward the excess top.

On the contrary, noise regions are characterized by the lack of "relevant" sets of pixels sharing common properties. "Relevant" will be further defined in section 3 by means of Bidimensional L-systems (BLSs).

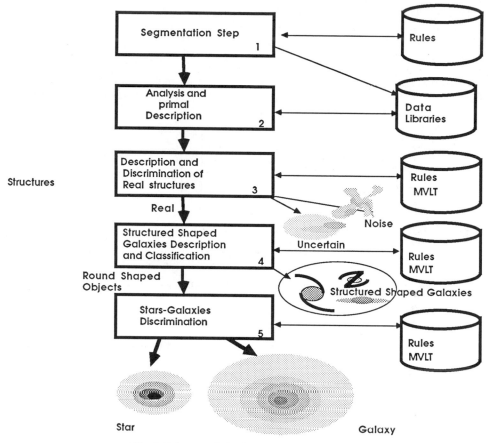

Fig.1 Scheme of the interpretation activity

## 3. CANDIDATE STRUCTURE IDENTIFICATION: SEGMENTATION

In the segmentation step (step 1 in diagram of Fig 1) pixels which are candidate to belong to a structure of interest are selected. In this phase our approach is conservative: the set of candidate structures is so defined as to try to minimize the loss of meaningful pixels, although some noise structures are selected.

This procedure is performed by defining a proposition P which must be satisfied by the attributes either of pixels or of structures. Pixels which satisfy the proposition are set to 'visible' state ('1'), the others to 'invisible' state ('0'). The result of this operation is a binary image.

In the astronomical case the segmentation is performed by detecting all the excesses from the background which are plausibly associated to signal.

Their idenfication is made possible by the definition of a set of "positive" structures and a set of "negative" ones. The former identifies maximum and slope regions, while the latter describes unstructured zones (noisy regions), valleys and ridges. The definition of these structures is accomplished by associating to each pixel of the image the slope code which resumes the local variation of the colour, and then specifying, by means of BLSs, the two sets, as derived by the astrophysicist's experience.

For each pixel P :(x, y, c) a 'slope code' SC is defined:   $SC = \sum ( c_j > c ) \times 2^j$ where $c_j$ is the colour of the j-th neighbour of pixel P, ordered as shown in fig.2.

247

The slope code constitutes a subset of the alphabet V which is used by a set of BLSs identifying the set of 'positive' and 'negative' structures.

A bidimensional L-system $L_b$ is a 4-tuple $L_b = <V, Ax, R_b, I>$ where:

V is an alphabet: in our case the union of the set $V_1$ of the slope codes,

$V_1 = \{0,...,255\}$ and $V_{2;} = \{256,257,258,259,260,261,300,301,302,303,400,401,402\}$ .

Ax is the set of axioms: it is a non-empty set of bidimensional words on V. A bidimensional word on V is a matrix MxN (with $M,N \in N$) whose elements are in V.

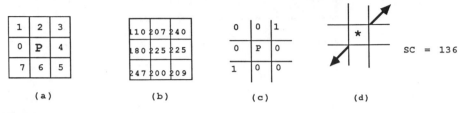

|   |   |   |
|---|---|---|
| 1 | 2 | 3 |
| 0 | P | 4 |
| 7 | 6 | 5 |

(a)

|   |   |   |
|---|---|---|
| 110 | 207 | 240 |
| 180 | 225 | 225 |
| 247 | 200 | 209 |

(b)

|   |   |   |
|---|---|---|
| 0 | 0 | 1 |
| 0 | P | 0 |
| 1 | 0 | 0 |

(c)

(d)    SC = 136

(a) Ordered neighbourhood of pixel P
(b) An example of 3x3 raw image
(c) Application of |j > c| to central pixel in (b)
(d) Graphical representation of Slope Code (SC = 136) associated to pixel P in (c)
➡ direction of colour's growth in the raw image (b)
✶ position of central pixel P

Fig.2 The Slope Code

$R_b$ is a set of bidimensional rules on V: it is a collection of couples $<\alpha, \beta>$ (denoted as $\alpha \rightarrow \beta$) where $\alpha, \beta$ are bidimensional words on V.

I is the set of metarules specifying the direct generation process and the halt rule.

The parallel bidimensional direct generation process is stated as following: a bidimensional word $\eta$ over V generates directly a bidimensional word $\omega$, in symbols $\eta == > \omega$ if each subword $\eta_k$ of $\eta$ is re-written into a subword $\omega_k$ of $\omega$, and $<\eta_k, \omega_k> \in R_b$

Parallel bidimensional generation $=*=>$ is defined as usual as the reflexive transitive closure of $==>$ .

Given a bidimensional L-system $L_b$, the language recognized by the L-system is the following set of bidimensional words over V:

$$\mathscr{L}(L_b) = \{\omega : \omega = *=> \alpha \in Ax\}$$

where $\omega$ indirecly generates an axiom instead of the contrary, because the bidimensional L-systems are here used as recognition devices (Salomaa,73).

The sets of structures which empirically were found of interest are detected and grouped into candidate or background structures by a set of 10 L-systems (Accomazzi et al,88). A candidate structure is a hill in the image surface, i.e. a slope surrounding a maximum plain. A background structure is a region characterized by a random distribution of growth directions. In these regions, therefore, many single valleys or ridges exist, separated by unrelevant pixel hills. Single valleys and ridges will be defined by a suitable BLS. 'Unrelevant' means that such hills are filamentous, i.e. connected pixels which have no adjacent slope or maximum neighbours.

In this way we eschew the use of any threshold on the image intensity, thus avoiding the possible loss of faint objects; overlapping objects are separated by the identification of valleys and ridges.

Only three L-systems are here described due to the lack of space: the BLS for minimum and flex plain identification, for single valley and ridge detection, and the one for east-west valley and ridge identification; their application is shown on the image of Fig.3.a. Due to the fact that the alphabet is common, as the metarules are, each BLS is described by the set of rules $R_b$ and of its axioms.

248

| | |
|---|---|
| $(\begin{smallmatrix} a_0 & 0 \end{smallmatrix}) \dashrightarrow (\begin{smallmatrix} a_0 & 400 \end{smallmatrix})$ | $(\begin{smallmatrix} 400 & 0 \end{smallmatrix}) \dashrightarrow (\begin{smallmatrix} 400 & 400 \end{smallmatrix})$ |
| $(\begin{smallmatrix} 0 & a_4 \end{smallmatrix}) \dashrightarrow (\begin{smallmatrix} 400 & a_4 \end{smallmatrix})$ | $(\begin{smallmatrix} 0 & 400 \end{smallmatrix}) \dashrightarrow (\begin{smallmatrix} 400 & 400 \end{smallmatrix})$ |
| $(\begin{smallmatrix} a_1 & b \\ c & 0 \end{smallmatrix}) \dashrightarrow (\begin{smallmatrix} a_1 & b \\ c & 400 \end{smallmatrix})$ | $(\begin{smallmatrix} 400 & b \\ c & 0 \end{smallmatrix}) \dashrightarrow (\begin{smallmatrix} 400 & b \\ c & 400 \end{smallmatrix})$ |
| $(\begin{smallmatrix} b & \\ 0 & a_3 \end{smallmatrix}) \dashrightarrow (\begin{smallmatrix} b & \\ 400 & a_3 \end{smallmatrix})$ | $(\begin{smallmatrix} b & 400 \\ 0 & c \end{smallmatrix}) \dashrightarrow (\begin{smallmatrix} b & 400 \\ 400 & c \end{smallmatrix})$ |
| $(\begin{smallmatrix} 0 & b \\ c & a_5 \end{smallmatrix}) \dashrightarrow (\begin{smallmatrix} 400 & b \\ c & a_5 \end{smallmatrix})$ | $(\begin{smallmatrix} 0 & b \\ c & 400 \end{smallmatrix}) \dashrightarrow (\begin{smallmatrix} 400 & b \\ c & 400 \end{smallmatrix})$ |
| $(\begin{smallmatrix} b & 0 \\ a_7 & c \end{smallmatrix}) \dashrightarrow (\begin{smallmatrix} b & 400 \\ a_7 & c \end{smallmatrix})$ | $(\begin{smallmatrix} b & 0 \\ 400 & c \end{smallmatrix}) \dashrightarrow (\begin{smallmatrix} b & 400 \\ 400 & c \end{smallmatrix})$ |
| $(\begin{smallmatrix} a_2 \\ 0 \end{smallmatrix}) \dashrightarrow (\begin{smallmatrix} a_2 \\ 400 \end{smallmatrix})$ | $(\begin{smallmatrix} 400 \\ 0 \end{smallmatrix}) \dashrightarrow (\begin{smallmatrix} 400 \\ 400 \end{smallmatrix})$ |
| $(\begin{smallmatrix} 0 \\ a_6 \end{smallmatrix}) \dashrightarrow (\begin{smallmatrix} 400 \\ a_6 \end{smallmatrix})$ | $(\begin{smallmatrix} 0 \\ 400 \end{smallmatrix}) \dashrightarrow (\begin{smallmatrix} 400 \\ 400 \end{smallmatrix})$ |

where b, c $\in V_1$ and $a_i \in A_i$, where i = 0,... 7

$A_i$ is the set of slope codes characterizing elements with a colour growth in direction i + 4 (mod8), together with SC = 0. APL $\alpha$ - $\omega$ notation for this set is:

$$0,,((1+2\times\iota 128 \neq B)\times B)\circ.+\iota B \leftarrow 2*8 \mid I+4$$

1 - BLS for minimum and flex plain identification.

Maxima, minimum and flex plains (respectively marked with symbols + and . in Fig.3.b) are identified by SC = '0'. This L-system translates into '400' the minimum and flex codes. The set of axioms is the set of images on $V' = V_1 \cup \{ 400 \}$.
The set of rules is presented in table 1

2 - BLS for single ridge and valley (RiVa) detection.

The RiVa codes (RV) are a subset of alphabet V and denote those pixels whose growth directions cluster into two or more connected sets.
The set of axioms is the set of images on $V'' = ( V' \cup \{ 261 \} ) - RV$.
The set of rules defining such structures is:

$R_2$:      ( a ) --> ( 261 )      where    a $\in$ RV.

3 - BLS for multiple RiVa identification.

Three L-systems are required for the identification of structures which separate ascending regions (multiple RiVa); they are composed of two or more adjacent pixels whose codes indicate separate directions of growth in the colour intensity.

3.1 - BLS for identification of east-west RiVas.

East-west RiVas are bidimensional words of dimensions 1xn where 2 $\leq$ n $\leq$ N.

Such structures are characterized by horizontal sequences of code SC = '400' on the same raw of the digital image (i.e. plain regions) lying between two extremes having a SC belonging to sets EAST, WEST respectively, where:
EAST = { 1, 2, 3, 128, 129, 131 }
WEST = { 8, 16, 24, 32, 48, 56 }
The set of axioms is the set of images on $V''' = V'' \cup \{ 401,256 \}$.
The set of rules is:

         ( a   400 ) --> ( a   401 )
         ( 401   400 ) --> ( 401   401 )     where    a $\in$ EAST
         ( 401   b ) --> ( 256   b )                  b $\in$ WEST
         ( 401   256 ) --> ( 256   256 )

after the application of such rules, these multiple RiVas are characterized by horizontal sequences of SC = '256' (carrying the information of east-west valley finding), between SC values a and b, elements of sets EAST, WEST. These latter SC are not re-written since they can be part of North-South RiVas to be investigated further on.

In Fig.3.c single and multiple RiVas are marked.

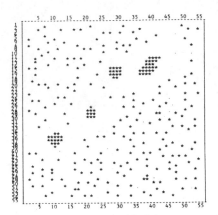

Fig.3.a A subfield of the Plate 157J of the ESO Sky Survey digitized by ESO Garching microdensitometer.

Fig.3.b Structures identified by means of the first BLS on the field shown in Fig.3.a
'*' identifies maxima,'.' identifies minima and flex

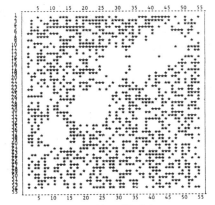

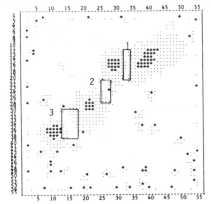

Fig.3.c Structures identified by means of the second and third BLS on the field shown in Fig.3.a '*' identifies single and multiple RiVas

Fig.3.d Structures identified by means of the last BLS on the field shown in Fig.3.a
'*' identifies maxima, '.' identifies slopes
In windows 1,2, and 3 it can be noticed how the applied BLSs yield a separation between close objects

Analogous BLS are defined for North-South and double RiVas. The BLSs allow the recognition of different kinds of structures, some of which are assigned to the set of positive structures (slopes and maxima planes, Fig.3.d) and others to the set of negative ones (RiVas, single minima, unrelevant pixel hills).

The three last BLSs are called for in order to associate the remaining structures (among which minimum and flex plains) to either the negative or positive ones, by means of contextual rules. For example, a flex plain surrounded by positive structures becomes part of them.

At Last, positive structures are set to visible state, negative ones are set to invisible state, so completing the evaluation of the proposition P of the segmentation.

250

## 4. CANDIDATE STRUCTURES DESCRIPTION AND CLASSIFICATION

Every structure obtained from the segmentation step (candidate structure) can be associated either to a faint object, a star, a galaxy, or noise. These structures are described identifying suitable substructures, their features, and evaluating their properties (Fig.1 step 2). The description is thereafter evaluated to obtain a plausible verdict on the nature of the corresponding source in the real scene. Candidate structures which do not belong to the class of galaxies or stars are plausibly classified as 'Real' if they may be considered the track of a faint object, 'Uncertain', or 'Noisy' otherwise (see Fig.1, phases 3-4-5).

The description to be evaluated (named measured description) is obtained by a process of successive refinement of the rough description. The whole process mimics the heuristic activity of the astronomer, when interpreting analog images; it is defined plausible (Prade,85) because the same set of properties may be mapped into a different classification, in relation with the context.

The rough description is the unordered set of triples (c,x,y) associated to a particular subset of visible pixels called extreme contour pixels in which a change in direction takes place (Merelli et al,85). The "c" in the triple is a suitable code obtained by indexing the eight neighbours of a pixel as in Fig.2 and computing:

$c = \sum V_i \times 2^i$  where $V_i = 1$ for visible neighbours and $V_i = 0$ for invisible neighbours.

It has been demonstrated (Merelli et al,85) that a binary image can be fully reconstructed and described from its rough description.

From the rough description, more specialized descriptions may be derived by the use of a Pattern Directed Inference System (PDIS) (Watermann and Hayes-Roth,78) described in Cugini et al (1982) which is governed by suitable sets of rewriting rules described in (Balestrieri et al,83).

Once obtained the measured description, the classification stage begins. The classification procedure is organized in two steps. First the meaning of each observed property of a structure is established as if it were the only known property of the structure with reference to the real scene. This first step is performed using labeling functions (l-functions) which take into account the imprecise meanings associated to the data.

As far as this point is concerned, each attribute of a feature (a structure has got features, a feature has got attributes, each attribute has got properties or values which can be numeric or not) spans over a given range. Different values in this range may however be equivalent with respect to a specific interpretation. Thus properties are mapped into labels by means of labeling functions, where labels reflect the interpretation verdicts.

Formally, an L-function maps a set of properties $P = \{ p_1, p_2, ..., p_n \}$ into a set of labels $L = \{ l_1, l_2, ..., l_m \}$, where $m \leq n$. The set P is partitioned into equivalence classes. To each class a label is associated to denote the meaning of the class in the interpretation process. This partition is performed exploiting the image model. The use of a whole set of defined l-functions maps the structures' measured description into a new one, the s-description, with a score associated to each attribute. The s-description is now used as input to a context evaluation function represented in the form of multivalued logical tree (MVLT) (Michalsky,80) which supplies as output the label of the class to which the structure is assigned (Garribba et al,85).

The processes of label-function and MVLT definition are empiric and subject to the uncertainties of the experts' reasoning.

### 4.1 Application to the astronomical structures

The general method is here tailored to meet the requirements of astronomical data, and interpretation goals, that is:
(a) identifying the tracks of Real sources (Real objects) (see Fig. 1, phase 3)

251

(b) classifying the Real objects whose tracks are well-resolved (i.e. not faint) as stars and galaxies (see Fig. 1, phase 4,5).

To pursue these objectives, a suitable set of attributes and related properties meaningful in the astronomical experiment must be selected. The following example shows how the attributes and properties are chosen, and related l-functions and MVLT are built.

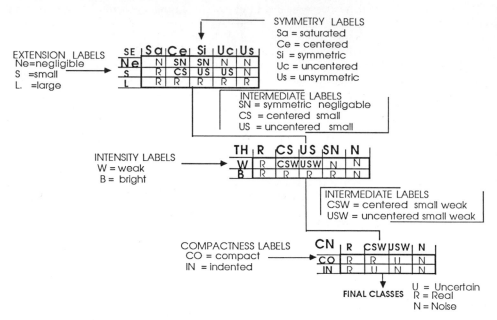

Fig.4. The MVLT combines the attribute properties to classify the candidate structures as Uncertain, Real, Noise. The SE, TH, CN operators are given in the form of tables

The astrophysicist judges as real structures both the intense and extended ones, while his verdict is often doubtful if the structures are faint and small ones, as these last properties are comparable with those of spurious tracks generated by the grain patterns on the photographic plates.

Area and Intensity are in this way recognized as discriminant morphometrical properties. The study of their dependence from the plate characteristics and digitization parameters determines a set of thresholds, scores, which identify plausible classes. Two thresholds Thr1 and Thr2 are identified so as to label the area of an excess into three classes of extension, which were found useful by the astrophysicist. The definition of such thresholds exploits knowledge about:

1) The plate: plate scale (sc), seeing (see), exposition time (et), granularity (gr)
2) The digitization process: pixel size (pixs), quantization range (qr)
3) The distribution $\varphi$ of the structures' extension in the binary image.

A structure is so classified as Negligible, Small or Large by the following l-function:

$$\text{LEXT} = (\ \text{Negligible} \quad (N) \quad \text{if} \qquad\qquad \text{area} \ \le \ \text{Thr1}$$
$$\text{Small} \qquad\ (S) \quad \text{if} \quad \text{Thr1} \ < \ \text{area} \ \le \ \text{Thr2}$$
$$\text{Large} \qquad\ (L) \quad \text{if} \quad \text{Trh2} \ < \ \text{area}\ )$$

The classification into weak and bright objects is based on a threshold $\text{THR} = g(D(CS))$ estimated on colour distribution of pixels belonging to excesses $D(CS)$. A second l-function is defined so as to identify two classes (Weak and Bright) of intensity:

$$\text{LINT} = (\ W \quad (\text{weak}) \qquad \text{for} \quad I \ \le \ \text{THR}$$
$$B \quad (\text{bright}) \qquad \text{for} \quad I \ > \ \text{THR})$$

252

where I is the intensity value of the absolute maximum pixel in the structure in hand; THR is computed as the mean value of the pixel colour distribution estimated on the set of candidate objects incremented by 2 times the distribution variance.

Given an excess, the couple of these two properties computed by the two l-functions is one of its simple s-descriptions. This s-description may be used to identify noisy structures by the following rule: an excess which is Large or Bright is classified as Real track, while if it is Negligible and Weak it is directly assigned to the class of Noise.

This tool is not powerful enough to allow a good classification of small and weak tracks. To manage these cases, two more attributes are introduced which were empirically obtained by the analysis of the results of the classification of the astronomer. These attributes are Symmetry and Compactness. The symmetry of an excess is a structural attribute whose value resumes some significant configurations of the maximum pixel neighbourhood. Symmetry is estimated through a score, a label for which only order can be defined and not distance. Four degrees of symmetry have been defined and here are described starting from the most symmetric one and ending with the least symmetric.

Sa (Saturated): a blob with at least two maximum pixels;
Ce (Centered): the maximum pixel has no or only one invisible neighbour in an odd position;
Sy (Symmetric): the maximum pixel has more than one invisible neighbour in odd position;
Uc (Uncentered): the maximum pixel is also a contour pixel with only one invisible neighbour;
Us (Unsymmetric) the maximum pixel is a contour pixel with more than one invisible neighbour.

| NSC | BX | BY | ARE | PER | PE/A | LSYM | LEXT | LCON | LINT | CLAS |
|---|---|---|---|---|---|---|---|---|---|---|
| 2 | 30 | 21 | 119 | 41 | .34 | SA | L | CO | B | REAL |
| 6 | 26 | 33 | 10 | 9 | .90 | US | N | IN | W | NOISE |
| 7 | 27 | 7 | 13 | 11 | .85 | US | N | IN | W | UNCERTAIN |
| 9 | 30 | 12 | 5 | 5 | 1.00 | US | N | IN | W | NOISE |
| 10 | 32 | 35 | 7 | 7 | 1.00 | US | N | IN | W | NOISE |
| 15 | 37 | 10 | 66 | 28 | .42 | SA | L | CO | B | REAL |
| 16 | 42 | 16 | 7 | 6 | .86 | US | N | IN | W | NOISE |
| 18 | 35 | 35 | 7 | 7 | 1.00 | US | N | IN | W | NOISE |
| 21 | 36 | 41 | 6 | 6 | 1.00 | US | N | IN | W | NOISE |
| 22 | 40 | 48 | 7 | 6 | .86 | US | N | IN | W | NOISE |
| 24 | 44 | 6 | 7 | 6 | .86 | US | N | IN | W | NOISE |
| 27 | 47 | 8 | 6 | 6 | 1.00 | SA | N | IN | W | NOISE |
| 30 | 49 | 52 | 5 | 5 | 1.00 | US | N | IN | W | NOISE |
| 32 | 49 | 12 | 5 | 5 | 1.00 | US | N | IN | W | NOISE |
| 35 | 50 | 26 | 8 | 7 | .88 | US | N | IN | B | NOISE |
| 36 | 51 | 18 | 6 | 6 | 1.00 | US | N | IN | W | NOISE |
| 37 | 53 | 37 | 7 | 7 | 1.00 | US | N | IN | W | NOISE |
| 42 | 53 | 32 | 5 | 5 | 1.00 | US | N | IN | W | NOISE |
| 44 | 8 | 52 | 6 | 6 | 1.00 | US | N | IN | W | NOISE |
| 47 | 49 | 38 | 7 | 6 | .86 | SA | N | IN | W | NOISE |
| 53 | 4 | 45 | 7 | 7 | 1.00 | US | N | IN | W | NOISE |
| 55 | 14 | 41 | 189 | 56 | .30 | SA | L | CO | B | REAL |
| 58 | 12 | 3 | 5 | 5 | 1.00 | SA | N | IN | W | NOISE |
| 59 | 17 | 28 | 74 | 34 | .46 | SA | L | CO | B | REAL |

NSC = Connected set identifier  
BX  = X coordinate of Baricenter  
BY  = Y coordinate of Baricenter  
ARE = Area  
PER = Perimeter  
PE/A= Perimeter-Area ratio  
LSYM=Symmetry labels  
Sa=Saturated  
  Ce=Centered  
  Si=Symmetric  
  Uc=Uncentered  
  Us=Unsymmetric  

LEXT=Extension labels  
  L=Large  
  S=Small  
  N=Negligible  

LCON=Compactness labels  
  Co=Compact  
  In=Indented  

LINT=Intensity labels  
  B=Bright  
  W=Weak  

CLAS=Final Classes

Fig.5. Synthesis of Description and Classification verdicts for the candidate Structures in Fig.3.d

The compactness of an excess is a morphological attribute defined as the ratio between PErimeter and Area PE/A; this value is judged by the following l-function:

LCOM = ( Co   (compact)     for  PE/A   > T

In   (indented)     for  PE/A   $\leq$ T )

where T is an empirically estabilished threshold.

The s-description of an excess is now represented by four scores which are feeded to the MVLT of Fig.4 which yields the final classification Real/Uncertain/Noise. Fig.5 shows the results of the classification to the segmented image in Fig.3.d.

Once the tracks of real objects have been identified, objective (b) must be pursued: first recognition of structured galaxies takes place, then discrimination between stars and galaxies follows. To this end each Real object structure is analysed to find out the presence of some characteristic features like a strong ellipticity, spiral arms or bars as described in (Balestrieri et al,83) following the definition of Sandage (1975). The details of the process will be described in a further paper.

## REFERENCES

Accomazzi A., Bordogna G., Mussio P., Rampini A., 1988, Plausible interpretation of astronomical digital images, Technical Rep. 4.88 IFCTR-CNR, Milano.

Balestreri M., Della Ventura A., Fresta G., Mussio P., 1983, A proposed pictorical language for describing astronomical objects, in: "Image processing in Astronomy", G. Sedmak, M. Capaccioli, R.J. Allen eds., Trieste.

Bianchi S., Gavazzi G., Mussio P., 1982, Un metodo per la Classificazione Automatica di Oggetti in Immagini Astronomiche, Mem S.A.It., vol. 53, n.1

Cugini U.,Dell'Oca M.,Merelli D.,Mussio P., 1982, A Computer aided system for interactive definition of digital image interpretation, Proc. II Conference on Image Analysis and Processing, Selva di Fasano, Italia

Dacey M. F., 1970, The syntax of a triangle and some other figures, Pattern Recognition 2, 11

Garribba S., Guagnini E., Mussio P., 1985, Multiple-Valued Logical Trees: Meaning and prime implicants, IEEE Trans. on Reliability vol. R-34, 5

Godwin J. G., Metcalfe N., Peach J. V., 1983, The Coma cluster-I, M.N.R.A.S. 202, 113

King I. R., Raff M. R., 1977, Publ. of the Astronomical Society of the Pacific 89, 120

Kron R. G., 1980, Photometry of a complete sample of faint galaxies, Astroph. J. Supp. Series, 43, 305

Merelli D., Mussio P., Padula M., 1985, An approach to the definition, description, and extraction of structures in binary digital images, Computer Vision, Graphics and Image Processing, 31

Michalski R. S., 1980, Pattern Recognition as Rule-Guided Inductive Inference, IEEE Trans. Patt. Anal. Mach. Int. vol PAMI-2, 4

Mussio P., 1980, Design of a Pattern-Directed System for the interpretation of digitized astronomical images, in "Data Analysis in Astronomy", V.Di Gesu', L.Scarsi, P.Crane, J.H.Friedman and S.Levialdi eds., Plenum Press

Oegerle W. R., Hoessel J. G., Ernst R. M., 1975, The nearby Abell Clusters. I. Abell 168, The Astron. Journ. 91, 4

Prade H., 1985, A computational approach to approximate and plausible reasoning with application to expert systems, IEEE Trans. Patt. Anal. Mach. Intell., vol. PAMI-7

Salomaa A., 1973, "Formal languages", Academic Press, New York

Sandage A., 1975, in "Galaxies and the Universe". Stars and Stellar Systems,vol 9

Sedmak G., 1986, in "Image Processing for Astronomy", Mem. SaIt vol.57, n.2

Watermann D. A., Hayes-Roth F., 1978, "Pattern Directed Inference Systems", D. A. Watermann and F. Hayes-Roth eds., New York, Academic Press

# PARALLELISM IN LOW-LEVEL COMPUTER VISION - A REVIEW

Vipin  Chaudhary  and  J. K.  Aggarwal

Computer and Vision Research Center

The University of Texas at Austin, Austin, Texas 78712

## ABSTRACT

In  this  paper  we  review  various  parallel  algorithms  and architectures  used  in  Computer  Vision.  The  problem  of  visual recognition  is  divided  into  three  conceptual  levels  -  low-level, intermediate-level  and  high-level  There  are  few  conceptual  difficulties in  parallelizing  low-level  vision  and  most  of  them  have  been parallelized.  However,  not  much  work  has  been  done  in  parallelizing intermediate  and  high-level  vision.  We  present   parallel  algorithms  for low-level  vision.

## INTRODUCTION

Computer vision tasks require an enormous amount of computation, especially  when  the  data  is  in  image  form,  demanding high-performance  computers  for  practical,  real-time  applications. Parallelism appears to be the only economical way to achieve the level of performance  required  for  vision  tasks.  Researchers  in  human  and machine  vision  share  the  belief  that  massive  parallel  processing characterizes early vision.

Parallel processing is the concurrent manipulation of data elements belonging  to  one  or  more  processes,  solving  a  single  problem. Pipelining  and  data  parallelism  (referred  to  as  parallelism)  are  two methods used to   achieve concurrency .   Pipelining is characterized by overlapping  computations  to  exploit  temporal  parallelism,   while parallelism  is  the   use   of  multiple  computing  resources  to  achieve spatial  parallelism.  The  first  few  high-performance  computers  used pipelining; explicit parallelism followed.

255

In past, there have been various arguments against the merits of high-level parallelism. Notable of those is Grosch's law, which states that the speed of computers is proportional to the square of their cost. This has been totally disproved by the advent of commercially available parallel machines which have a much higher throughput compared to the serial machines of similar cost. Another observation called the Minsky's conjecture states that the speedup achievable by a parallel computer increases as the logarithm of the number of processing elements, thus making large-scale parallelism unproductive. This conjecture has also been disproved by experimental results, although the speedup achievable is a function of the algorithm and the target architecture. Amdahl's law states that a small number of sequential operations can effectively limit the speedup of a parallel algorithm. If f ($0 \leq f \leq 1$) is the fraction of sequential operations, then the maximum speedup S achievable by a parallel computer with p processors performing the computation is (Hwang and Briggs [1])

$$S \leq 1 /(f + (1-f)/p).$$

Amdahl's law sets an upper limit on the maximum speed-up theoretically achievable, although the actual speed-up achieved is dependant on the machine used. Algorithms may have to be significantly altered to adapt them to a specific architecture or totally new algorithms may have to be developed.

There have been several methods of classifying computer architectures (Quinn [2]). We follow a simpler classification based on processor organization and three models of computations; pipelined computers, processor arrays, multiprocessors and multicomputers. A pipelined computer performs overlapped computations to exploit temporal parallelism. An array processor uses multiple synchronized arithmetic logic units to achieve spatial parallelism. A multiprocessor system achieves asynchronous parallelism through a set of interactive processors with shared resources. Although these three parallel approaches to computer system design are not mutually exclusive they are reasonably distinct.

In this paper we follow the standard usage in computer vision by dividing the problem of visual recognition into three conceptual levels (Ballard and Brown [3]); low, intermediate and high. Low level (or early) vision is characterized by the local nature of its computations. There are few conceptual difficulties in designing parallel algorithm for these tasks and several existing systems do many of these tasks in parallel. Intermediate level vision involves segmentation – a reduction of the incoming visual information to a form that will be effective for the recognition step of high level vision. High level vision involves the "cognitive" use of knowledge. In general, parallelism is not immediately evident in both, intermediate and high-level vision tasks.

The remainder of this paper discusses various low-level

algorithms that have been implemented in parallel. The algorithms include Smoothing, Convolution, Histogram generation, Clustering, Hough transforms, Thinning, Normalization, Contour extraction, Geometric algorithms, etc.

## LOW-LEVEL VISION ALGORITHMS

In general, low-level vision tasks require computations corresponding to each pixel of the image. This is indicative of the amount of computation required. Fortunately, these computations are usually highly regular, in that the same computations are performed for all portions of the image. We discuss some of these operations in this section.

### Smoothing

Smoothing operations are primarily used for diminishing spurious effects (usually noise). Given an N X N image $f(x, y)$, the mean filtered image $g(x, y)$ is obtained by averaging the grey level values of the pixels in a predetermined neighborhood $h(x, y)$ of $(x, y)$

$$g(x, y) = (1/H) \ \Sigma \ f(x, y)$$

where  H is the total number of points in $h(x, y)$

Similarily, the median filtered image is obtained by taking the median of the grey values of the pixels in the neighborhood $h(x, y)$ of $(x, y)$

$$g(x,y) = \text{Median } f(x,y)$$

where the unary operation Median evaluates the median of $f(x, y)$ in the neighborhood $h(x, y)$ of $(x, y)$.

These operations are parallelized by dividing the N X N image into n equal subimages , possibly overlapping, of size (N X N)/n each. The n processors perform the above operations on corresponding subimages.

For the shared memory SIMD computer, each processor accesses its allocated subimage in an order which voids using overlapping or communication with adjacent processors for calculations at the boundaries of the subimages. Similarily, for other SIMD computers with the various processor organizations and multicomputers, the image is divided into overlapping subimages and each processor works on its subimage.

Reeves and Rostampur [6], discuss the Balanced Binary Filter and Median Filter on a simulated Binary Array Processor (BAP). In addition, a

257

set of preprocessing algorithms are described which are designed to register two images of TV-type video data in real time. The memory cycle cost of a general BAP for some characteristic operations have been presented based on the 1-bit wide data path constraint.

## Convolution

Convolution is often expressed as $g(x,y) = c(x,y) * f(x,y)$ where $f(x,y)$ is the input image, $g(x,y)$ the output image, $c(x,y)$ the convolution function (window) and $*$ indicates convolution operation. This operation is easily parallelized by dividing the image into subimages, possibly overlapping, as in smoothing and each processor convolving on its corresponding subimage.

The parallel algorithms for SIMD are implemented similar to smoothing. Templates are generated for the convolving function and it is applied by each processor to its corresponding subimage. Literature about implementation of convolution is abundant. Kung and Song [7], present the 2-D convolution algorithm on a 2-D systolic array. The single basic cells are interconnected to form a 2-D systolic array. Cells are implemented bit serial due to pin requirements. If the memory speed is higher than the cell speed, then 2-d structure is used, else 1-D structure is used.

Kung and Picard [8], introduce the multi-dimensional convolution algorithm based on hardware pipelines. The 1-D systolic array is used, each of which has a multiply-add unit, and the staging registers for input and output. The arithmetic rate of one cell is greater than 18MOPS.

Kung et al. [9], describe a systolic array for the computation of n-dimensional (n-D) convolutions for any positive integer n. The systolic array for 1-D convolution which uses pipelined arithmetic units is described and it is shown that the systolic array can be extended to handle convolutions of any dimensionality and then the total amount of memory required by the systolic array is optimal. The systolic array in this paper uses a second level of pipelining by allowing the processing elements themselves to be pipelined to an arbitrary degree. Also, noteworthy is the fact that the array suggested by them can execute n-D convolutions with only as many cells as there are non-zero elements of the kernel. Finally, if the number of cells in the array are insufficient for a particular problem, the problem can be decomposed by passing the signal through the array multiple times, each time with a successive segment of the kernel.

Lee and Aggarwal [10], present a parallel 2-D convolution scheme for a mesh connected array processor consisting of the same number of simple processing elements as the number of pixels in the image. For most windows considered, the computation steps required is the same as that of the coefficients of a convolution window. The proposed scheme can be easily extended to convolution windows of arbitrary size and shape. 1-D systolic scheme has been used to implement 2-D convolution on the mesh structure. The computation has been carried out along a path called the convolution path which is the Hamiltonian path ending at the center of the window, the length of which is equal to the number of window coefficients. The simple architecture and control strategy make the proposed scheme suitable for VLSI implementation.

Maresca and Li [11], present a generalized convolution algorithm to process the regular convolution and other morphological operations uniformly for the mesh-connected computers through a snake-sweeping mechanism. This generalized convolution algorithm is not restricted to any particular kernel size. Also, it handles images of any size on any size mesh array and the image/processor mapping is only limited by the local memory of each processing element.

Giordano *et al.* [12], suggest an architecture for a high speed pyramidal convolver which would permit parallel computation to be carried out at different resolutions. The basic element of the convolution board is a programmable VLSI component and several such components can be connected in a semi-systolic way in order to achieve the desired throughput rate. The main application for the proposed architecture is for fast edge detection based on zero-crossings from the second derivative of Gaussian filtered images.

## Histogram generation

A histogram of the grey levels of an image provides a global description of an image. The algorithms developed for histograms are machine dependent. Conceptually, the image is divided into equal subimages and histograms generated for each subimage, which are then combined in a second pass to form the histogram for the entire image.

In case of a shared memory SIMD computer, the image is split into equal subimages and a global shared variable accessable to all processes is updated. In this case, mutual exclusion of the variable histogram is to be guaranteed, which is the bottleneck to the efficiency of the algorithm. Most computers have explicit mutual exclusion implementing commands such as locking a variable, unlocking a variable, semaphores, etc. (e.g. m_lock() and m_unlock() functions in Sequent Balance).

Ramamoorthy *et al.* [13], divide the image into n strips, and each processor calculates the partial histogram independently. They next accomplish the merging of the n partial histograms by recursive doubling. Suppose that the histogram consists of B bins. Initially, processors $P_{2l+1}$; $l=0..(n/2-1)$ accumulate the sums for the $B/2$ least significant bins of the partial histograms contained in their own as well as the memory of their neighbours to the right $P_{2l+2}$. Similarily, processors $P_{2l+2}$ ; l=0..(n/2-1) accumulate the remaining B/2 bins located in their as well as the memory of their neighbours to the left i.e. $P_{2l+1}$. Next, processors $P_{4l+1}$ ; l=0..(n/4-1) transfer the least significant bins of the partial histograms to their neighbouring processors $P_{4l+2}$ while processors $P_{4l+4}$ transfer the B/2 most significant bins to processors $P_{4l+3}$ ; l=0..(n/4-1). At this point, processors $P_{4l+2}$ and $P_{4l+3}$ ; l=0..(n/4-1) contain partial B-bin histograms, and the process is repeated. The final completed histogram can be found in the processor $P_{n/2}$.

On the Connection machine Little et al. [14], each processor determines whether its left neighbour is less than itself. Each processor where this holds sends its cube address to the location $H_k$, in the histogram table, resulting in a cumulative frequency distribution table, easily converted to a histogram. Computing a histogram in m buckets by counting involves stepping from $0 \leq i \leq m$, selecting processors where intensity=i, and counting the number of selected processors ($H_i$), using m global counting operations. When m is less than 64, this can be more efficient than sorting. Finally, when only one value in the distribution is needed, say, finding the $k^{th}$ percentile, the percentile can be found in a binary search, using O(log m) global counting operations.

## Clustering

Cluster analysis is indispensible in exploratory pattern analysis, especially when very little prior information about the data is available. It is also vital in unsupervised pattern recognition and image segmentation applications. The squared-error clustering technique is the most popular one and being iterative, requires substantial computation.

Ni and Jain [15], have used the recent advances in VLSI microelectronic technology to implement the squared-error clustering directly in hardware. A two-level pipelined systolic pattern clustering array is proposed by them. The memory storage and access schemes are designed to enable a rythmic data flow between processing units. Each processing unit is pipelined to further enhance the system performance. Detailed architectural configuration, system performance evaluation , and simulation experiments are presented. The systolic pattern clustering array can off-load the host from the time-consuming pattern clustering tasks. The host will provide the initial cluster center. Given a number of cluster centers, the systolic pattern clustering array will result in a partition, which minimizes the squared error, of the patterns among clusters. Depending on the heuristic used, the host may add, delete or merge the clusters and initialize another cluster task. The processing time for one pass of the new cluster centers is dominated by the memory cycles required to fetch the pattern matrix once. Simulation results show an estimation of a potential performance gain of 1300 over the serial processor.

Li and Fang [16], also use the squared error clustering approach. For a clustering problem with N patterns, M features and K clusters, the processing time complexity for one pass on a serial computer is $O(N*M*K)$. They present a parallel algorithm for the above on an SIMD computer with $N*M$ processors and a hypercube interconnection network. Each processor has a local memory of size K. The time complexity is reduced to $O(K* \log_2 N + K* \log_2 M)$.

## Hough   Transform

The detection of straight lines in the digitized images is a common problem in image processing. Hough transform is a technique by which straight line segments are extracted by detecting global consistencies in the image data. The technique is designed to detect collinear sets of edge pixels in an image by mapping the pixels into a parameter space (the Hough space) defined in such a way that collinear sets of pixels in the image give rise to peaks in the Hough space. Although Hough transform can be used for any distribution of pixels most parallel implementations of this technique are used for detection of straight lines.

Siberberg[17], has implemented the Hough transform in a binary image on the Geometric Arithmetic Parallel Processor (GAPP). The GAPP is a SIMD machine in which the processors are arranged to form an array of arbitrary size. The problem has been split into two:  the computation of the values in the accumulator array, and next, determination of the maximum valued cell in the array. Since the size of the GAPP array is not large enough to match the image data, the divide-and-conquer strategy is used.

Ibrahim  *et al.* [18], implement the Hough transform on a fine-grained tree-structured SIMD machine. The machine, called NON-VON, is a massively parallel machine. The novelty of the algorithm is the use of duplication of data and the avoidance or delay of the communication of intermediate results. As usual, the communication across the tree is the bottleneck. Sanz *et al.*[19], also use Hough transform for line detection. Impressing is the fact that projection data constitutes a powerful description of images for computer vision.

Systolic arrays algorithms have been proposed by Chuang and Li [20], to compute the   conventional   and   the   modified   Hough transform. The disadvantages of the conventional Hough transform are: it fails to check the connectivity of the edge pixels, the pixel count does not necessarily reflect  the  length  of  a  line segment and  it cannot detect collinear line segments. Their modified Hough transform takes care of the above listed disadvantages. It is implemented in three steps: the distances of the edge pixels on a candidate line segment from a reference point are computed and stored – this reduces memory requirement, the distances are sorted to establish connectivity of the pixels and the systolic array processor filters out the noisy components and detect the collinear line segments.

Chandran and Davis [21], implement the Hough transform on two different types of machine: Butterfly and NCUBE. The algorithm involves computing the Hough array and then extract the peaks. These processes are done in a sequence but the emphasis is on allocating processors optimally. On the Butterfly, each processor is allocated an angle, as it is the unit of computation. As a result, P processors evaluate the first P tasks in parallel and then process the next P tasks. Each processor thereby does not work on the task corresponding to contiguous angles. In the next situation, rows are allocated from the input image such that the processors do not work on contiguous rows, but the adjacent elements in a row of the input matrix are all physically adjacent as the block transfer of physically adjacent memory cells from memory is efficient on the Butterfly. This prevents contention which occurs when more than one processor is trying to access the input image at the same time. The next step of merging the individual Hough arrays computed by the various arrays into a global Hough array requires atomicity of access which is accomplished by locking mechanisms available in Chrysalis (the Butterfly Operating System). This approach reduces the efficiency significantly. Some strategies have been used to reduce this contention, one of which is to allocate alternate rows to processors. The processor allocation for peak detection is done by splitting the Hough array into P strips; the adjacent strips sharing a common row. Each processor computes the first k elements in its data and these individual k peaks are combined in a tree-like fashion to produce the final desired peaks.

Little et al. [14], use the Connection machine to determine the Hough transform. They parametrize lines by the normal angle, i, and the perpendicular distance from the origin, j. The Hough transform table will be stored in a matrix of processores, indexed by (i, j). They compute the Hough transform in i separate stages; each computes the values for a particular angle q(i). For each angle they broadcast cosq(i) and sinq(i). Each processor computes the scalar product of its (x, y) address with the normal described by the broadcast pair. Each processor then knows its (i, j). Then, count votes by sending a 1 from each active pixel, summing at the destination processor. If one were simply to send each vote to the table, there would be many collisions, especially when (i,j) does in fact represent a line. This can be slow when the number of collisions is large ($\geq 64$). To avoid this, they reconfigure the votes for each step, so that scan operations can be used to sum the votes. Each processor computes its location on lines normal to the current angle; the location is read from bottom to top along a line, and then continues to the next line. This assigns each pixel a unique address, in which all pixels with a particular j have addresses forming a contiguous block. Each processor sends its value to the processor at its address, in one router cycle (with no collisions). The votes then lie, in linear order in the machine, by their

262

positions on the normal lines. Pixels at the beginning of one of the normal lines set a segment-bit. Then a plus-scan operation, using segments, accumulates the sum of votes in a block. One send operation collects the values into the histogram. This suffices to construct a column (i, j), $0 \leq j \leq$ max, of the histogram. The procedure described above uses unique addresses for the linearization step. The routing hardware incurs little penalty for having up to 32 collisions per destination. By assigning, to pixels on a line, random vote destinations in a block, the number of collisions is minimized. For each i, a pixel computes a location based on j and a random value, using a linearized normal lines as above. For a fixed j, pixels send their votes to a contiguous block of processors. Block sizes are calculated to reduce the average number of collisions. The votes are summed, using a built-in capability of the router. Then the votes are tallied using a plus-scan operation with segments. This uses a simpler address computation than the deterministic method.

## Thinning

Kuehn et al. [22], present two image thinning algorithms : a "peeling algorithm using Arcelli's masks and the other a hybrid scheme that combines a "distance-measurement" and "peeling" algorithm to achieve a better average performance. The "distance-measurement" thinning algorithm determines each pixel's distance from the edge of the line to which it belongs. Pixels with locally maximum distance are retained in the thinned line and; non-local maximum pixels are removed. "Peeling" is another thinning algorithm in which pixels are "peeled" (removed) from line edges until a one-pixel-wide line remains. For a fixed image size, the processing time for peeling is proportional to the line-width wheras the distance algorithm requires fixed time, but both the procedures leave certain extraneous pixels. They discuss the parallel implementation of the above two techniques on SIMD and MIMD computers and also try to evaluate the preferred architecture: SIMD, MIMD or both. They also implement the "hybrid" algorithm that uses both the techniques and conclude that without a priori knowledge of the image chararcteristics to dictate the use of a certain algorithm, the hybrid approach is preferred.

A faster version of the above parallel algorithm has been discussed in Lu and Wang [23]. They claim that the improved algorithm overcomes some of the disadvantages encountered previously by preserving necessary and essential structures for certain patterns which should not be deleted and maintains a speedup of 1.5 to 2.3 compared to the previous work.

An algorithm for parallel thinning on recoded binary images has been discussed by Favre and Keller [24]. This algorithm is isotropic and performs a one-pixel thinning along the object boundaries per scan. It is based on syntactic rules applied to the 3 X 3 neighborhood but enough

263

information on connectivity is further added from outside by allowing five different codes for the pixels: rim points, interior points, core points and skeleton points for the objects, and the background points. Recoding accumulates sufficient contextual information from a 5 X 5 neighborhood to operate the algorithm within a 3 X 3 window. Initial recoding is achieved in three scans using parallel computation of conditional expressions combined with minimum or maximum evaluation on the 3 X 3 window. This overhead is compensated by the parallelism of the thinning step, all four cardinal operations being processed at the same scan. The skeletons produced show a thickness less than or equal to two outside the branching parts. One-pixel thickness may be reached by arbitrarily choosing two or more orientations of thinning by an algorithm working on each of them separately. The complexity of the operation is linearly related to the largest thickness observed on the initial objects. The important point here is the isotropic thinning step as it divides roughly by four the number of iterations necessary for other algorithms. Thinning is handled by structural information. Conditions for shrinking are made on local image descriptions. The local structure of the codes is determined through a binary cascade of tables. To each structure corresponds a rewriting rule. A two-scan postprocessing is needed to obtain a skeleton with linear structure.

## Normalization/Vectorization/Restoration

Image normalization is an important function and is frequently used in image recognition tasks. In general, normalizing the image refers to the process of creating a description that is invariant to the position, orientation, and size of the object in the image. It is especially useful in case of real time object recognition as the object may appear in any position at any scale and orientation in the image. Also, it is efficient to store just the normalized image. Lee *et al.* [25], [26], propose a parallel processing structure and its control scheme for fast binary image normalization for a mesh connected array processor. They use a 4-neighbor-connected mesh of simple processing elements (PEs), each of which has a queue dedicated to the temporary storage of data. The normalization process essentially requires parallel pixel mapping, after which the problem reduces to routing pixel data through the mesh. A simple control scheme which utilizes a store and forward mechanism is described in the paper. The normalization process is decomposed into three procedures: translation, scaling and rotation. In each procedure, the mapping of a pixel from its original position to its destination is controlled by the repetitive application of basic flow controls. Normalization is completed with parallel boundary reconnection and filling algorithms. The proposed algorithm is demonstrated by computer simulation with three different objects and the results are shown in figures. For a N X N image size with each PE assigned a pixel, the overall complexity of the algorithm reduces from $O(N^2)$ by sequential methods, to $O(N)$.

Murary *et al.* [27], consider how the image restoration technique of Geman and Geman [28], which involves searching for the maximum a posteriori distribution of an image modelled as bounded Markov random fields using simulated annealing, can be approximated on a parallel SIMD processor array, the IDL Distributed Array Processor (DAP). The problems associated with MRF (Markov Random Fields) distribution and synchronous updating schemes have been avoided by simultaneously changing only elements that do not interact directly. In the current version the potential speedup over an equivalent serial processor is equal to half the number of processors in the machine (50% speedup).

Many computer image processing algorithms require that objects in images be represented by sets of straight line segments (vectors) as it is conceptually more familiar, is commonly used in display devices and requires significantly less storage space. The vectorization algorithm discussed by Kuehn *et al.* [22] assumes that a thinned binary image is available. The algorithm consists of three main elements: line end/junction identification and line following, topological reconstruction, and iterative polygonal approximation. Line ends/junctions are identified by applying a template at each 1-valued point. A data structure which describes the graph model (vertices and edges) of the lines in the image is constructed. Finally, sets of vectors that approximate the lines in the image are obtained. Linear speedup is obtained so long as the subimage size does not become too small in comparison to the average length of lines in the image. Also, for a given machine size, arbitrarily breaking lines into vectors at PE subimage boundaries results in better speedups than those obtained for the more complex approach of iterative polygonal approximation across subimage boundaries.

## Contour Extraction

Tuomenoksa *et al.* [29], develop and analyze parallel algorithms for edge-guided thresholding and contour tracing. Edge-guided thresholding uses adaptive thresholding to allow contour extraction where gray level variations would not allow global thresholding to be effective. The parallel scenario was found to embrace both SIMD and MIMD subtasks, involves significant PE-to-PE data transfer, and contains both nearest neighbor and non-adjacent PE communication patterns. They suggest desirable system architectural features, including SIMD/MIMD capability with dyanamic mode switching, dedicated PE-to-PE communication support hardware, and arbitrary PE-to-PE interconnection capability.

Guerra [30], presents a parallel dynamic programming algorithm for the extraction of a smooth curve from a digitized image.The algorithm is implemented on a systolic architecture with a limited number of processors , independent of the image size. Wu *et al.* [31], derive some contour properties of curves in parallel by a string or cycle of automata in linear time. The curves are represented by chain-codes. The properties extracted include the intersection of two contours, the straightness of a line, the union or intersection of two contours and polygonal approximations of a contour.

Bertolazzi and Pirozzi [32], investigate the problem of recognizing a curve in a noisy environment using a multiprocessor system. They present a dynamic programming parallel algorithm. It is shown that under hypothesis of unbounded parallelism and using a simple parallel architecture, the algorithm converges in a number of steps which is proportional to the length of the curve. The extension of the algorithm to the case of bounded parallelism is also presented. Dynamic programming technique is normally computationally expensive, but parallelizing the problem makes it attractive for practical implementation.

## Geometric Algorithms

Geometric algorithms includes a variety of algorithms, such as distances, convexity, convex hulls, diameters, etc. Miller and Stout [33], present an $O(n)$ parallel algorithm for internal distance, convexity, and external distance on a n X n mesh connected computer (MCC). The well known MCC algorithms such as rotating data within a row (column), passing a row (column) through the MCC, rotating data in snake-like order, sorting, random access read, random access write and component labeling are used for the geometric algorithms.

Kumar and Raghavendra [34], describe parallel geometric algorithms for finding convex hull and nearest neighbor on the enhanced MCC. The two-dimensional enhanced MCC with multiple broadcasting has been proposed, and each PE has six links instead of four. Finding the convex hull and the nearest neighbor take $O(N^{1/4})$ time, whereas finding extreme points of figures takes between $O(N^{1/4})$ to $O(N^{1/2})$ time depending on the type of input images(N X N). The authors [35], have also developed algorithms for median row, convex polygon and the nearest neighbor, all in $O(N^{1/6})$ time on a multidimensional architecture with efficient local and global communication features. They have used multiple shared buses where the global information flow is not large, but the drawback is the long wires which increases the delay.

Dubitzki et al. [36], have used quadtrees to represent regions in binary images. By using the roped quadtree (a graph formed from a quadtree by connecting the nodes representing equal sized blocks in the image) network as a cellular computer, the perimeter and the Euler number (genus) of a binary image as well as the quadtree distance transform are computed in O(tree height) = O(log image-diameter) time. The area and the centroid of the image are computed in O(tree height) time without roping.

266

Systolic arrays have been used by Guerra [37], to calculate the distance transform. An iterative algorithm that converges in two iterations is presented. The following local operation is applied sequentially

$$n_{ij} = \min\nolimits_{Q*ij} \{n_{ij} , n_{rs} + d_{ij,rs}\}$$

where $d_{ij,rs}$ is the distance between the pixels $n_{ij}$ and $n_{rs}$, and $Q*ij$ is the subset of $Qij$ consisting fo neighbors $n_{rs}$ of $n_{ij}$ which have already been transformed in the fixed sequence. The hardware implementing the systolic algorithm is an array of linearly connected cells. Each cell communicates only with its (two) nearest neighbors and the flow of data through the array is similar to that in case of two-dimensional convolution.

## SUMMARY

This paper has reviewed low-level computer vision algorithms implemented on parallel computers. Algorithms for a given problem vary with the architecture of the computer it is implemented on. There seem to be no conceptual difficulties in parallelizing of low-level vision algorithms. However, one cannot make a similar observation for intermediate and high-level vision tasks.

## ACKNOWLEDGEMENTS

It is our pleasure to acknowledge suggestions and criticism by M. H. Sunwoo. This research was supported in part by IBM.

## REFERENCES

[1]   K. Hwang and F. A. Briggs, *Computer Architecture and Parallel Processing*, McGraw-Hill, New York, 1984.
[2]   Michael J. Quinn, *Designing Efficient Algorithms for Parallel Computers*, McGraw-Hill Book Company, 1987.
[3]   D. Ballard and C. Brown, *Computer Vision*, Prentice Hall, 1982.
[4]   L. C. Higbie, The OMEN Computers: Associative array processors, *COMPCON 72 Digest, IEEE*, New York.
[5]   K. E. Batcher, The STARAN Computer, In *Infotech State of the Art Report: Supercomputers*, vol. 2, C. R. Jesshope and R. C. Hockney, eds. Infotech, Maidenhead, England.
[6]   A. P. Reeves and A. Rostampour, Computational Cost of Image Registration with a Parallel Binary Array Processor, *IEEE Trans. on PAMI*, 4, NO. 4, July 82.
[7]   H. T. Kung and S. W. Song, A Systolic 2-D Convolution Chip, *CAPAMI*, 85.
[8]   S. Y. Lee and J. K. Aggarwal, Parallel 2-D Convolution on a Mesh

Connected Array Processor, *IEEE Trans. on PAMI*, Vol. 9, No.4, July 87.

[9] A. Giordano, M. Maresca, G. Sandini, T. Vernazza, D. Ferrari, A Systolic Convolver for Parallel Multi resolution Edge Detection, *IEEE Proc. of CVPR,86.*

[10] H. T. Kung and P. L. Picard, Hardware Pipelines for multi-dimensional Convolution and Resampling, *CAPAMI, 81.*

[11] H. T. Kung, L. M. Ruane and D. W. L. Yen, Two-level pipelined systolic array for multidimensional convolution, *Image and Vision Computing*, Vol. 1,No.1, Feb 83.

[12] Massimo Maresca and Hungwen Li, Morphological Operations on Mesh Connected Architecture : A Generalised Convolution Algorithm, *IEEE Proc. on CVPR*, 86.

[13] D. V. Ramanamurthy, N. J. Dimopoulos, K. F. Li, R. V. Patel and A. J. Al-Khalili, *IEEE Procs. on CVPR*, 1986.

[14] J. J. Little, G. Blelloch and T. Cass, Parallel Algorithms of Computer Vision on the Connection Machine, *International Conference on Computer Vision*, 87.

[15] Lionel M. Ni and Anil K. Jain, A VLSI Systolic Architecture for Pattern Clustering, *IEEE Trans. on PAMI*, Vol. 7, No.1, Jan 85.

[16] Xiaobo Li and Zhixi Fang, Parallel Algorithms for Clustering on Hypercube SIMD Computers, *IEEE Proceedings of CVPR*, 86.

[17] T. M. Siberberg, The Hough Transform on the Geometric Arithmetic Parallel Processor, *CAPAMI*, 85.

[18] Hussien A. H. Ibrahim, John R. Kender and David Elliot Shaw, The Analysis and Performance of two Middle-level Vision tasks on a Fine-Grained SIMD Tree Machine, *IEEE Proc. on CVPR*, 85.

[19] Jorge L. C. Sanz and Its'hak Dinstein, Projection Based Geometrical Feature Extraction for Computer Vision: Algorithms in Pipeline Architectures, *IEEE Trans. on PAMI*, 9, No. 1, Jan 87.

[20] H. Y. H. Chuang and C. C. Li, A Systolic Array Processor for Straight Line Detection by Modified Hough Transform, *CAPAMI*, 85.

[21] Sharat Chandran and Larry S. Davis, The Hough transform on the Butterfly and the NCUBE, *CAR-TR-226, CS-TR-1713, Sept 86, Center for Automation Research, University of Maryland, College Park, MD 20742.*

[22] James T. Kuehn, J. A. Fessler and H. J. Siegel, Parallel Image Thinning and Vectorisation on PASM, *IEEE Proc. on CVPR*, 85.

[23] H. E. Lu and P. S. P. Wang, An Improved Fast Parallel Thinning Algorithm for Digital Patterns, *IEEE Proc. on CVPR*, 85.

[24] A. Favre and Hj. Keller, Parallel Syntactic Thinning by Recoding of Binary Pictures, *Computer Vision, Graphics, and Image Processing* 23, 1983.

[25] S. Y. Lee, S. Yalamanchili and J. K. Aggarwal, Parallel Image Normalization on a Mesh-Connected Array Processor, *Pattern Recognition*, 20, Vol. 1, 87.

[26] S. Y. Lee, S. Yalamanchili and J. K. Aggarwal, Parallel Image Normalization, *IEEE Proc. on CVPR*, 85.

[27] D. W. Murary, A. Kashko and H. Buxton, A Parallel approach to the Picture Restoration Algorithm of Geman and Geman on an SIMD

machine, *Image and Vision Computing*, 4, NO. 3, Aug 86.

[28] S. Geman and D. Geman, Stochastic relaxation, Gibbs distributions, and the Bayesian restoration of images, *IEEE Trans. on PAMI*, Vol. 5, 1984.

[29] D. L. Tuomenoksa, G. B. Adams, H. J. Siegel and O. R. Mitchell, A Parallel Algorithm for Contour Extraction : Advantages and Architectural Implications, *IEEE Proc. on CVPR*, 83.

[30] C. Guerra, A VLSI Algorithm for the Optimal Detection of a Curve, *CAPAMI*, 85.

[31] A. Y. Wu, T. Dubitzki and A. Rosenfeld, Parallel Computation of Contour Properties, *IEEE Trans. on PAMI*, May 81.

[32] P. Bertolazzi and M. Pirozzi, A Parallel Algorithm for the Optimal Detection of a Noisy Curve, *Computer Vision, Graphics, and Image Processing* 27, 1984.

[33] R. Miller and Q. F. Stout, Geometric Algorithms for Digitized Pictures on a Mesh-Connected Computer, *IEEE Trans. on PAMI*, March 85.

[34] V. K. P.Kumar and C. S. Raghavendra, Image Processing on Enhanced Mesh Connected Computers, *CAPAMI*, 85.

[35] V. K. P. Kumar and C. S. Raghavendra, An Enhanced Mesh Connected VLSI Architecture for Parallel Image Processing, *IEEE Proc. on CVPR*, 85.

[36] T. Dubitzki, A. Y. Wu and A. Rosenfeld, Paralle Region Property Computation by Active Quadtree Network, *IEEE Trans. on PAMI*, Nov 81.

[37] Concettina Guerra, Systolic algorithms for Local Operations on Images, *IEEE Trans. on Computors*, Vol. c-35, No.1, Jan 86.

# PERFORMANCE OF THE MESH ARCHITECTURE IN THE ANALYSIS OF SPARSE IMAGES

A. Machi'[1], V. Di Gesu'[1,2], and F. Lupo[1,2]

(1) Istituto di Fisica Cosmica ed Informatica, CNR
    Palermo, Italy

(2) Dipart. di Matematica ed Applicazioni
    Univ.di Palermo, Italy

## ABSTRACT

Aim of the paper is to show a parallel cluster algorithm on a mesh architecture useful in the analysis of *sparse images*. The algorithm computes the 1NN and an approximate version of the MST of a weighted undirected graph, which represents the image "on" pixels. Some indexes of performance are given on the basis of an experimental evaluation.

Key words: *parallel processing, mesh architectures, clustering, sparse images.*

## 1. INTRODUCTION

*Sparse images* [1,2,3] are binary pictures, $\mathbb{I}$, with 'on' pixels scattered on a structured background. The 'on' pixels may be considered as nodes of a random graph in iconic representation with arc weights equal to the distance between endpoints.

Their analysis is of great interest in a wide range of applications (X and gamma ray astronomy, biomedical imaging, restoring of damaged images,...). Graph-theoretical methods are widely used for this purpose.

For example, if the image points are connected through a Minimum Spanning Tree (MST), the number of connected components $Nc(\phi)$

271

obtained by cutting the MST edges having weight greater than some threshold $\phi > 0$ gives several informations about the spatial distribution of the 'on' pixels and guides the choose of the relevant segments (clusters) in $\mathbb{I}$.

Whenever the image size is greater then $10^4$ pixels and the number of the 'on' pixels exceeds 5% of the total size parallel algorithms are suitable in order to reduce the computation burden.

In the literature several parallel algorithms on graphs have been studied, referring to virtual machine architectures[4,5]; adjacency matrices are commonly used for graph representation.

In the paper a different computational paradigm is used representing always graphs through binary images. It is shown that even if the algorithms may be classified as *not local* nevertheless the paradigm adopted mantains the whole analysis in the algorithm domain where mesh architectures show better performance[6].

A parallel implementation of a segmentation method, based on the computation of the MST of a *sparse image* $\mathbb{I}$, is described. Its implementation on the first level of the **PAPIA** pyramid machine[7] is shown. Experimental complexity and performance evaluations are given and compared to serial implementation.

## 2. BASIC DEFINITIONS

An undirected, simple and weighted graphs is a triple $G = <N,E,W>$, here $N$ is the set of nodes, the cardinality of which is denoted by $|N|$, $E$ is the set of arc (an arc between the couple of nodes $v$ and $v'$ will be denoted by $[v,v']$), $W$ is a weight function $E -> R^+$, where $R^+$ is the set of positive real numbers. In order to light the notation we indicate by $W(v,v')$ the weight of the arc $[v,v']$. If the nodes of $G$ are elements of a $d$-dimensional space $X$, then $W$ could be the Euclidean distance.

For each node $v \varepsilon N$, $N_v$ denotes the set of nodes linked to $v$. A natural order may be defined between the nodes of $N_v$:

$$\forall v',v'' \varepsilon N_v \quad v' <= v'' <=> W(v,v') <= W(v,v'')$$

The first $K$ nodes of $N_v$ are the $K$-Nearest Neighbours linked to $v$. The $KNN(G)$ is a subgraph of $G$, such that each node $v \varepsilon N$ is linked with its $K$ nearest nodes; for $K = 1$ we get the $1NN(G)$. The $MST(G)$ is a spanning tree of $G$, such that the sum of the weight of its arcs is minimal.

## 3. COMPUTATIONAL MODEL

The following computational model has been adopted:

- Graph nodes and arcs are always represented by binary images.
- The arc paths follow the city_block topology (4-connection) and the weight assigned to each arc is the city_block norm:

$$\text{dist } (P,P') = | P_x - P'_x | + | P_y - P'_y |$$

- A massive parallel processor with a mesh architecture (cellular automaton) operating in Single_Instruction_Multiple_Data_stream fashion (SIMD) is assumed for algorithm development.
- The processor mesh size is equal to the image size and each processing element (PE) of the automaton is assigned to one image pixel.
- The PAPIA pyramid machine instruction set is used for the simulated performance study.

PAPIA is a pyramid S.I.M.D. machine composed by a number of arrays of decreasing size piled one over the other. Each array layer is a mesh and each PE is connected to its four neighbours (N,S,E,W). Each processor has one ALU and a set of boolean registers for arithmetic and logical operations, for masking the processor activity, for I/O and for Near Neighbour PE register access. Two variable length shift registers for arithmetic and a bit-addressable local memory of 256 bits are available to each PE.

## 4. THE PARALLEL ALGORITHM

Five main steps are required for the complete image analysis process.

**Step 1** : After image loading, the region of interest (ROI) is selected, by means of an internal propagation of the image edges until the minimum rectangle surrounding the graph nodes is obtained. The ROI mask, so obtained, is then used to restrict the active area in the next steps (S2-S4).

**Step 2** : One approximation of the 1NN is obtained by iteration[2]. Nodes still isolated propagate in four different directions looking for neighbours (fig. 1a). Nodes reached from propagation waves backtrack to the original ones to mark valid paths (fig. 1b). An extra forward propagation plus peeling is applied to refine valid edges. Cycles are randomly generated in this process (fig. 1c).

**Step 3** : The clusters in the 1NN graph are connected by using again an iterative algorithm as in S2. One cluster is randomly selected and its nodes are propagated. When clusters are reached, they are joined. An extra shrinking is required to reduce paths to thickness 1. The process

273

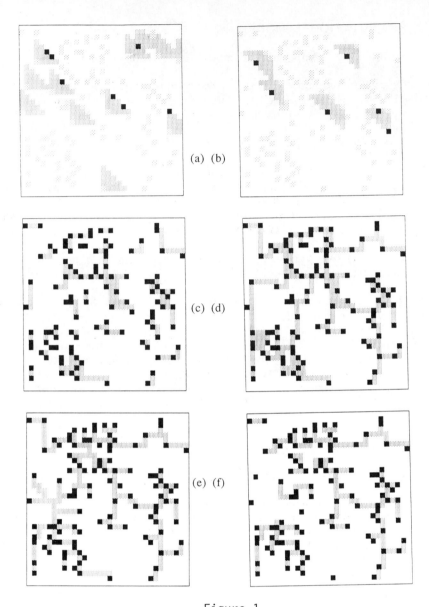

Figure 1.

a) Step 2 : One cycle of *forward* propagation ( *nodes reached* in dark grey)
b) Step 2 : One cycle of *backward* propagation (*originating nodes* in dark grey)
c) Output from Step 2 . 1NN graph with cycles.
d) Output from Step 3 . Link of 1NN components with cycles.
e) Output from Step 4 . Approximated Minimum Spanning Tree.
f) Clusters from MST cutting branches longer than 4 units.

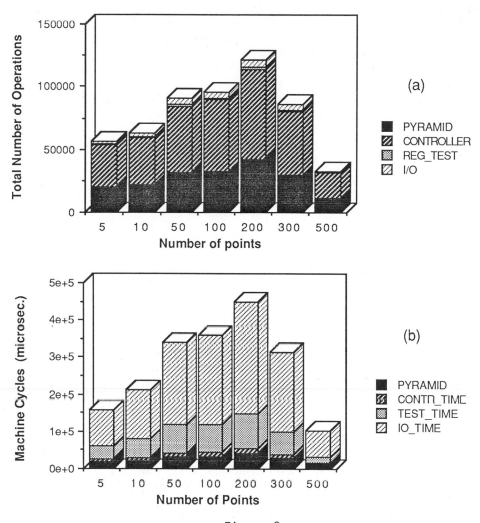

Figure 2.

a) Operation type complexity distribution
b) Operation type time distribution.

loop is repeated until all clusters are connected. Cycles are, again, randomly generated (fig. 1d).

**Step 4** : The graph previously obtained is reduced to a Spanning Tree opening the cycles (fig. 1e). The ST is not exactly Minimum because the arcs to be cut in each cycle are randomly selected. However it remains still a good approximation, because the number of cycles is limited in respect to the total number of edges.

**Step 5** : In order to compute the values of the function $Nc(\phi)$ the image points are propagated over the MST until the MST is completely covered. At each propagation step the number of connected components is obtained recognizing and counting *fingerprint* pixels and applying the Euler formula:

$$\mathbf{N.\,of\ Components} = \mathbf{Npixels}\,(\,{}^{1\,0}_{0\,0}\,) + \mathbf{Npixels}\,(\,{}^{1\,0}_{0\,1}\,) - \mathbf{Npixels}\,(\,{}^{1\,1}_{1\,0}\,)$$

On the host computer a uniformity test[1] is then applied to the distribution $Nc(\phi)$, a threshold $\phi_\tau$ is chosen and the relevant clusters are obtained applying step 5 using the chosen distance $\phi_\tau$ ( fig. 1f).

## 5. EXPERIMENTAL PERFORMANCE EVALUATION

A good definition of parallel computation is still subject of study; Turing machines are not satisfactory theoretical models, and the relation between problems and their parallel solution is not always well understood. Moreover the complexity and performance of not local algorithms, as the one shown, are highly dependent from the node distribution over the input image.

To evaluate experimentally the algorithm performance, the various steps were coded in PMACRO[8] and run on a simulator of the PAPIA machine. Using the simulator facilities the number of mesh operations **PYR**, controller internal operations **CONTR**, controller instruction **IO** and register test **TEST** were separately counted.

Conservative unit time cycles of 1, .3, 50, and 50 microseconds were assumed respectively for the **PYR, CONTR, TEST** and **IO** operations. The array and image linear size **D** was kept equal to 32. The following quantities were computed for the evaluation of the performance:

- The Total number of elementary Operations, **TO**, as a function of the image density (number of points).
- The Total number of machine Cycles, **TC**, as a function of the image density (number of points).

276

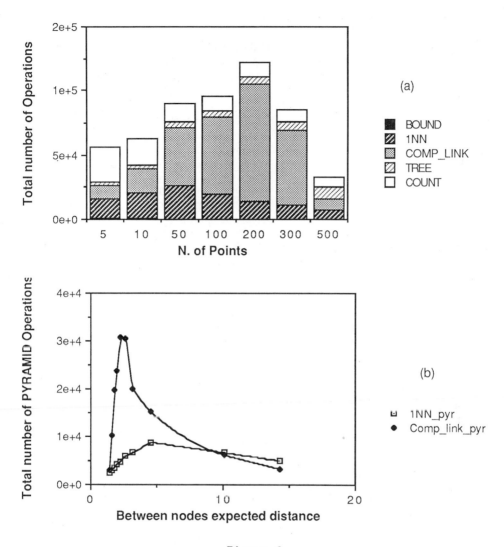

Figure 3.

a) Contributes to the Total number of Operations from the various steps.
b) Dependence of Steps 2 & 3 complexity from the expected distance between nodes.

The results obtained are shown in Fig. 2 a,b. The number of operations **TO** is dominated by the **PYR** and **CONTR** but the maximum contribution to the total time **TC** is given by the exchange of information between the controller and the mesh.

If the mesh is driven by a microprogrammable controller holding a few scalar registers (as the PAPIA machine[9]) transactions with the scalar CPU may be avoided and the relevant contribution is given just by the **PYR** operations.

Fig 3a shows the contribution of each algorithm step to the parameter **TO.** The results show that Steps 2 and 3 are the heaviest ones. Their complexity versus the expected internode distance is shown in Fig 3b. Here the expected internode distance is defined as $\mathbf{dist} = \mathbf{O}\left(\frac{\mathbf{D}}{\sqrt{\mathbf{N}}}\right)$

Both the **1NN** and the **LINK** steps are dependent on two factors: the first constant (peeling and shrinking), the second variable (propagation).
At low densities (high expected distance) the distributions of both neighbour distance and direction is not homogeneous; a limited number of points are collected at each iteration and the constant factor is predominant. At higher densities both distances and directions are homogeneously distributed, the parallelism increases and it follows that the propagation factor predominates.

The maximum in the **TO** distribution for the **LINK** step is obtained at higher densities because there are less components to link in the **1NN** than points to connect in the input image.

For both steps an asymptotic dependence of $\mathbf{O}\left(\frac{\mathbf{D}}{\sqrt{\mathbf{N}}}\right)$ is obtained for high node densities ( > 20% ).

A comparative performance evaluation has been carried out by comparing the parallel algorithm timings with the optimal serial implementation (complexity $\mathbf{O}$ ($\mathbf{N}log\mathbf{N}$), which requires about 300 seconds for N=1000 on a Vax 11/750 [1]. The following quantities have been calculated and reported in Fig. 4a,b:

- Absolute Speed_up Factors (**ASF**) : Serial Time / Parallel Time.
- Relative Speed_up Factor (**RSF**) : ASF/ Number of processors utilized.

The results of Fig.4b show that also the relative speed-up factor is greater than one for point-density greater than 20%.

## 6. FINAL REMARKS

A single link algorithm has been implemented using cellular logic operations and counting on a mesh architecture. An asymptotic

278

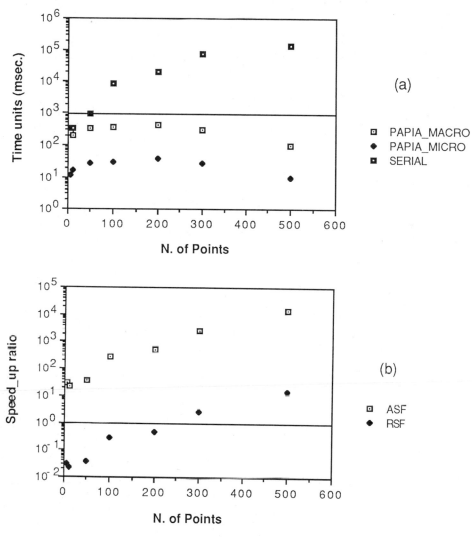

Figure 4.

a) Parallel versus serial time comparison
b) Absolute and Relative speed–up Factors

complexity of $O\left(D/\sqrt{N}\right)$ has been obtained for image point densities greater than 20%. The experimental simulation shows that a good performance is reached by using the PAPIA instruction-set, so that real_time processing may be obtained using microprogramming techniques.

The speed_up increases with the point density and it is advantageous even if the number of processors utilized is taken in account.

The algorithm to compute the **MST** is *not_local*, however a good performance is reached by using an *homogeneous* SIMD mesh of processors, if local approximation is made. Further effort will be made to study the theoretical speed-up and the level of the approximation.

## REFERENCES

1. G. De Biase, V. Di Gesù, B. Sacco, "Detection of Diffuse Clusters in Noise Background", Pattern Recognition Letters, 4 , pp. 39-44, 1986.

2. A. Machì, V. Di Gesù, "Two Parallel Algorithms for the Analysis of Random Images" in A. K. Jain ed.: "Real Time Object Measurement and Classification", Springer Verlag, 1987.

3. V. Di Gesu' and M.C.Maccarone, "An Approach to Random Images Analysis", in Image Analysis and Processing II, V.Cantoni, V.Di Gesu' and S.Levialdi (eds.), Plenum Press, 1987.

4. M. J. Atallah, S. R. Kosaraju, "Graph Problems on a Mesh-connected Processor Array", Journal of ACM, vol 31, n.3 Jul 1984, pp. 649-667.

5. R. Miller, Q. Stout, "Pyramid Computer Algorithms for Determining Geometric Properties of Images", Proc. Symp. Comp. Geom. June 1985, ACM 263-271.

6. M. Duff ed., "Intermediate Level Image Processing", Ac. Press, 1986.

7. V. Cantoni et al., "The PAPIA Image Analysis System", 2nd Intern. Techn. Symposium ANRT-SPIE, Cannes, Nov. 1985.

8. G. De Gaetano, V. Di Gesù, G. Gerardi, A. Machì, D. Tegolo, " Low lelel Languages for the PAPIA Machine", in V. Di Gesù, L. Scarsi et al eds. "Data Analysis in Astronomy II", Plenum Press, 1986.

9. G. Gerardi, " The PAPIA Controller: Hardware Implementation", in V. Cantoni & S. Levialdi eds. "Pyramidal Systems for Computer Vision", Springer Verlag, NATO ASI Series F, 1986.

# CRYSTAL GAZING VI.

# UPDATING COMPUTER TECHNOLOGY PROJECTIONS*

*Donald C. Wells*

*National Radio Astronomy Observatory* [†]
*Edgemont Road, Charlottesville, VA 22903-2475 USA*

> "Prediction is very difficult, especially of the future."
> (Neils Bohr)[1]

## INTRODUCTION

Present-day computer procurements need to be *planned* on the basis of *predictions* of *future* technology. This is because present CPUs, peripherals and software will have to *coexist* and *interconnect* with future CPUs, peripherals and software.

This paper updates predictions made in a tutorial session entitled "Computer System Technology Projections" during the second Erice Workshop in 1986[4] to the state-of-the-art in 1988. It also includes some new predictions and a discussion of a procurement strategy which is robust in the face of computer system diversity and rapid evolution.

## UPDATING THE 1986 PREDICTIONS

The viewgraph of predictions made in 1986 is shown in Figure 1. These predictions were intended to have a lifetime of more than five years and so it is no surprise to see that after only two years they all still appear to be quite valid. However, some minor revisions can now be made, and the details of confirmation of several of the items are informative.

### COMPUTER CAPACITIES

For several years the price/performance ratio of state-of-the-art scientific computers has been halving about every 18 months. In June 1988, when this paper was delivered, a Cray-1

---

* *crystal gazing:* the practice of staring into a crystal ball in order to arouse visual perceptions, as of distant happenings, future events, etc.[2][3]

[†] NRAO is operated by Associated Universities, Inc., under contract with the U. S. National Science Foundation.

**In order of decreasing certainty:**

- computer capacities will continue historic rate of increase; for example, Cray-1 power will cost less than US$1M by 88; system designers should not think small!
- UNIX will be the dominant OS, although frequently overlaid with special environments; system designers must start their UNIX projects *now.*
- Multi-window/Icon/Menu-based graphics will be pervasive; this is sure to affect command language designs.
- Optical disks will be distribution & interchange medium; 9-tk are nearly obsolete, assume larger capacities in future.
- Worldwide high bandwidth standard networks will exist; system designers should plan for remote computing.
- It will be harder and harder to avoid ADA (ADA is really a project-management tool!) Fortran must evolve rapidly or become obsolete.
- "Artificial research assistants" will be feasible with AI; this will permit new command languages, pattern recognition algorithms, data management, etc.

Figure 1. Predictions for the Early 90's [as of April 1986[4]]

equivalent machine had not yet been delivered but was still expected by the end of 1988. It will be the Convex C-240, a second-generation mini-supercomputer with 4 CPUs, each of which is $\approx 2.5\times$ as powerful as the Convex C-1 of 1985 and costing about US$1M. Several other vendors (e.g,, Alliant) are expected to match the Convex price and performance by 1989.

In the spring of 1988 a new class of machines was introduced, the "mini-supercomputer workstations" (also called "graphics supercomputers"). These machines have nearly the MFLOPs power of the first-generation mini-supercomputers at entry prices under US$100K.

Some astronomy organizations tell the author that they cannot afford these new machines, and that they can only buy conventional workstations. The author disagrees: the same organizations bought VAXen a few years ago at higher prices than the new high-powered workstations! Of course, if we wait for four or five years we can probably buy such power for conventional-WS prices, but by then we will probably be able to buy Cray-class power for mini-super WS prices. At some point one has to just decide to spend the money on the state-of-the-art of the moment!

Another phenomenon that will become important during the next few years is the exploitation of central compute servers with cheap WSes over LANs. However, for image processing local I/O devices are important for ultimate performance.

Recently an astronomer suggested to the author that software productivity gains will soon be more important than hardware performance gains. The author disagrees: for all of the decade of the 80's hardware price/performance gains have dominated computing technology and there is no sign of an end to this trend.

## UNIX

The standardization process is accelerating. The "Posix" (U.S.) and "X-Open" (European) efforts have made steady progress; U.S. and European governments are likely to write these standards into procurements during the next few years. Finally, at long last, the Berkeley and Bell dialects of Unix are being merged! But in reaction to this event, seven other vendors (Apollo, Bull, DEC, HP, IBM, Nixdorf, Siemens) have formed the OSF (Open Systems Foundation). The formation of OSF is a clear sign of the universal acceptance of Unix as the operating system of the future.

## WINDOWS, ICONS, GRAPHICS

Polygon rendering hardware will become common in all classes of WSes by the early 90's. The new mini-supercomputer workstations carry this to an extreme: they offer very nearly the highest polygon rendering performance available. Probably this capacity will become important for some astronomy imaging problems during the next few years—the 3-D spectral-line cube problem is a likely candidate.

The recent widespread acceptance of the X-11 portable window interface standard means that it will be the "look and feel" of scientific workstations in the 90's. The fact that it was designed to support remote computing will stimulate widespread use of the technique. The portability of the standard means that application programmers will build it into a vast array of applications.

## ARCHIVE/INTERCHANGE MEDIA

Both optical disks (14-inch and CDs) and helical-scan cassettes (VHS, 8 mm, 4 mm) will be used for both archiving *and* interchange. It is difficult to predict at this time which one of these could become the universal data interchange format which will replace classical 0.5-inch 9-track tape. Maybe there will be more than one, but the author speculates that perhaps the 4 mm DAT (Digital Audio Tape) cassettes have the best potential to become the new de facto data interchange medium.

## HIGH-BANDWIDTH NETWORKS

The standard protocol for worldwide networking after about 1991 will be the ISO/OSI suite, but TCP/IP will continue in heavy use in the U.S. until about 1995.

It is now clear that Sun's NFS (Network File System) will be the dominant standard for transparent remote file access, and that the ANSI FDDI (Fiber Distributed Data Interface) will be the standard for high-performance LANs (100 Mb/s).

The U.S. NSFnet strategy for a national research Internet based on TCP/IP protocols is working; about 200 universities will be on the national net by the end of 1988. The transcontinental backbone bandwidth is being raised to 1.5 Mb/s by stages and the routers are being engineered to operate at even higher bandwidths later on. NSFnet hopes to raise its backbone bandwidth to 45 Mb/s by 1991; this will be a good match for FDDI campus LANs for those campuses that happen to be on the backbone net. The regional nets are now operating at 56 Kb/s and some will be raised to 1.5 Mb/s by 1990.

Recently the government of the F.R.G. authorized the operation of the DFN (Deutsche Forschungs Netz); this probably implies that a European research net similar to the NSFnet will be operating by 1990.

## LANGUAGES

Ada still looks important for the 90's, but the C++ language will compete with it because Sun and AT&T are building their new Unix kernel with this object-oriented superset of C. If software engineering tools are offered for C++ comparable to those being offered for Ada then the author believes that C++ will be widely used.

The author no longer believes that Fortran has any advantage over C for scientific programming. The growing availability of scientific programming libraries in C testifies that the author is not alone in this opinion.

## ARTIFICIAL INTELLIGENCE

The AI papers presented at this 1988 Erice workshop demonstrate that astronomy is progressing satisfactorily with AI development, and that the 1986 prediction of AI in the 90's for astronomy is correct.

## FOUR NEW PROJECTIONS

In Figure 2 we tabulate details of four new predictions. Many astronomy projects are already using DBMS's for internal record-keeping in software systems. So far comparatively few astronomers have any experience with commercial DBMS technology, and few of them have used these systems for scientific data analysis. This situation will change over the next few years. The Structured Query Language SQL is now an ANSI standard and will clearly become the standard interface to DBMSs.

At the moment it appears that the Postscript page description language will dominate in the laser printer market for many years to come.

CASE (Computer-Aided Software Engineering) technology has become increasingly important in industrial projects and it is beginning to creep into astronomy projects. In CASE the software is data for tools which manipulate code objects. The intent is improve the efficiency of programmers in manipulating large software packages. The author's 1986 argument that Ada will be important was based on the projection that the best CASE environments will be associated with the Ada language, but CASE tools are beginning to appear for other languages and software systems.

```
┌─────────────────────────────────────────────────────────────────┐
│                                                                   │
│   Data Base Management Systems                                    │
│                                                                   │
│      • DBMS will be widely used in astronomy                      │
│      • SQL will be the standard interface                         │
│                                                                   │
│   Laser Printers                                                  │
│                                                                   │
│      • Postscript will be the standard interface                  │
│                                                                   │
│   Computer-Aided Software Engineering                             │
│                                                                   │
│      • CASE will be used in large software projects               │
│      • it will reduce costs and improve documentation             │
│                                                                   │
│   Canonical Architecture                                          │
│                                                                   │
│      • almost all future computers will use 8/16/32-bit           │
│        twos-complement integers and 32/64-bit IEEE floating       │
│        point, without byte-swap, and 8-bit ASCII characters,      │
│        which DCW calls the "canonical architecture."              │
│      • sharing of large binary files will be most efficient if NFS│
│        is used; this implies need for canonical architecture      │
│      • implies proprietary architectures may be excluded in some  │
│        future procurements because they inhibit file sharing      │
│      • argument is still true even for Sun's XDR (External Data   │
│        Representation) protocol because of performance reasons    │
│                                                                   │
└─────────────────────────────────────────────────────────────────┘
```

Figure 2. Additional Predictions for the Early 90's

## THE CANONICAL ARCHITECTURE

Almost all new computers designed in the 1980's have compatible data formats. This process has proceeded to the point that we can consider the standard combination to be a canonical architecture. This has profound implications for binary file sharing using the NFS protocol, especially between compatible supercomputers and workstations. Protocols for transmission of arbitrary binary structures have been devised (most notably Sun's XDR protocol), but they will suffer from inefficiency compared to NFS and from the fact that they are not yet generally implemented in compilers. The author expects that some organizations will soon write binary compatibility requirements into procurement specifications in order to facilitate distributed operations in a multi-vendor environment.

Note that only exact images of Fortran COMMON blocks or C global structures can be shared. These are nearly guaranteed to have compatible formats in spite of other differences in compiler implementation because the Fortran EQUIVALENCE-statement rules and the C union-statement rules constrain the compiler implementors to produce compatible structures.

**"Unix":**

- Posix/X-Open

**Window Standards:**

- X-11
- (NeWS & Display-Postscript?)

**Graphics Standards:**

- GKS
- PHIGS?

**Canonical Data Architecture:**

- 8/16/32-bit integers
- 32/64-bit IEEE floating point
- ASCII, non-byte-swapped

**I/O Busses:**

- VME
- (plus NuBus & AT-Bus)

**Device Interfaces:**

- SCSI
- IPI

**LANs:**

- Ethernet (Thick & Thin)
- FDDI

**Network Protocols:**

- ISO
- TCP/IP
- NFS

**DBMS:**

- Relational & SQL
- Standard SQL Server?

Figure 3. Open-System Interfaces

## THE OPEN-SYSTEM-INTERFACE STRATEGY

How do we use our projections to plan our procurements in order to achieve reasonable compatibility in a diverse and evolving environment? The author advocates an "open-interface strategy" for procurements. The point is that the connections between systems and components occur at *interfaces*, and we want to procure systems and components which have interfaces which have the greatest likelihood of becoming *open* industrial standards, de jure standards if possible, de facto at least. In Figure 3 a set of currently recommended interfaces is tabulated. In the real world of procurements one cannot afford to be pedantic and un-compromising about these interfaces—they are merely guidelines to steer the pattern of procurements in a desirable direction. The general rule is to avoid proprietary interfaces as much as possible in order to maximize freedom in future procurements. This principle is destined to become *much* more important as the Unix phenomenon comes to dominate the market more and more, and computers begin to be commodity items, almost interchangeable.

## REFERENCES

[1] Thomas A. Bass 1985, "The Eudæmonic Pie", Vintage Books (Random House), New York. The Bohr quotation is taken from the title page of Chapter 1. This is a very entertaining book about microcomputers and roulette and a variety of other subjects.

[2] Definition of *crystal gazing* from: **Random House Dictionary of the English Language**, 1971, The Unabridged Edition.

[3] Previous papers in the author's "Crystal Gazing" series include:

I. "Gazing Into the Crystal Ball", pp. 490-494 in **Image Processing in Astronomy**, ed. G. Sedmak, M. Capaccioli, R.J. Allen, Osservatorio Astronomico di Trieste, the proceedings of an international workshop held June 4-8, 1979, in Trieste, Italy.

II. "Implications of Advances in Digital Data Storage Technology", pp. 345-359 in **Astronomical Microdensitometry Conference**, ed. D.A. Klinglesmith, NASA Conference Publication 2317, the proceedings of a workshop held May 11-13, 1983, at Goddard Space Flight Center.

III. "Image Processing Technology for the NNTT", a report and invited talk prepared for delivery before the NNTT Scientific Advisory Committee, 29 September 1983.

IV. "Workstations in Astronomy—Image Processing and Data Analysis", invited talk at the "Workshop on Applications of Supermicrocomputer Workstations in Physics and Astronomy", held 20-22 January 1984 at the Univ. of North Carolina, Chapel Hill, NC.

V. "Computer System Technology Projections", pp. 153-170 in **Selected Topics on Data Analysis in Astronomy**, ed. V. Di Gesù, L. Scarsi, P. Crane, World Scientific, Singapore, 1987, the general lectures given at the Second Workshop on Data Analysis in Astronomy held April 20-27, 1986, in Erice, Italy.

[4] see paper V of [3].

# New Frontiers in Astrophysics

# COSMIC BACKGROUND RADIATIONS

George F. Smoot

Space Sciences Laboratory and Lawrence Berkeley Laboratory
University of California, Berkeley, California 94720

ABSTRACT     The photon is the best probe we have of the distant regions of the universe. There are two regions in the electromagnetic spectrum where one can expect to see far back into the universe. In both of these windows there are backgrounds which are now being explored and studied. These observations have led to suprisingly strong statements about the nature of our universe.

## 1. INTRODUCTION

Astronomers have long studied distinct luminous objects in the sky. Most of these have been point-like and until recently astronomers divided all objects into two categories, stars or nebulae.

### 1.A. Diffuse Astronomy

Diffuse astronomy is much different from standard astronomy. Instead of looking for point sources or bright objects, one looks at structures that are large in scale and indistinct. In diffuse astronomy one attempts to look beyond foreground objects to observe the large scale structure of the universe. Our present observations already indicate that it is difficult to find significant features and understand their implications. That is why the discovery of a connection between galactic clusters or of a large void makes headlines. There is one salient observation of diffuse astronomy: it is easy to observe that the moonless night sky is surprisingly dark away from the Milky Way.

The astronomer Heinrich Olber (1823) among others recognized that a dark night sky was a significant measurement of the cosmic background and was in conflict with the then-accepted theories that the universe was infinite and static. Olber's Paradox: If the universe were static and infinite, then no matter in which direction one observed, eventually the line of sight would terminate

291

on a star so that the night sky would be as bright as the sun. Olber's Paradox stated more mathematically is: Consider any large spherical shell centered on the earth. Within this shell, the amount of light produced by stars can be calculated by knowing the average cosmic density of stars. Then consider a shell at twice the radius. Within this larger shell, the stars are only one quarter as bright, but there are four times as many of them, so on average they add as much light to the night sky. For each radius doubling, the night sky doubles in brightness. Following this argument to larger and larger radii, the night sky increases in brightness indefinitely, until the entire sky is apparently covered with stars.

If one tries to absorb the light with an intervening material, as the more nearby stars eventually do to more distant stars' light, then one finds that the intervening material is eventually heated to the temperature of the stars and then radiates like the stars.

## 1.B. Cosmic Backgrounds

We can test and bolster this conclusion by checking the brightness of the night sky over all wavelengths including the visible. Figure 1 is a plot of diffuse photon flux as a function of frequency. It shows primarily upper limits but there is evidence for four cosmic background radiations: (1) a cosmic radio background (CRB), (2) the cosmic microwave background (CMB), (3) the cosmic x-ray background (CXB), and (4) a cosmic gamma-ray background (CGB).

The cosmic radio background is detected in the region between 1 MHz and 1 GHz (0.3 to 300 m wavelength). It is difficult to detect since the flux is dominated by a foreground radiation from our galaxy composed of synchrotron and HII emission. The CRB seems readily explained as the summation of this kind of radiation from all galaxies and radio sources in the universe.

The CMB is the dominant cosmic background and its existence is well-established. It is thought to be the relic radiation of a hot dense phase of the early universe. The CMB will be treated in detail below.

The x-ray region contains another well established and partially understood cosmic background. The CXB has a non-thermal spectrum that extends well into the gamma-ray region; its intensity is nearly the same in all directions. The origin of the CXB is not yet determined conclusively. The CXB may well be the sum of radiation from a number of discrete sources, just as the CRB is the sum of radiation from the radio galaxies and quasars. This is controversial and a possible connection to the CMB is considered in the conclusion of the CMB spectrum discussion.

In the gamma-ray region the data are not conclusive. Observations have not yet achieved the sensitivity to find extragalactic sources that would account for the observed flux. The Gamma Ray Observatory scheduled to be launched in 1990 is likely to provide us much new information about this region.

From the low levels of these measured fluxes and Olber's Paradox we can conclude that the universe is not infinite and static. There is now a gener-
292

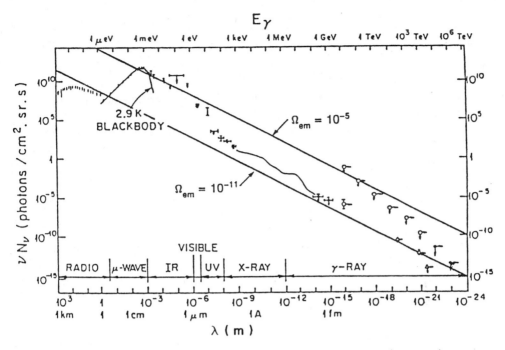

Figure 1. Diffuse Photon Flux vs. Frequency (Energy)

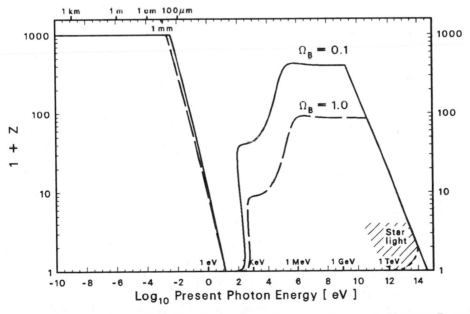

Figure 2. Redshift of Optical Depth Unity vs. Present Photon Energy

ally accepted class of cosmological models called the Big Bang theories that describe the evolution of the universe.

## 1.C. Cosmology - The Hot Big Bang

The standard cosmological model (the Hot Big Bang) holds that in the past the universe was much hotter and denser than it is in the present epoch. The universe is expanding throughout all space and thus has been cooling, according to the usual rules of physics. The universal expansion is demonstrated by Edwin Hubble's discovery of the general recession of galaxies. The distance and differential velocities of this galactic recession give a time scale to the history of the universal expansion of about $10^{10}$ years. If we extrapolate the universal expansion backward, we find a point where all the matter and energy, such as the galaxies and CMB we observe presently, would be so crowded together that the density would be singular (approaching an infinitely large value). As a reference in the theories we call this point/instant the beginning or the singularity. This point occurs about $10^{10}$ years ago. (According to Steven Hawking the measured isotropy of the CMB allows us to extrapolate back to an epoch at which the CMB energy density ensures that the universe must have passed through a singularity.)

Assuming a uniform, homogeneous, hot, expanding universe we can extrapolate the evolution of the universe forward from a fraction of a second after the singularity, and through calculations using known physics and parameters predict (1) the form of the universal expansion, (2) the synthesis of light atomic elements in an epoch occurring a few minutes after the singularity, and (3) the existence and properties of the the the CMB (Weinberg 1977, The First Three Minutes, Basic Books).

With this model we can also explain Olber's Paradox, since the universe is not static nor has it existed for an infinite time. The Hot Big Bang model predicts that there was an epoch, roughly $10^5$ years after the singularity, when the sky was as bright as the sun (a few thousand degrees Kelvin), but the universal expansion has since cooled the CMB radiation to about 3 K at present. This epoch occurred when the CMB photons last scattered on their travels through the universe. In the intervening time, roughly $10^{10}$ years, the universe has expanded by roughly a thousand fold, and galaxies, stars, planets, and life have formed.

Figure 2 is a plot showing how far back into the universe one can see as a function of frequency assuming the Big Bang theory is correct. For this particular plot it is assumed that the density of the universe is equal to the critical density. The critical density defines the boundary between having sufficient or insufficient matter density for gravitation to stop the universal expansion and cause a collapse. ($\rho_{critical} = 3H_o^2/8\pi G$) This is the density predicted by the Inflationary Hot Big Bang models and is aesthetically favored by many. As we can see from this plot there are two windows in which we can hope to make cosmological observations of the structure of the universe at great distances. It is not totally fortuitous that the CMB lies in the deepest of these windows.

# 2. THE COSMIC MICROWAVE BACKGROUND

The nature of the CMB and its dominance over other diffuse radiation fields, as well as the deep transparency of the universe to its photons, make it a uniquely important cosmological and astrophysical probe. The CMB automatically registers the large-scale structure of the universe. Empirical evidence concerning the spectrum and angular distribution of this radiation can set limits on processes and perturbations that may exist. These departures from an idealized universe are what produced the world we observe today.

In the context of the Big Bang theory the origin of the CMB is intrinsic to the nature of the evolving universe.

## 2.A. Spectrum

In the Big Bang theory the CMB has a thermal origin and it carries much of the entropy of the universe. In the idealized case the spectrum of the CMB is Planckian, reflecting its thermal origin and the close coupling of matter and radiation in the early universe. In this context the CMB is formed very early in the history of the universe; however, even if one wants to start the Big Bang cold, it is straightfoward to show that the bulk of the CMB must have been formed within about the first hour after the beginning. All measurements to date tend to support the thermal nature of the CMB.

### 2.A.1 Measurements and Potential Distortions

Recent measurements of the cosmic microwave background (CMB) temperature by several authors (Crane *et al.* 1986; Johnson and Wilkinson 1986; Matsumoto *et al.* 1988; Meyer and Jura 1985; Peterson, Richards, and Timusk 1985; Sironi *et al.* 1987; Smoot *et al.* 1987) have greatly reduced the uncertainty in the spectrum of the radiation at wavelengths between 50 cm and 0.05 cm. Danese and De Zotti (1978, 1980) have analyzed previous CMBR temperature measurements for the presence of distortions due to Compton scattering (scattering of photons by free electrons) or other processes; Hayakawa *et al.* (1988) and Smoot *et al.* (1988) have updated their work.

Precise measurements of the CMB spectrum are the most powerful probe available of the history of the universe in the period before galaxies and clusters of galaxies formed. The CMB spectrum provides a record of the thermal history of the universe; in particular, the spectrum of the CMB contains information about energy-releasing processes that may have distorted the CMB spectrum from its initial blackbody distribution (Danese and De Zotti 1977, 1978, 1980, and 1982). Such distortions remain until the present, unless there is sufficient time for bremsstrahlung and Compton scattering to relax the spectrum to a blackbody. A number of cosmologically revealing spectral distortions could remain today.

A Compton distortion is the most likely spectral deviation from a Planckian (blackbody) spectrum. If an energy source heats the ionized intergalactic matter, the hot electrons scatter low energy photons to higher energy, making the CMB cooler at frequencies below the peak and hotter above the peak (wavelength $\lambda < 0.1$ cm). This process must occur in X-ray emitting galactic clusters and is called the Sunyaev-Zel'dovich effect. The S-Z effect is usually investigated by looking at the spatial variation between a galactic cluster and a neighboring blank field since galactic clusters are relatively small and well defined and the effect is small ($\leq 1$ mK in the Rayliegh-Jeans region). However, if there is a hot intergalactic plasma, observing the distorted spectrum may be the only means to detect it. A detection of this distortion measures the Thomsonization parameter $y$,

$$y = \int \frac{kT_e}{m_e c^2} n_e c \sigma_T dt$$

which is the number of Thomson (low-energy Compton) scatterings times the dimensionless electron temperature, $kT_e/m_e c^2$, which is also the mean fractional energy (or equivalently frequency) shift per scattering for an electron plasma of temperature $T_e$.

If the matter of the universe is heated with extra energy before recombination ($z \gtrsim 10^3$), there will still be cooling below the peak and heating at high frequencies but bremsstrahlung will produce additional low-frequency (or long-wavelength) photons. Thus the brightness temperature will be higher at low and high frequencies but cooler in the middle frequency range. For energy release between redshifts of about $4 \times 10^4$ and $10^6$ the number of Thomson scatterings is sufficient to bring the photons into thermal equilibrium with the primordial plasma but the bremsstrahlung and other radiative processes do not have sufficient time to add enough photons to recreate a Planckian distribution. The resulting distribution is a Bose-Einstein spectrum with a chemical potential, $\mu$, that is exponentially attenuated at low frequencies, $\mu = \mu_o e^{-2\nu_0/\nu}$. The Planckian spectrum is the special case of the Bose-Einstein with zero chemical potential.

The energy injection discussed above causes cooling for the frequencies below the peak with the maximum decrease in the range from about 3 to 10 cm wavelength (10 to 3 GHz). This corresponds to a photon underpopulation (positive chemical potential) in that frequency region. If a process were to add photons in this range, one would observe a bump (negative chemical potential). There are a number of potential processes that could add photons to the CMB. Through the 10 micron silicate and other emission features, the very early existence of dust could produce an apparent excess in the CMB spectrum (Rees 1978). Another possible mechanism is the cosmological production of a weakly interacting particle, such as an unstable massive neutrino, that decayed into a photon and other daughters. The decay photons would add a bump to the spectrum. Even though the photons could have a sharply defined frequency, they would be smeared in frequency by the thermal energy of the parent, the varying redshifts at the times of the decays, and successive Thomson scatterings with the plasma.

At redshifts greater than a few times $10^6$, the combined action of bremsstrahlung and radiative Compton scattering maintains a tight coupling between the matter and the radiation field. Photon production can take place on time scales short compared to the expansion time, and any non-blackbody spectral features are quickly erased. (Illarionov and Sunyaev 1975; Danese and De Zotti 1982; Lightman 1981).

At redshifts smaller than $\sim 10^5$, the electron density is no longer high enough for Thomson scattering to establish a Bose-Einstein spectrum. The spectrum assumes a mathematically more complex form, but its main characteristics are an increased brightness temperature in the far Rayleigh-Jeans region due to bremsstrahlung emission by relatively hot electrons, a reduced temperature in the middle Rayleigh-Jeans region where the photons are depleted by Thomson scattering, and a high temperature in the Wien region, where the Thomson-scattered photons from long wavelengths have accumulated.

Neither Thomson scattering nor bremsstrahlung takes place in a neutral medium, so in simple cosmological models, the spectrum at recombination is more or less preserved until the present, though redshifted by a factor of a thousand or so. More complex cosmological models may call for substantial energy injection from galaxy or quasar formation which could reionize the intergalactic or intercluster medium, creating conditions somewhat like those at redshifts between $10^3$ and $10^5$ but with a lower electron density and a higher ratio of electron temperature to radiation temperature. The effect of these conditions on the CMB spectrum would again be to raise the brightness temperature in the Wien region and at very long wavelengths.

Soon after the Big Bang the matter and radiation are in close thermal contact, and the CMB spectrum is Planckian. At $z < z_T \approx 10^6$, Thomson scattering is the most effective mechanism of interaction between the matter and the radiation field, transferring energy to and from existing photons rather than through photon production and absorption, so a net transfer of energy results in distortions in the radiation spectrum. The mean rate at which a photon can gain energy through Thomson scattering is given by the parameter $\alpha_0(z)$, defined as

$$\alpha_0(z) = \frac{kT_e(z)}{m_e c^2} \sigma_T n_e(z) c$$

where $T_e(z)$ is the electron temperature at redshift $z$, $n_e(z)$ is the electron

297

density at $z$, and $\sigma_T$ is the Thomson-scattering cross-section. Note that $\alpha_0$ is simply the product of the ratio of electron kinetic energy to rest mass (which determines how efficiently Thomson scattering transfers energy) and the rate at which the photon undergoes Thomson scatterings; this quantity is proportional to the pressure of the medium $(n_e k T_e)$.

The Thomsonization parameter discussed above is the integral of $\alpha_0$ with respect to time, and is a measure of the fractional change in photon energy caused by Thomson scattering between some initial time $t$ and the present. This integral is defined as

$$
y(z) = -\int_0^{t(z)} \alpha_0 dt
$$

$$
= -\int_0^{z(t)} \alpha_0(z) \frac{dt}{dz} dz
$$

(Zel'dovich and Sunyaev 1969). The value of $y$ is unity at a redshift $z_a \approx 1.8 \times 10^4 \hat{\Omega}_b^{-1/2}$ where $\hat{\Omega}_b = \Omega_b(H_0/50 \text{km/sec/Mpc})^2$, $\Omega_b$ is the ratio of the baryon density to the critical density, and $H_0$ is the Hubble parameter (Danese and De Zotti 1980). If energy injection occurs at a redshift $z_h \gtrsim z_a$, Thomson scattering causes the spectrum to approach a Bose-Einstein distribution with a non-zero chemical potential $\mu$.

Energy injection at redshifts between $z_a$ and $z_T$ gives rise to a blackbody spectrum at long wavelengths (due to bremsstrahlung), a Bose-Einstein spectrum at short wavelengths (due to Thomson scattering), and a transition region in between. The Bose-Einstein chemical potential $\mu_0$ is proportional to the ratio of the injected energy to the energy previously present in the radiation field:

$$
\mu_0 \approx 1.4 \frac{\delta E}{E_R} \tag{1}
$$

where $E_R$ is the energy density of the unperturbed radiation field and $\delta E$ is the energy added to the radiation field (Chan and Jones 1975).

If energy injection occurs at a redshift $z_h$ smaller than $z_a$, Thomson scattering may still affect the radiation spectrum even though it cannot establish a Bose-Einstein distribution. The Thomson-distorted spectrum is given by (Danese and De Zotti 1978) and is characterized by a parameter $u$ which is defined by the equation:

$$
u = -\int_0^{z_h} \alpha_0(z) \frac{T_e - T_R}{T_e} \frac{dt}{dz} dz = -\int_0^{z_h} \frac{k(T_e - T_R)}{m_e c^2} \sigma_T n_e(z) c \frac{dt}{dz} dz ,
$$

$T_e$ and $T_R$ being respectively the electron and radiation temperatures (Illarionov and Sunyaev 1974). Note that for $T_e \gg T_R$, $u \approx y$. The value of $u$ is determined by $\delta E/E_R$, the fractional energy added to the CMB:

$$
\begin{aligned}
\frac{\delta E}{E_R} &= e^{4u} - 1 \\
&\approx 4u \qquad (u \ll 1)
\end{aligned} \tag{2}
$$

(Sunyaev and Zel'dovich 1980). Thomson scattering has the effect of depressing $T_B$ by an amount $2uT_R$ at frequencies $x < 1$ and sharply enhancing it beyond. Zel'dovich et al. (1972) have shown that the amount of the temperature rise is

$$\Delta T = 7.4\, T_R u \ .$$

## 2.A.2 Fits to Observations

The combined results of recent measurements at wavelengths longer than 0.1 cm (Table 1) yield a CMB temperature of $2.74 \pm 0.02$ K and fit a blackbody spectrum with a $\chi^2$ of 24.0 for 17 degrees of freedom. Including the new results of Matsumoto et al. (1987), the data set is inconsistent with a blackbody spectrum.

As well as checking the results for overall consistency with a blackbody spectrum, one can also analyze the measurements for possible distortions. When only the CMB measurements in Table 1 at wavelengths longer than 0.1 cm are fitted to equation (3), the resulting values are:

$$\hat{\Omega}_b = 1.0 \quad T = 2.76 \pm 0.02\ \mathrm{K} \quad \mu_0 = (4.2 \pm 4.9) \times 10^{-3} \quad \chi^2 = 23.14$$
$$\hat{\Omega}_b = 0.1 \quad T = 2.76 \pm 0.02\ \mathrm{K} \quad \mu_0 = (2.4 \pm 3.5) \times 10^{-3} \quad \chi^2 = 21.98$$
$$\hat{\Omega}_b = 0.01 \quad T = 2.76 \pm 0.02\ \mathrm{K} \quad \mu_0 = (2.0 \pm 1.5) \times 10^{-3} \quad \chi^2 = 21.76$$

for 19 degrees of freedom and the errors are for the 68% confidence level.

Because the value of $z_h$ affects the contribution of bremsstrahlung to the low-frequency portion of the spectrum, $z_h$ and $\hat{\Omega}_b$ must both be specified for the model for the best-fit values of $T_R$ and $u$. Table 2 lists the best-fitted values of $T_R$ and $u$ over a range of redshifts and densities. The values are derived from all the measurements in Table 1. The best fit values of $u$ indicate a Thomsonization distortion at the 12-$\sigma$ level with the entire significance due to the data of Matsumoto et al. (1987); without their data the best-fitted value is negative but consistent with zero.

## 2.A.3 Impact on CMB-Production Models

The measured values of $\mu_0$, which are all consistent with zero, can be used in conjunction with equation (2) to set upper limits on the energy transferred to the radiation field at redshifts between $z_a$ and $z_T$. The complete data set yields the 95% confidence level limits:

$$\hat{\Omega}_b = 1.0 \quad \frac{\delta E}{E_R} \leq 10.0 \times 10^{-3}$$
$$\hat{\Omega}_b = 0.1 \quad \frac{\delta E}{E_R} \leq 6.7 \times 10^{-3}$$
$$\hat{\Omega}_b = 0.01 \quad \frac{\delta E}{E_R} \leq 3.6 \times 10^{-3} \ .$$

Similarly, the measured values for $u$ restrict the energy release that could have occurred at more recent times. Depending on the density parameter and

## TABLE 1

### Recent Measurements of the Cosmic Background Radiation Temperature

| References | Wavelength (cm) | $\nu$ (GHz) | $T_{CBR}$ (K) |
|---|---|---|---|
| Sironi *et al.* 1987 | 50.0 | 0.6 | $2.45 \pm 0.7$ |
| Levin *et al.* 1987 | 21.2 | 1.41 | $2.28 \pm 0.39$ |
| Sironi and Bonelli 1986 | 12.0 | 2.5 | $2.79 \pm 0.15$ |
| De Amici *et al.* 1987 | 8.1 | 3.7 | $2.58 \pm 0.13$ |
| Mandolesi *et al.* 1986 | 6.3 | 4.75 | $2.70 \pm 0.07$ |
| Kogut (Smoot *et al.* 1987 ) | 3.0 | 10.0 | $2.61 \pm 0.06$ |
| Johnson and Wilkinson 1986 | 1.2 | 24.8 | $2.783 \pm 0.025$ |
| De Amici (Smoot *et al.* 1985*b*) | 0.909 | 33.0 | $2.81 \pm 0.12$ |
| Bersanelli (Smoot *et al.* 1987 ) | 0.333 | 90.0 | $2.60 \pm 0.10$ |
| Meyer and Jura 1985 | 0.264 | 113.6 | $2.70 \pm 0.04$ |
|  | 0.132 | 227.3 | $2.76 \pm 0.20$ |
| Crane *et al.* 1986 | 0.264 | 113.6 | $2.74 \pm 0.05$ |
|  | 0.132 | 227.3 | $2.75^{+0.24}_{-0.29}$ |
| Peterson, | 0.351 | 85.5 | $2.80 \pm 0.16$ |
| Richards, and | 0.198 | 151 | $2.95^{+0.11}_{-0.12}$ |
| Timusk 1985 | 0.148 | 203 | $2.92 \pm 0.10$ |
|  | 0.114 | 264 | $2.65^{+0.09}_{-0.10}$ |
|  | 0.100 | 299 | $2.55^{+0.14}_{-0.18}$ |
| Matsumoto *et al.* 1987 | 0.116 | 259 | $2.795 \pm 0.018$ |
|  | 0.0709 | 423 | $2.963 \pm 0.017$ |
|  | 0.0481 | 624 | $3.150 \pm 0.026$ |

before this time are erased since Thomson scattering is essentially isotropic. At the present time most anisotropy measurements have produced only upper limits. There are only two anisotropies that are known to exist and one as yet unconfirmed report. These measurements and upper limits are shown in Figure 5.

The angular variations in temperature of the cosmic microwave background convey valuable information on the large scale structure of the universe. Although careful and precise searches have been made for CMB temperature fluctuations on all angular scales from a few arc seconds to 180 degrees, the only discovered departures from isotropy seem to be the dipole anisotropy thought to be due to the observer's motion relative to the cosmic background and the fluctuations observed toward clusters of galaxies caused by the Sunyaev-Zel'dovich effect.

The current upper limits on the CMB anisotropies on all angular scales significantly constrain cosmological theories. At present the lack of observed anisotropy and spectral distortion has caused cosmologists to embrace more exotic theories including the inflationary scenarios for the early universe and dark matter models where the bulk of the mass in the universe is carried by weakly interacting massive particles (WIMPs).

**Primeval Anisotropies and Dark Matter:** Some of the strongest evidence for the existence of dark matter in the universe is the currently observed isotropy of the CMB. On angular scales $\geq 2°$, which corresponds to the causal horizon at the epoch of last scattering ($z \approx 10^3$), the observed isotropy can be understood only by invoking either special initial conditions or inflationary cosmology. Inflation models unambigously predict that the density of the universe is the critical density ($\Omega = 1$), and thus that the universe is dominated by dark matter. Independent of the assumption of inflation, the observed isotropy on angular scales of 10 minutes to 2° requires $\Omega \geq 0.3$ in order to reconcile the homogeneity of the matter distribution at $z \approx 10^3$ with the amplitude of the large-scale structure observed today. Taken together with the upper limit on the density of baryonic matter ($\Omega_B \leq 0.2$), which is derived from Big Bang nucleosynthesis, the isotropy of the CMB leads one to conclude that at least some of the matter in the universe is not only dark but also is not baryons.

The presence of dark matter reduces, but does not eliminate, the predicted anisotropy in the CMB. Most dark matter models predict anisotropies within an order of magnitude of current upper limits. As sensitivity improves, measurements of the CMB isotropy thus promise to shed light on the nature of the dark matter, particularly if anisotropies can be detected and studied in detail. Various classes of dark matter can be distinguished from one another by the amplitude, power spectrum, and morphology of the anisotropies they produce in the CMB. Cosmic strings, for example, would produce a network of step function anisotropies, while the large-scale structure associated with weakly interacting massive particles would produce a gaussian distribution

301

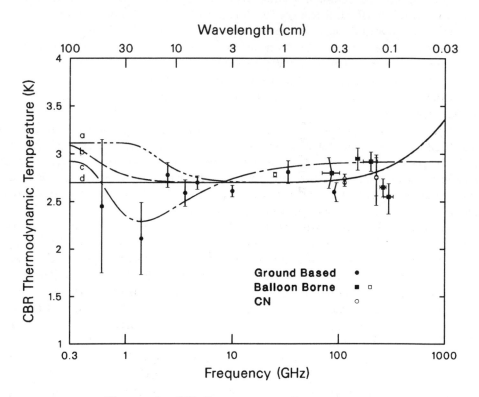

Figure 3. CMB Temperature Measurements
with 95% C.L. Distortions

(a) $\hat{\Omega}_b = 0.1$, $T_R = 2.795$ K, U = 0.01196, $z_h = 40000$;

(b) $\hat{\Omega}_b = 1.0$, $T_R = 2.809$ K, U = 0.01282, $z_h = 1000$;

(c) $\hat{\Omega}_b = 0.1$, $T_R = 2.780$ K, $\mu_o = 0.0056$;

(d) $\hat{\Omega}_b = 1.0$, $T_R = 2.805$ K, $\mu_o = 0.0141$.

TABLE 2

Best-Fitted Values of $T_R$ and $u$ from All Measurements[a]

| $\hat{\Omega}_b$ | $z_h$ | $T_R$ (K) | $u$ | Maximun $T_R$ (K) | Maximum $u$ | $\chi^2_{best}$ |
|---|---|---|---|---|---|---|
| 1.0 | $4 \times 10^4$ ($z_a$) | 2.805 | $1.78 \times 10^{-2}$ | 2.826 | $2.09 \times 10^{-2}$ | 79.49 |
|  | $1 \times 10^4$ | 2.807 | $1.95 \times 10^{-2}$ | 2.828 | $2.28 \times 10^{-2}$ | 55.49 |
|  | $4 \times 10^3$ | 2.811 | $2.00 \times 10^{-2}$ | 2.830 | $2.34 \times 10^{-2}$ | 45.28 |
|  | $1 \times 10^3$ | 2.814 | $2.01 \times 10^{-2}$ | 2.833 | $2.34 \times 10^{-2}$ | 38.68 |
| 0.1 | $1 \times 10^5$ ($z_a$) | 2.812 | $2.01 \times 10^{-2}$ | 2.832 | $2.34 \times 10^{-2}$ | 40.43 |
|  | $4 \times 10^4$ | 2.813 | $2.01 \times 10^{-2}$ | 2.833 | $2.34 \times 10^{-2}$ | 39.82 |
|  | $1 \times 10^4$ | 2.813 | $2.01 \times 10^{-2}$ | 2.833 | $2.34 \times 10^{-2}$ | 39.16 |
|  | $4 \times 10^3$ | 2.813 | $2.01 \times 10^{-2}$ | 2.833 | $2.34 \times 10^{-2}$ | 38.88 |
|  | $1 \times 10^3$ | 2.814 | $2.01 \times 10^{-2}$ | 2.833 | $2.34 \times 10^{-2}$ | 38.68 |

[a] The maximum $T_R$ and $u$ are the maximum values of these parameters on the $\chi^2_{best} + 4$ ellipse. This ellipse corresponds to about the 95% confidence level. The $\chi^2$ is computed with 19 degrees of freedom.

the redshift of energy release, these results indicate an energy transfer to the CMB of approximately 8%.

One can do a series of parameter fits using various assumptions about the density of the universe, the baryon (electrons for Thomson scattering) density, and the epoch and nature of the energy release. These families of parameter limits can then be converted into limits on energy release in those situations. One can then make a series of curves showing maximum allowable fractional energy release, $\delta E/E_R$, as a function of redshift for various densities.

The fractional energy release limits can then be used to set limits on processes of cosmological interest. Examples of these are: (1) The spectrum of adiabatic density perturbations (Sunyaev and Zel'dovich 1970) (2) The spectrum of primordial turbulence and vorticity (Illarionov and Sunyaev 1974; Chan and Jones 1976) (3) Annihilation of matter and antimatter in the early universe (Stecker and Puget 1972; Sunyaev and Zel'dovich 1980) (4) Energy release by evaporating primordial black holes or unstable (decaying) particles (Dolgov and Zel'dovich 1981; Silk and Stebbins 1983) (5) an improved estimate of the average photon density of the Universe.

From the constraints on the chemical potential, the energy in turbulence on scales which are currently 30 kpc to 4 Mpc is less than one per cent of that in the CMB. This limit is sufficient to rule out turbulence and vorticity as the drivers of galactic formation. Similar arguments can be made regarding adiabatic perturbations and the limits are borderline. However, the small-scale anisotropy measurements place an even more restrictive limit, if re-ionization of the universe did not happen sufficiently early to scatter the CMB and erase the distortions.

The annihilation of matter and antimatter is constrained similarly by the limits on energy release provided by the limits on $y$ and the chemical potential $\mu$. If the Grand Unification and Inflation theories were well established, then the excess of matter over antimatter and the possibility of residual antimatter might be of no concern. For redshifts between $10^6$ and $4 \times 10^4$ the amount of energy released in annihilation is less than 1% of that in the CMB. After a redshift of about $4 \times 10^4$ the limit is about 10%.

The decay of massive particles can also produce a non-Planckian spectrum. Silk and Stebbins (1983) have shown that one can use the limits on CMB spectral distortions to rule out weakly interacting particles which have lifetime, $\tau$, between about 0.1 years and the Hubble time ($10^{10}$ years) and masses, $m$, between about $10^{-6}$ and $10^6$ eV, with the exception of a small range along $m^2\tau = 10^{10}(\text{eV})^2$ years which is hidden in the interstellar background.

Upper limits on $\delta E/E_R$ derived from $\mu_0$ and $u$ are summarized in Figure 4. Some care must be taken in applying these limits to astrophysical processes, since the energy generated by these processes may not be readily transferred to the CMB. Figure 3 shows the data in Table 1, along with five sample distortions.

The situation after recombination is not so simple. Because the neutral matter interacts only very weakly with the radiation, the kinetic energy of its bulk motion does not cause distortions in the CMB spectrum. On the other hand, any significant release of thermal energy by the matter would ionize some fraction of it once again. If the ionized matter is not strongly clumped, the reionized medium interacts with the radiation as before. At redshifts greater than $\sim 8$, the time scale on which the matter is cooled by Thomson scattering is shorter than the expansion time of the universe (Sunyaev and Zel'dovich 1980), so most of the excess thermal energy in the reionized medium is taken up by CMB photons, causing a distortion. In this case, the observational bounds on $u$ limit the thermal energy released to reionized matter between $z \approx 8$ and $z \approx 1500$ to less than 10% of the energy in the CMB. For energy release after $z \approx 8$, the same 10% limit applies, but in this case it refers to the energy actually transferred to the CMB, which will, in general, be less than the energy released to the matter. Similarly, if the electrons in the reionized medium are relativistic, the distorted spectrum will have a steeper slope at high frequencies, rather than that shown in curves a, b, and d of Figure 3.

Much of this analysis breaks down if a large fraction of the matter in the universe is highly clumped or bound up in an early generation of stars. In that case, Thomson scattering may not be able to transfer the heat efficiently from the matter to the CMB. The form and extent of the resulting distortions depend on the details of the model. One such class of models—the hypothesis

304

that some or all of the CMB is thermal radiation from warm dust produced by Population III stars at $z \gtrsim 10$—has been suggested by Rees (1978), Negroponte et al. (1981), Wright (1982), and others to explain the spectral distortion reported by Woody and Richards (1979, 1981). Although the details of such models have to some extent been tailored to fit the Woody-Richards distortion, significant departures from a blackbody curve are almost inevitable because of the spectral characteristics of the carbon or silicate materials that make up the grains. The absence of significant tail-off in the Rayleigh-Jeans region of the CMB spectrum is a heavy blow against such theories. To devise a plausible dust-emission model that gives the observed spectrum over a hundredfold range in wavelength may well prove an impossible task, although Hawkins and Wright (1987) and Wright (1987) have attempted to do so using long thin needles and fractal dust grains. It is more straightforward to explain only the high frequency excess by a background of dust emission. A good fit can be obtained by a model of emission from dust heated by very early stars (Hayakawa et al. 1988).

An important implication of this spectral distortion limit is the constraint it places on the contribution that a hot intergalactic medium (IGM) could make to the diffuse X-ray background (called CXB above). Such a hypothesis has been contemplated (Cowsik and Kobetich 1972) and found to be consistent with the X-ray data (Field and Perrenod 1977; Sherman 1980; Marshall et al. 1980), although a test of this hypothesis requires a careful (and somewhat uncertain) subtraction of the contribution from discrete X-ray sources. The residual spectrum may be too flat to be simply explained as having thermal bremsstrahlung origin (Giacconi and Zamorani, 1987). In a recent analysis, Guilbert and Fabian (1986) restate that one can obtain an accurate fit to the X-ray observations, but that the energy required to heat the IGM is a severe problem. Nonetheless, if the presumed IGM were to contribute significantly to the CXB, it must have $\hat{\Omega}_b \gtrsim 0.22$ and must have been heated within the redshift interval $3 \lesssim z_h \lesssim 6$ to an electron energy $kT_e \approx 36(1 + z_h)$keV.

A hot IGM would distort the CMB spectrum by Thomson scattering. Guilbert and Fabian predict that the expected Rayleigh-Jeans temperature change, $\Delta T = 2uT_R$, should be about 0.07 K so that $u \sim 0.013$ is implied. The best-fitting value for $3 \lesssim z_h \lesssim 6$ is $u = 0.020 \pm 0.002$ at the 95% confidence level. In fitting the results of measurements to the type of spectrum resulting from interactions with a hot IGM, data in the wavelength range from 200 to 300 microns are of particular interest. Emission from interstellar dust is not yet dominant in this region, and the predicted hot IGM flux remains significant due to a high-frequency tail (see, e.g., Matsumoto et al. 1987, Figure 2; and Guilbert and Fabian 1986, Figure 4). It is, perhaps, of interest to note that the predicted flux in Figure 4 of the latter paper is somewhat higher than the Matsumoto et al. value at 262 microns. However, in that particular model the required value of $\hat{\Omega}_b$, 0.27, is relatively high, and a nominal value of 3 K was

305

used for the CMB temperature. We conclude that the current CMB data are not inconsistent with a thermal origin for the CXB.

To interpret the apparent distortion as due to Thomson scattering raises difficult questions with regard to energetics. The high energy required to heat the gas is a basic difficulty if the energy is produced by baryonic matter. (This difficulty is due essentially to the low baryon/photon ratio of the universe.) This difficulty might be avoided if non-baryonic dark matter were the source of the energy. Ostriker and Thompson (1987) have recently estimated that energy deposition of superconducting cosmic strings can cause a distortion of the CMB spectrum at a level corresponding to $u = 1 - 5 \times 10^{-3}$, where the range of values depends on the mass scale of the strings. The low predicted value of $u$, compared to the value derived from spectrum measurements, is probably not a difficulty in view of qualitative uncertainties in the cosmic string scenario.

The potential implications of this distortion call for confirmation of the Matsumoto et al. (1987) results and determination of the epoch at which the distortion occurred. An alternate explanation of the Matsumoto et al. data (Hayakawa et al. 1988) is that dust was produced by a very early generation of stars. That early cosmic dust was heated by starlight and reradiated into the submillimeter. The amount of dust necessary is not excessive ($\Omega_{dust} \sim 10^{-5}$; however, it must have been generated at a relatively early time ($z \geq 6$). This occurence is earlier than stars and galaxies have been thought to form. However, the recent discovery of a galaxy at a redshift greater than 3 may be followed by the discovery of mature galaxies at even greater redshifts and the introduction of a paradigm calling for earlier star and galaxy formation. This will, of course exacerbate the already difficult problem of successfully modelling galaxy formation.

## 2.B. Photon Statistics

If a radiation is thermal in origin, its photons should be distributed according to a planckian distribution, and the distribution, its mean and r.m.s. can be easily computed from basic principles. This provides a means to investigate the statistical properties of the photon stream constituting the CMB in regards to fluctuations in numbers of photons per quantum state in a manner similar to the Hanbury Brown and Twiss stellar interferometer. (Hanbury Brown, R., and Twiss, R.Q., 1954, vol 45, 663.) A result disagreeing with predicted fluctuations would contradict a thermal origin of the CMB.

### Thermal Radiation Statistical Distribution – **Planck Radiation Law**

Photons in a thermal distribution obey a special case of Bose-Einstein particle statistics originally derived by Planck at a turning point in physics. In his explanation, Planck laid the foundation for quantum theory in which the (Planck's) constant h = $6.6256 \ 10^{-34}$ joule-seconds is the fundamental unit of action just as the electronic charge e = 1.601 $10^{-19}$ coulomb is the fundamental unit of charge. Planck asserted that an atomic resonator (matter) could absorb or emit only discrete amounts of energy which were proportional to its oscillator frequency, $E = h\nu$.

The processes that lead to thermal (blackbody) radiation are stochastic. This means that the number of photons (amount of radiation) emitted as a function of time is a random variable whose distribution is set by the physics of the processes involved.

For thermal equilibrium situations there is relative probability of a system being in a given state compared to that of being in another state

$$Boltzmann\ factor = Relative\ Probability = e^{-\Delta E/kT}$$

where $\Delta E$ is the energy difference between states. Following the quantum theory as expressed by Planck the energy of a state with $n$ photons of frequency $\nu$ is $E_s = nh\nu$ so that the relative probability of a state with $n$ photons to a state with 0 photons is

$$rp(n) = e^{-nh\nu/kT}.$$

We can normalize by summing all the relative probabilities from $n$ equals zero to infinity:

$$Normalization^{-1} = \sum_{n=0}^{\infty} rp(n) = \sum_{n=0}^{\infty} e^{-nh\nu/kT} = \frac{1}{1 - e^{-h\nu/kT}}.$$

This is just the sum of a geometric series

$$\sum_{n=0}^{\infty} a^n = \frac{1}{1-a}$$

with $a = e^{-h\nu/kT}$. Now we can compute the normalized probability distribution for the number of photons in a given state for a system in thermal equilibrium at thermodynamic temperature T:

$$p(n) = (1 - e^{-h\nu/kT})e^{-nh\nu/kT} \qquad Photon\ Occupation\ Number\ Probability$$

This is a probability distribution with a well defined mean and clearly a large dispersion especially for cases when $h\nu/kT << 1$.

Not all processes produce photon distributions of this type. An important example is the maser (microwave laser) which depends upon stimulated emission from what are called population inversions or non-equilibrium distributions of energy states in atoms. There are well-known masers that occur naturally including CO masers operating in the Martian atmosphere as well as interstellar masers in large molecular clouds. In a maser the photon occupation number distribution is very different. For a well-behaved maser the distribution is very narrow. For a thermal distribution the state with zero photons is the most likely or frequently occurring state while for the good maser the most frequent state is the one containing the mean number of photons. Thus a good maser would have the same mean number of photons as a matching thermal source but would have a very much smaller variation in the number occupying successive states.

When considering a statistical distribution one often uses the various moments to characterize the distribution with pragmatically useful parameters. In this case we are especially interested in the first moment of the distribution - the average radiation emitted, which is a measure of mean power and the second moment, which is representative of the mean square fluctuations in the power.

### 2.B.1 First Moment

We can calculate the first moment of the thermal radiation probability distribution which is the average photon occupation number (mean number of photons in a quantum state characterized by frequency $\nu$).

$$< n_\nu > = \sum_{n=0}^{\infty} n p(n) = (1 - e^{-h\nu/kT}) \sum_{n=0}^{\infty} n e^{-nh\nu/kT}$$

$$= (1 - e^{-h\nu/kT}) \sum_{n=0}^{\infty} n (e^{-h\nu/kT})^n = (1 - e^{-h\nu/kT}) \frac{e^{-h\nu/kT}}{(1 - e^{-h\nu/kT})^2}$$

$$< n_\nu > = \frac{1}{(e^{h\nu/kT} - 1)} \qquad Planck\ Mean\ Photon\ Occupation\ Number$$

### 2.B.2 Second Moment

The second moment of the photon statistical distribution is a measure of the mean square variation of the quantum state, s, occupation number.

$$< \Delta n_s^2 > = < (n_s - < n_s >)^2 > = < n_s^2 > - < n_s >^2 .$$

For the Planckian distribution the mean square variation in the photon occupation number is

$$< \Delta n_\nu^2 > = \sum_{n=0}^{\infty} (n^2 - < n >^2) p(n) = (1 - e^{-h\nu/kT}) \sum_{n=0}^{\infty} (n^2 - < n >^2) e^{-nh\nu/kT}$$

$$< \Delta n_\nu^2 > = < n_\nu >^2 + < n_\nu > \qquad Planck\ Occupation\ No.\ Variance.$$

The first part, the $< n_\nu >^2$ term, is due to the coherence, the fact that photons in the same state can interfere constructively or destructively; the $< n_\nu >$ term is just what we expect for a Poisson distribution of discrete objects with expected number $< n_\nu >$. This is the quantum noise, the equivalent of electron shot noise.

## 2.B.3 Photon Statistics and Detection Efficiency/Sensitivity

Photon noise on detectors arises from the random fluctuations in the rate of absorption of photons. The process of absorption or amplification of photon signal depend upon the coherence of the incident photons. Due to the quantum processes involved there is a minimum noise added by an amplifier or detector that corresponds to a single quantum. In most systems this is currently not a serious concern since almost no amplifier is quantum limited and only a few detectors are. However, some bolometers and photodetectors are near the quantum limit and the photon collection system can contribute a significant degradation to the detector performance due to the degradation of photon coherence by beam attenuation and by partial absorption by the detector (Mather, J. 1982, Applied Optics, **21**, 1125). The effect is directly related to the effective photon occupation number for absorption of the photons in the detector.

## 2.C Polarization of the CMB

Any anisotropy intrinsic to the CMB which was established prior to the last scattering will result in a net linear polarization due to the polarization dependence of Thompson (low-energy Compton) scattering. The linear polarization is of special interest in that Rees (1968, Ap. J. **153**, L1.) first pointed out that it preserves information about the expansion and anisotropies through the last few scatterings. This means that even if some primordial anisotropies have been washed out by a reionization of the intergalactic medium or the last moments of decoupling the linear polarization allows us to look back a couple of more optical depths than temperature anisotropy measurements. As a result the primordial perturbations might produce polarizations of the CMB anisotropies as large as 25%. The actual degree of polarization in an anisotropic universe is sensitive to the ionization history and the fractional hydrogen content (Negroponte and Silk, 1980) and typically lies in the range of 1 to 100% of the temperature anisotropy. Polarization measurements of the CMB can potentially provide a unique signal of cosmological anisotropy.

Nanos (1979, Ap. J. **232**, 341-347) and Lubin, Melese, and Smoot (1983, Ap. J. **273**, L51-L54) have placed 95% confidence level limits of 0.2 mK ($\Delta P/T \leq 7 \times 10^{-5}$) and 20 mK on any linear or circular polarization components respectively. These data thus give support to the thermal origin of the CMB.

## 2.D Anisotropy

In the standard hot Big Bang model at a redshift of about a thousand the primordial plasma of electrons and ions combined to neutral form, and the universe became optically thin to the CMB. Any inhomogeneity that was present after this time produced an anisotropy in the CMB. Anisotropies present from characterized by a power law spectrum.

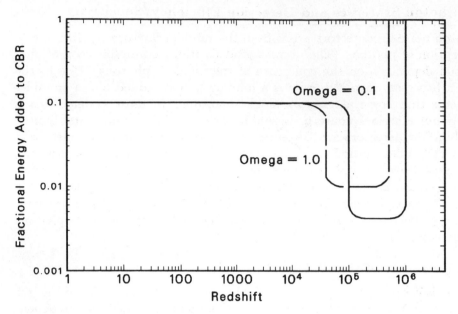

Figure 4. 95% Confidence Limits on Fractional Energy Release

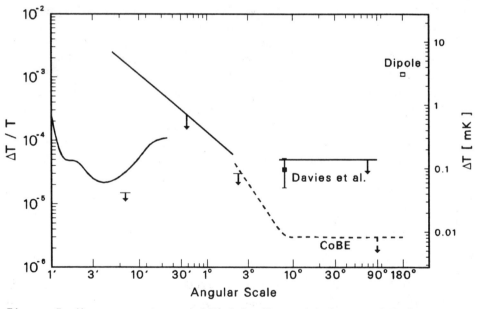

Figure 5. Measurements and 95% C.L. Upper Limits on CMB Anisotropy

Hot and cold dark matter models can be distinguished from one another because free-streaming of the hot dark matter produces relatively smaller fluctuations on scales less than 10 Mpc. This produces a break in the power spectrum of CMB anisotropies at about 5 arcminutes. Because each model must be normalized to the amplitude of density perturbations observed for scales less than 10 Mpc in the current epoch, the predicted amplitude of anisotropies on larger angular scales are quite different for hot and cold dark matter models. Unfortunately, the power spectrum of primeval anisotropies will be difficult to observe on angular scales smaller than about 10 arcminutes. Compton scatterings, the primary interaction between the CMB and plasmas, is nearly isotropic and thus destroys the anisotropy on scales smaller than the width of the last scattering surface. The standard models predict that the width of the last scattering surface at $z \approx 10^3$ corresponds to angular scales of about 10 arcminutes. The amplitude of primeval anisotropies on smaller scales will be diluted.

Anisotropies on angular scales greater than about 10 arcminutes will remain undiluted and reflect the structure of the universe on very large scales and at times when linear perturbation theory still applies. Anisotropies thus provide information about the origins and evolution of structure in the universe which is complementary to that which can be obtained by studying the distribution of galaxies at the current epoch. Such information will be essential to any complete understanding of the nature of dark matter and the role it plays in the formation of structure in the universe.

**Medieval Anisotropies** Other interpretations of the CMB anisotropy exist. If the universe were re-ionized at an early enough epoch ($z \geq$ ), the optical depth to Compton (Thomson) scattering would exceed unity.

Any energy release capable of re-ionizing the universe (for example, supernova explosions from an early generation of massive stars) may also play a critical role in the formation of structure on the scale of galaxies, clusters, and superclusters. In this case, neither the matter distribution that we observe today nor the isotropy of the CMB reflect the primeval matter distribution, and the connection to dark matter is radically altered. Dark matter might provide an energy source for the re-ionization (e.g. the decay of massive particles) or might be contained in the relics of an early generation of stars that fueled the re-ionization. In addition the re-ionization will leave its imprint in both the spectrum and anisotropy of the CMB.

Even if an energy release at intermediate redshifts does not re-ionize the universe sufficiently to wash out the primeval anisotropies, it may create medieval anisotropies that we must distinguish from the primeval anisotropies. Medieval anisotropies will reflect the distribution of matter at the epoch energy release, and may provide important clues to the formation of structure in the later stages of the evolution of the universe. Medieval anisotropies could be created by Compton scattering of the CMB or by the emission from cosmic dust that is heated by the energy release. Emission by cosmic dust is expected to dominate the CMB at submillimeter wavelengths (Negroponte, 1986, MN-RAS, **222**, 19, McDowell, 1986, MNRAS, **223**,763) and to produce significant anisotropies at millimeter and submillimeter wavelengths (Bond, Carr, and Hogan, 1986, Ap. J., **306**, 428). A suprisingly large cosmic submillimeter background which might be a result of dust emission at high redshift was reported by Matsumoto *et al.* (1988) and discussed in the spectrum analysis.

Medieval anisotropies complicate the study of primeval anisotropies, but are extremely interesting in their own right and crucial to understanding the dark matter problem. While the reported detection of a cosmic submillimeter background requires confirmation, such a background has a sound theoretical grounds and may well provide us with information on intermediate redshifts. When a CMB anisotropy is detected, we must be able to distinguish between the various possible origins. This requires an eventual measurement at several wavelengths from centimeter to submillimeter.

### 2.D.1 Dipole or First Order Anisotropy

The velocity of the solar system and thus our galaxy and its neighbors produce a first order CMB anisotropy due to the doppler shift. Thus a careful measurement of the CMB dipole anisotropy can be used to determine our peculiar velocity with respect to the distant matter in the universe and thus give us insight in cosmic dynamics and corrections in determining the Hubble constant.

*Table 3 Summary of measured dipole anisotropy amplitude*

| Research Group | Dipole Amplitude(mK) | Right Ascension(hr) | Declination (degrees) |
|---|---|---|---|
| *Berkeley* | $3.46 \pm 0.17$ | $11.3 \pm 0.1$ | $-6.0 \pm 1.4$ |
| *Princeton* | $3.18 \pm 0.17$ | $11.2 \pm 0.1$ | $-8 \pm 2$ |
| *Moscow* | $3.16 \pm 0.12$ | $11.3 \pm 0.15$ | $-7.5 \pm 2.5$ |

The discovery of the dipole anisotropy and its implication that our galaxy was moving some 550 km/sec relative to the CMB and thus the distant matter in the universe eventually spawned the industry of galactic flows around our local part of the universe and the successful search for the "Great Attractor". The current results yield a galactic velocity of $540 \pm 50$ km/sec towards $\alpha = 10.8 \pm 0.3$ hours and $\delta = -25 \pm 5°$ or $l^{II} = 272 \pm 5°$ and $b^{II} = 30 \pm 5°$ assuming a solar velocity of 295 km/sec towards $l^{II} = 97.2°$ and $b^{II} = -5.6°$.

### 2.D.2 Quadrupole or Second Order Anisotropy

The limits on the quadrupole anisotropy are particularly powerful in analyses of anisotropy and model constraints. The quadrupole (2nd order spherical harmonic) anisotropy is of intrinsic interest, for its amplitude is independent of how the universe has evolved during the interval from decoupling ($z = 1000$) to the current epoch. Limits on the quadrupole amplitude constrain the allowed rotation of the universe, the long wavelength gravity wave density, anisotropic expansion, etc. The current limits on the density of gravity waves and the rotation of the universe are significant constraints for cosmological models.

### 2.D.3 Higher Order Anisotropies

The formation of structure in the universe is as yet one of the major unsolved problems in cosmology. One reason that it remains so are the severe limits set by the general isotropy of the CMB.

There is a report of a detection of anisotropy at an angular scale of 8° by Davies *et al.* (Nature 1987), which has not yet been confirmed. Because of the low frequency of the observation (10.4 GHz) and the importance of finding anisotropy, a number of groups including Lasenby and Davies are making observations to check this finding.

The only other anisotropies reported and confirmed as those related to the Sunyaev-Zel'dovich effect briefly mentioned in the spectrum discussion. This is a small anisotropy resulting from the scattering of the cool CMB photons to higher energies by the hotter electrons in galactic clusters. One can predict the magnitude of the effect from the observed x-ray intensity assuming the x-rays are produced by thermal bremsstrahlung and assuming reasonable parameters for the properties of the galactic cluster. The measured anisotropies are roughly consistent with the model prediction but are not yet sufficiently precise to be used as a diagnostic probe of galactic and cosmic parameters. (See for example: Birkinshaw *et al.* 1984; Nature **309**, 34.)

From the moment of its discovery researchers have investigated the isotropy of the CMB. After 23 years of investigation we primarily have upper limits and have yet to find features in the maps that we recognize as cosmological effects. This has been disappointing to the observers and is beginning to discomfort the theorists. At first there were simple and plausible ways for theorists to lower the level of the original fluctuations expected to be at the $10^{-3}$ level expected from the existence of galaxies and clusters. Since then the levels have had to be pushed down by about $10^2$. This has forced more exotic theories into being including the need for Inflationary models, dark matter, cosmic strings, and other devices to generate galaxies and still not distort the CMB noticeably.

Inhomogeneities in the large-scale geometry and distribution of matter in the universe inevitably lead to the generation of large-angular-scale anisotropy of the CMB. The observed isotropy of the CMB thus implies a high degree of isotropy and homogeneity for the universe.

## 3. FUTURE WORK AND CONCLUSIONS

While the photon is the best probe that we have of the distant universe and astronomers have long studied distant objects, diffuse astronomy is just now moving from the pioneering age to a phase of systematic study. Thus there are a number of large projects for a study of diffuse and distant objects and their relationship to cosmology in process at the present time.

### 3.A. COBE - NASA Cosmic Background Explorer Satellite

The Cosmic Background Explorer (COBE) satellite to be launched in 1989 is designed to map the spectrum and angular distribution of diffuse radiation from the universe over the wavelength range from 1 micron to 1 cm (Mather, 1982, Optical Engineering, **21**, 769). It carries three instruments: (1) a set of differential microwave radiometers (DMR) operating at 3.3, 5.7 and 9.5 mm wavelengths, (2) a far infrared absolute spectrophotometer (FIRAS) covering 100 microns to 1 cm, and (3) a diffuse infrared background experiment covering 1 to 300 microns. The planned mission is one year of observations from a polar orbit aligned on the day-night terminator. The instruments are located inside a large conical shield, which protects them from the earth, the sun, and the telemetry emissions. The spacecraft rotates at about 0.8 Hz and orbits the earth about every 90 minutes at a 900 km altitude. The earth's quadrupole moment precesses the plane of the spacecraft orbit at a rate that keeps the orbit-sun-earth geometry roughly constant. The motion of the earth along its path around the sun causes the satellite to scan the entire sky every six months.

**COBE DMR** The purpose of the DMR (differential microwave radiometers) is to map the CMB and detect anisotropies. The radiometers operate at frequencies of 31.5, 53 and 90 GHz, chosen to minimize interference from known astronomical and man-made sources of radiation, e.g. galactic synchrotron and free-free (HII) emission at the longer wavelengths and interstellar dust at shorter wavelengths. The rms noise sensitivity of the radiometers is estimated to be 50, 20, and 30 mK/Hz$^{1/2}$ for each frequency respectively. If the measurement goes as expected the measured temperature difference for a pixel the size of the antenna's 7° field of view will have rms noise of less than 0.2 mK corresponding to $\Delta T/T \leq 7 \times 10^{-5}$. The sensitivity to measuring large angular scale anisotropies will be proportionally better.

**COBE FIRAS** The FIRAS (far infrared absolute spectrophotometer) will measure the spectrum of the CMB, the interstellar dust, any submillimeter background, and any other sources in the wavelength range from 100 microns to 1 cm. Its primary purpose is to detect and measure CMB spectral distortions.

The FIRAS is a cryogenically cooled Michelson interferometric spectrometer. It is rapidly scanned and uses a polarizer as a beam splitter. It is cooled to 1.6 K and is symmetrized so that both inputs and outputs are used. For maximum sensitivity to distortions of the CMB spectrum, the instrument second input is a reference blackbody whose temperature is maintained near the CMB temperature. In addition, there is a full beam blackbody calibrator which can replace the beam from the sky with an accurately known blackbody spectrum.

**COBE DIRBE** The DIRBE (diffuse infrared background experiment) is a 10 band filter photometer covering the wavelength range from 1 to 300 microns. Its purpose is to detect and measure any extragalactic background radiations which may exist and to measure the foreground sources such as starlight, sunlight reflected and reradiated from the interplanetary dust, and thermal emission from interstellar dust heated by starlight.

The light from the sky is received by an off-axis Gregorian telescope with a primary aperture of 20 cm and about a 1° square field of view. The DIRBE sensitivity will be better than about $10^{-12}$ W cm$^{-2}$sr$^{-1}$Hz$^{-1/2}$. These limits will be achieved using photoconductors and bolometers cooled to about 1.6 K.

## 3.B. Relikt 2 and Aelita - Institute for Space Research Moscow

The Institute for Space Reserch in Moscow is planning two satellites to measure the cosmic backgrounds in the early 1990's. The Relict 2 mission will fly in 1992 and the Aelita mission in 1994. The Relict 2 mission will be competitive with the COBE DMR. The Aelita will carry a cryogenic submillimeter telescope and small angular scale anisotropy instruments and is well along in its planning, particularly the difficult cryostat development.

**RELICT 2** The Relict 2 satellite is planned for launch in about 1992, and will carry 5 channels of microwave radiometers. The frequencies are 22, 35, 59, 83, and 193 GHz. These radiometers will be cooled to 100 K by radiative coolers, which is facilitated by its orbit about Lagrange point L2, with a sun-oriented spin axis. As with Relict 1 (Strukov *et al.* 1987 *Sov Astron. Lett*, **13**, 65), Relict 2 has a gas jet attitude control which will move the spin axis every 10 days or so. Also as with Relict 1, the reference beam is along the spin axis, and the measuring beam perpendicular to the spin axis. The antennas will be of the corrugated horn design like the COBE DMR. Most of the Relict 1 problems with Moon and Earth emission will be negligible at that distance and orientation.

One area where this group has done quite well is in the use of degenerate parametric amplifiers. They quote a rms noise of 3 mK/Hz$^{1/2}$ at a frequency of 30 GHz using a receiver cooled to 77 K. The advantage of parametric amplifiers at higher frequencies is less, but they project cooled-mixer noise temperatures that are lower than COBE's. They have concentrated on getting broad bandwidth, with an 8 GHz IF frequency and GaAs FET preamps and HEMT preamps.

They are keenly aware of the importance of systematic error analysis after their experience with Moon and Earth interference, and in fact have asked how it is being handled for COBE. They are particularly interested now in the modeling of the Galaxy. The observing location for Relict 2 may well be better for anisotropy studies than COBE's. If their instrumentation works as designed, it is fairly likely the Relict 2 data will be as good or better.

**AELITA** The Aelita satellite is still being defined and is planned for launch around 1994. It has two telescopes planned: a cryogenic 1 meter submillimeter telescope, and a small angular scale anisotropy telescope. The cryogenic telescope is fairly far along in design. The small scale anisotropy experiment also has about a 1 meter aperture but with warmer receivers, similar in technology to those used on Relict 2. The proposed orbit is a 4 day long ellipse, with apogee of 200,000 km.

The cryogenic telescope design has a neon-cooled stage at 27 K for the telescope, a superfluid helium stage for the photometer, and a 0.3 K stage for the bolometers. They have a helium - 3 refrigerator that works against gravity, using capillaries. It has an absorption pump and a cycle time of 24 hours with a 100 microwatt load. The photometer design is a combination of dichroics for the long wavelength bands, with a movable filter wheel in front of the short wavelength detector. The present planned bands are centered on 1390, 700, 513, 460, 350, 280, 230, and 180 microns. Filters are 4-layer mesh filters, both inductive and capacitive, and spectral bandwidth is typically $\lambda/\Delta\lambda = 6$. Detectors are composite bolometers in spherical cavities, with cone-shaped absorbers to help improve the absorption efficiency, and NEP's of $10^{-15}$. There would be on-board microprocessors for signal analysis and instrument control.

The Soviets are open to cooperation in a variety of areas on the Aelita, in part becase such cooperation will support their case for completing it. They curently have a collaboration with the Roma Italia group.

### 3.C. IRTS - Japanese Space Agency

The IRTS (Infrared Telescope in Space) is scheduled for launch in 1993. It is a 15-cm telescope cooled to 2 K and is designed to have a mission life of two weeks. There are four focal plane instruments, all of which observe simultaneously. The instruments are a near and a mid-infrared spectrometer (NIRS & MIRS) with 32-element gratings, a far-infrared line mapper (FILM) with resolving power of about 500 designed to look at the OI and CII lines, and a far infrared photometer (FIRP) with 5 bands in the submillimeter. Prior to launch there will be two rocket flights, in February 1989 and February 1990, which will test prototypes of the focal plane instruments. The IRTS mission will cover about 10% of the sky and collect about 5% of the data of IRAS but with much greater sensitivity.

ACKNOWLEDGEMENTS. I would like to thank the Erice - "Ettore Majorana" Centre for Scientific Culture and particularly the Workshop on Data Analysis in Astronomy for inviting me to present this talk and participate in the workshop. This research was supported in part by NASA contract NAS5-27592, by National Science Foundation Grant No. DPP-87165548, and by the Department of Energy under Contract DE-AC03-76SF00098.

TABLE 4 IRTS Instruments

| | NIRS Nagoya | MIRS Tokyo/Ames | FILM ISAS/Berkeley | FIRP Berkeley/Nagoya |
|---|---|---|---|---|
| Type | Grating Spectrometer | Grating Spectrometer | Fabry-Perot Grating Spectrometer | Multi-Channel Photometer |
| Wavelength Range | 1.3 to 5 $\mu m$ | 7 to 14 $\mu m$ | 148 to 168 $\mu m$ 60 to 66 $\mu m$ | 100 to 800 $\mu m$ |
| Wavelength Resolution | $\Delta\lambda = 0.12\ \mu m$ | $\Delta\lambda = 0.2\ \mu m$ | $R = 500$ | $\Delta\lambda/\lambda = 0.2$ |
| F.O.V. | 8' x 8' | 8' x 8' | 8' x 20' | 30' FW circle |
| Detector | InSb 32 elements | Si:Bi 32 elements | Stressed Ge:Ga x5 | $^3$He bolometer x 5 |
| Sensitivity | $3 \times 10^{-13}$ W/cm$^2$ $\mu$msr ($\tau = 10$ sec, $3\sigma$) | $3 \times 10^{-13}$ W/cm$^2$ $\mu$msr ($\tau = 1$ sec, $3\sigma$) | $10^{-13}$ W/cm$^2$ sr ($\tau = 1$ sec, $3\sigma$) | $10^{-14}$ W/cm$^2$ sr ($\tau = 5$ sec, $2\sigma$) |
| Data Rate | 1.5 kbps | 500 bps | 640 bps | 640 bps |

# REFERENCES

Crane, P., Heygi, D. J., Mandolesi, N., and Danks, A. C. 1986, *Ap. J.,* **309,** 822.

Danese, L. and De Zotti G. 1977, *Riv. Nuovo Cimento* **7**, 277.

Danese, L. and De Zotti G. 1978, *Astron Astrophys.* **68**, 157.

Danese, L. and De Zotti G. 1982, *Astron Astrophys.* **107**, 39.

De Amici, G., Witebsky, C., Smoot, G. F., and Friedman, S. D. 1984, *Phys. Rev. D*, **29,** 2673.

De Amici, G., Smoot, G. F., Friedman, S. D., and Witebsky, C. 1985, *Ap. J.,* **298**, 710.

Friedman, S. D., Smoot, G. F., De Amici, G., and Witebsky, C. 1984, *Phys. Rev. D*, **29,** 2677.

Johnson, D.G., and Wilkinson, D.T., 1986, *Ap J. Lett.* **313**, L1-L4.

Mandolesi, N., Calzolari, P., Cortiglioni, S., and Morigi, G. 1984, *Phys. Rev. D*, **29,** 2680.

Mandolesi, N., Calzolari, P., Cortiglioni, S., Morigi, G., Danese, L., and De Zotti, G. 1986, *Ap. J.* **310,** 561.

Mather, J.C. 1982, *Optical Eng.,* **21**, 769.

Meyer, D.M., and Jura, M. 1984, *Ap J. Lett.* **276**, L1–L3.

Meyer, D.M., and Jura, M. 1985, *Ap J.,* **297**, 119.

Partridge, R. B., *et al.* 1984, *Phys. Rev. D*, **29,** 2683.

Partridge, R. B., *et al.* 1985, *Societa Italiana di Fisica* Conference Proceedings, **1**, 7.

Peterson, J.B., Richards P.L., and Timusk T. 1985, *Phys Rev Lett.,* **55**, 332.

Silk, J. and Stebbins, A., 1983, *Ap J.,* **269**, 1.

Sironi, G., Inzani, P., and Ferrari, A. *Phys. Rev. D*, 1984, **29**, 2686.

Sironi, G., *et al.* 1987, *13th Texas Symposium on Relativistic Astrophysics*.

Smoot, G. *et al.* 1983, *Phys. Rev. Lett.*, **51**, 1099.

Smoot, G. *et al.* 1985*a*, *Ap. J. Lett.*, **291**, L23–L27.

Smoot, G. *et al.* 1985*b*, *Societa Italiana di Fisica* Conference Proceedings, **1**, 27.

Smoot, G. *et al.* 1986, *Highlights of Astronomy*, Reidel IAU Conference Proceedings, **7**, 297-305.

Weiss R., Annual Rev. Astron. Astropys. 18, 489 (1980)

Witebsky, C., Smoot, G. F., De Amici, G., and Friedman, S. D., 1986, *Ap J.*, **310**, 1.

Witebsky, C., Smoot, G. F., Levin, S., and Bensadoun, M., 1987, *submitted to IEEE Antennas and Propagation*.

# COSMIC BACKGROUND EXPLORER (COBE) SATELLITE
# ANISOTROPY EXPERIMENT DATA ANALYSIS TECHNIQUES

S. Torres[1], J. Aymon[2], C. Backus[3], C. Bennett[4], S. Gulkis[5],
R. Hollenhorst[1], D. Hon[1], Q. H. Huang[1], M. Janssen[5],
S. Keihm[5], L. Olson[1], G. Smoot[2], and  E. Wright[6]


1. STX, Lanham, MD., 2. LBL, Berkeley, CA., 3. SAR, Lanham, MD
4. NASA-GSFC, Greenbelt, MD., 5. JPL, Pasadena, CA
6. UCLA, Los Angeles, CA

ABSTRACT - The COBE Differential Microwave Radiometer (DMR) experiment will measure the anisotropy of the cosmic microwave background radiation (CMBR). The initial phase of data analysis uses a 'sparse matrix' algorithm to convert the differential temperature data into sky maps. The sky maps are then fitted to a 'fast' multipole expansion in spherical harmonics. Since the CMBR anisotropy is very weak, powerful techniques are used to extract the angular unevenness of the sky from the low signal-to-noise data. Instrument signature and other systematic errors are subtracted by fitting models of these effects. Test results are presented.

## THE EXPERIMENT

The 3 K cosmic background radiation [1], first detected by Penzias and Wilson [2] in 1965, is believed to originate early in the evolution of the universe, and the photons have traveled freely for the past $10^{10}$ years. Thus, the information in the CMBR will help to answer fundamental questions regarding the origin, scale, and evolution of the universe. The CMBR will be measured by instruments carried aboard the Cosmic Background Explorer (COBE) satellite [3], to be launched by a Delta 3920A rocket in mid-1989. One of the instruments, the Differential Microwave Radiometer (DMR), will map the CMBR brightness and will determine whether the CMBR is equally bright in all directions or will discover the pattern of any detected anisotropy. Causes of anisotropy are the motion of the observer, long-wavelength gravity waves, shear, vorticity, cosmic strings, and primordial anisotropies. An orbit life of more than a year of uninterrupted observation will enable the DMR to detect celestial objects that are many times fainter than any yet seen. The DMR will map the sky at wavelengths of 3.3, 5.7, and 9.6 millimeters. The frequency dependence of the galactic microwave background signal will be used to distinguish the emission of our galaxy from the true CMBR.

The DMR employs two separate microwave receiver channels for each of the three frequencies. The antenna system of each DMR microwave receiver (Fig. 1) consists of a pair of conical, 7° field of view horns looking at two parts of the sky at a separation angle of 60°. To compare the signals from the two horns a Dicke switch alternately connects each horn to the electronics. The output signal is proportional to the difference in power received by the two horns. Three separate receiver boxes, one for each wavelength, are mounted near the center of the spacecraft and are protected by a sunshade. Spacecraft orbit and attitude will be selected so that neither the Sun nor the Earth will shine directly on the DMR. The entire spacecraft will rotate at about 0.8 rpm, sweeping the horns rapidly across the sky. The long wavelength (9.6 mm) receiver will be maintained near room temperature, but the other two receivers will cool radiatively to about 140 K for improved sensitivity.

Each instrument will be calibrated frequently during flight. Noise diodes will be commanded from the ground to radiate a known amount of power into each horn. In addition, the Moon will be observed frequently and will serve to verify the electronic calibrators and to confirm pointing directions of the horns. An independent absolute calibration can be derived from the dipole signal induced by the motion of the Earth through the CMBR.

The result of the data analysis will be the best sky maps ever made at the three DMR frequencies, with an uncertainty of only $3 \times 10^{-4}$ K at 9.6 mm and $1.5 \times 10^{-4}$ K at 5.7 and 3.3 mm, for each 7° pixel after one year. For measurements of large angular scale features, such as the difference between one half of the sky and the other, the sensitivity will be $10^{-5}$ K. This is a sensitivity of just one part in $3 \times 10^5$ of the total brightness of the CMBR and just one part in 50 million of the total receiver noise.

The data generated by the instruments will be recorded continuously on COBE and transmitted once daily to a ground receiving station. The COBE Science Data Room (CSDR) will provide each of the experimenters with direct access to all of the Observatory data for the purpose of data reduction, analysis, experiment management, and mission planning.


THE PIPELINE


In a 400 day mission DMR will generate 4-5 Gbytes of data, to be reduced to a manageable level and presented to investigators in a sufficiently preprocessed form suitable for scientific research. Other goals of the DMR software are to remove systematic errors from the data and to calibrate the signal correctly. Systematic errors can be produced by instrumental effects, interference from the COBE or other satellites, and a failure to calibrate fully.

The software written for this purpose consists of ~20,000 lines of executable

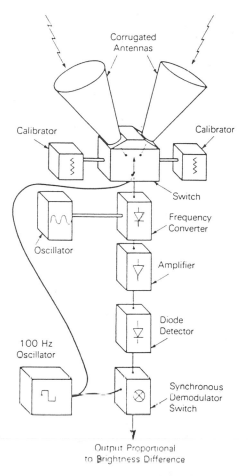

**Figure 1.** Schematic diagram of the Differential Microwave Radiometer (DMR).

FORTRAN code and relies heavily on the run-time library and advanced features of the VAX/VMS FORTRAN. The software process - called the DMR Pipeline - is partitioned into 20 programs that execute well-defined and independent functions. The interface between the programs as well as the final data bases are defined by means of the Record Definition Language, a software layer that accesses the VAX Common Data Dictionary. The programs, run in serial mode as illustrated in Fig. 2, produce intermediate results monitoring the performance of the programs and the quality of the data. The source code is managed by the DEC Code Management System (CMS) and built by the DEC Module Management System (MMS).

Pipeline input consists of the raw data, attitude information, and ground test data such as calibration information. The output comprises galactic emission maps, sky temperature maps, harmonic coefficients from the sky map fits, anisotropy measurements and power spectra, and parameters resulting from fits to cosmological models. COBETRIEVE, the data management system, maintains a catalog of all available data, permits loading and retrieval of archived data, and provides a history and pro-

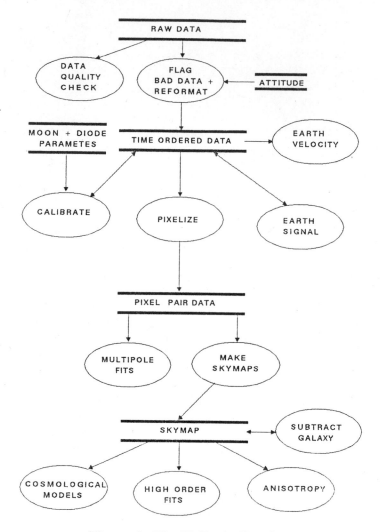

**Figure 2.** The DMR pipeline chart

tection for all transactions.

An interactive phase of the pipeline allows the user to inspect the data upon arrival and to flag bad segments of the time series. The rest of the pipeline runs in batch mode. The data reduction part runs daily with a required processing time of less than four hours. Data analysis is done weekly, monthly, or when sufficient data have been accumulated for a given type of analysis. A certification and history procedure is provided to release officially the product data bases to the National Space Science Data Center (NSSDC).

As seen in Fig. 2, the raw data (science and ancillary) are reformatted into a time

322

ordered data (TOD) file that contains additional information regarding calibration parameters, data quality flags, Moon proximity, and spacecraft attitude information. Subsequent pipeline programs fill in other fields of the data record. Before the data are passed to the TOD file, a fast search for spikes and abrupt baseline changes is performed. If such events are found, they are flagged, but left unchanged. The next step in the pipeline is to calibrate the data using information from an internal reference noise source. Once calibration and baseline parameters are found, attitude information is used to associate each measurement with one of the 6144 equal-area, quad-cube pixels into which the complete sky is divided (see Appendix). This pixelized information is written to a separate intermediate file, the Pixel Pair File, for use by the analysis programs. The instrument is designed so that two pixels are associated with each measurement, one pixel corresponding to each horn of the horn pair. The first analysis problem is to obtain a sky map from this differential information. Once a sky map solution is found, the data are fitted to a high order spherical harmonic expansion, and autocorrelation analysis is done. Since the DMR has two channels for each frequency, a criterion for the selection of the 'best' sky map or the 'best' average sky map is established by means of a set of figures of merit: the integral of the cross correlation surface, the peak and the width of the cross correlation, and the normalized sum of pixel ratios and pixel differences squared. A DMR simulator program relates noise levels and signal quality to these statistics and allows the determination of confidence limits for the rejection of a particular sky map.

## THE SIMULATOR PROGRAM

In order to test the efficiency and robustness of the algorithms, a simulator program generates data in the exact format of DMR experimental raw data but characterized by known and controlled parameters.

Simulating COBE orbit, attitude, and rate of data acquisition, the simulator scans the input sky map accepting sky temperatures for each antenna of the desired channel. The simulator then computes a temperature value for each horn by adding to the sky signal the secondary signals of Moon, Earth, Earth magnetic field, and instrument effects. Next, the simulator computes the difference between the signals of the two horns to obtain a differential signal. The outputs of the simulator are a raw data file, an attitude file and optionally a TOD file. Figure 3 shows 900 seconds of simulated output signal.

The output of the simulator, representing the signal that the DMR is expected to produce, is then processed through the pipeline. The user specifies a sky map and set of spherical harmonic coefficients. The pipeline data analysis produces an output sky map plus a set of spherical harmonic coefficients that can be compared with their corresponding inputs. This simulation technique verifies the efficiency of the algorithms in subtracting the systematic errors from the data and in deriving a calibration constant. In this manner one can learn which effects are significant and how these show up in the final sky map if they are not subtracted. Thus, processing simulated data with the pipeline evaluates the accuracy of pipeline programs.

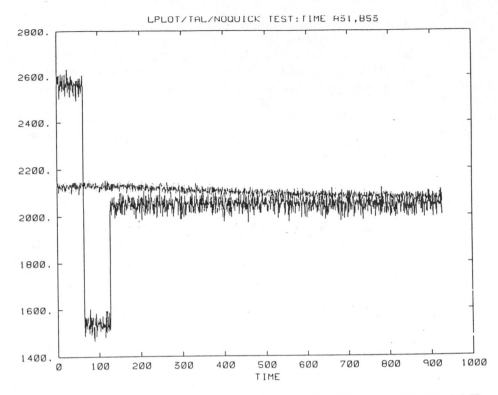

**Figure 3.** 900 seconds of simulated radiometer signal from one 31 GHz DMR channel with calibration pulses, 50 mK RMS noise, and one 53 GHz DMR channel with 20 mK RMS noise and baseline drift.

## SKY MAP SPARSE MATRIX ALGORITHM

Divide the Celestial Sphere into N cells (pixels) of equal area (Appendix), and denote the temperature of the $i$th pixel by $T_i$. A single measurement of the DMR represents a value proportional to the temperature difference of two pixels. An average of all the measurements associated with the same pair of pixels is represented as:

$$S_{ij} \sim < T_i - T_j > \sim < T_i > - < T_j >,$$

where $i$ and $j$ are determined by the orientation of the satellite and the opening angle between the horns. The object of the following algorithm is to deduce $T_i$ by solving a set of linear equations.

At a time $t$, the signal output from a DMR horn pair can be expressed as:

$$s(t) = g(t)[T_i - T_j](t) + z(t) + a_1 p_1(t) + a_2 p_2(t) + \ldots + a_q p_q(t) + \text{noise}$$

where $g$ is the gain, $z$ represents the radiometer offset plus any electrical bias, and the $p_i$ are possible systematic errors having coefficients $a_i$. Assume $z$ is slowly varying, and any systematic errors that are not modulated by the spacecraft rotation can also be represented by $z$. This term can be eliminated by applying a high-pass filter to the signal. The expression for the high passed signal can be written as:

$$\text{noise} = h(t) - V(t) \cdot A$$

where $h(t)$ is the high-passed signal, and the two vectors $V$ and $A$ are defined as:

$$V(t) = (0, 0, \ldots, 0, 1, 0, \ldots, 0, -1, 0, \ldots, 0, p_1(t), p_2(t), \ldots, p_q(t))$$

with the 1 and $-1$ occurring at positions (pixels) $i$ and $j$ respectively as a function of time, and:

$$A = (T_1, T_2, \ldots, T_i, \ldots, T_j, \ldots, T_N, a_1, a_2, \ldots, a_q).$$

To solve for $A$, minimize the sum over time of the square of the noise. The minimization yields the matrix equation:

$$(V \otimes V^T) \cdot A = h \cdot V.$$

A division of the sky into $N = 6144$ pixels yields 37,748,736 elements in the matrix. The fixed angular separation of the horns, however, reduces the number of non-zero entries substantially; the matrix is sparse. A noteworthy feature of the matrix is the character of the diagonal elements. These elements are equal to the sum over all the non-diagonal elements in their row. This singular matrix is solved numerically by the addition of a small arbitrary constant to the diagonal, and application of a Gauss-Seidel iteration. The result reflects ambiguity in the isotropic level (monopole) of the sky temperature which is not directly measured by the DMR.

Care has been taken to represent the sparse matrix by data structures that store only non-zero elements and allow for simple logical expressions to access and manipulate them. Each row of the matrix equation can have a small but somewhat arbitrary number of non-zero elements that are stored as a singly linked list and grow dynamically with the supply of measurements. The matrix is represented as an ordered array of linked-list pointers, one pointer for each row, each pointer containing the address of the first non-zero element (node) of its row. The contents of a node include the column number of the non-zero element, its value, and a pointer to the next node; the last node of a linked-list has a null-valued pointer.

The modified Gauss-Seidel algorithm applied to the linked-list representation executes efficiently on the VAX when provided with a sufficiently large amount of virtual memory in conjunction with a large 'working-set.' Most of the time the process is 'paging' through the linked lists. Optimization is merely a matter of providing a sufficient amount of memory to avoid 'paging' (~4 Mb) and letting the operating system (VMS) deal with memory management and I/O. The technique works well for very sparse matrices but becomes cumbersome for denser matrices. Other techniques, which are based on ordered contiguous storage, generally require more CPU effort to manage and access the storage area.

## MULTIPOLE FITS TO DIFFERENTIAL TEMPERATURE DATA

Anisotropy information can be extracted directly from a representation of the sky temperature distribution as a harmonic expansion on the surface of the unit sphere:

$$
S_{ij} = \sum_{l=0}^{N} \sum_{m=0}^{l} \{ a_{lm}^c [P_l^m(\theta_i) \cos m\phi_i - P_l^m(\theta_j) \cos m\phi_j]
$$

$$
+ a_{lm}^s [P_l^m(\theta_i) \sin m\phi_i - P_l^m(\theta_j) \sin m\phi_j] \} + \sum_{q=1}^{Q} a_q [(p_q(\theta_i, \phi_i) - p_q(\theta_j, \phi_j)]
$$

where $N (\leq 6)$ is the order of the expansion, the $p_q$ are $Q$ systematic error functions, the Legendre functions $P_l^m$ are evaluated at the directions of pixels i and j, and the coefficients $a$ are found by a least squares fit to the 172,800 temperature difference data points measured daily by the DMR. To optimize the fitting procedure, data points with the same pixel pair numbers are averaged. This reduces the sample to approximately 32,000 averaged points daily. Another technique to make the algorithm more efficient is to take advantage of the large virtual memory available in the VAX/VMS system so that the associated Legendre polynomials are computed only once and stored at the beginning of the program for each one of the 6144 directions in which the sky is pixelized. Standard recursive techniques are used to compute the polynomials. The most important systematic effects that may be reflected in the data and must be subtracted are: the coupling of the radiometer to the magnetic field of the Earth, the temperature-induced baseline drifts, the Moon microwave emission, and the residual Earth radiation that is diffracted over the ground shield.

The magnetic and temperature biases are subtracted by including the effect both in the sky map matrix algorithm and in the multipole fits described here. That is, the appropriate empirical and/or theoretical models are treated as basis functions in the least squares problem and coupling coefficients are then found. The disadvantage of this approach is that the models must be linear functions of the coefficients to be found by the fit. For that reason, the more complex effects such as the diffracted Earth and Moon emissions are treated separately with nonlinear fitting techniques.

To solve and invert the covariance matrix of the least squares problem, a method based on Cholesky factorization and iterative refinement is used [5]. The last section of this paper shows some results of the accuracy of the fits based on simulated data. A VAX-780 required approximately 12 minutes of CPU time to fit $\sim$ 30,000 pixel pairs to a spherical harmonic expansion of order 2 using the technique just described.

326

# HIGH ORDER MULTIPOLE FITS TO SKY MAPS

A least squares algorithm is also used to fit pixelized and time-integrated data in a sky map format to a two-dimensional spherical harmonic expansion:

$$T_i = \sum_{l=0}^{N} \sum_{m=0}^{l} P_l^m(\theta_i)[a_{lm}^c \cos m\phi_i + a_{lm}^s \sin m\phi_i].$$

Given the angular resolution of the DMR radiometers (7° FWHM), a spherical harmonic expansion of comparable resolution can include terms up to $N = 26$th order. That would require 729 basis functions for each pixel in the sky, a number that can not be efficiently accommodated in the available virtual memory. For this reason a technique to reduce the computations of the associated Legendre polynomials must be applied. Here, instead of using less accurate but faster techniques [6] to compute the coefficients of the harmonic expansion, the symmetry of the sky pixelization is exploited to evaluate the Legendre functions in the first octant only . The values of the functions in the other octants are the odd or even reflections of the functions in the first octant. The number of operations is reduced roughly by a factor of eight, and the full accuracy of the Legendre polynomial evaluation is retained. The timing of this algorithm is approximately $0.45 \times N^{3.43}$ CPU seconds in a VAX-780 for $7 < N < 26$ (where $N$ is the order of the fit). More results of simulated runs are presented in the test section at the end.


## ANISOTROPY STUDIES

Fast anisotropy tests can be applied to the DMR measurements by making use of the quad-cube organization of the sky map (see Appendix). Standard edge-detection techniques as well as two-dimensional Fourier analysis can be applied to each $32 \times 32$ element face. An appealing test is the two point (auto) correlation evaluated as a function of pixel angular separation from 0 to 180 degrees, or 0 to 64 pixels. Evaluating the angular separation of all pixel pairs is time-consuming but is performed only once. Calculation of the angular separation can be made faster by using precalculated values of colatitude and azimuth associated with the center of each pixel. The results can be stored in a table of dimension $= N \times (N + 1)/2$. For $N = 6144$, and using one byte storage, this table requires over 18 Mbytes of storage. The quad-cube formulation, however, provides an attractive compromise: it is possible to access the data at a lower resolution by slicing the quad tree at one step higher in the quad-tree structure. Each group of four adjacent DMR sky measurements can be accessed and averaged into a lower resolution 'super-pixel,' yielding a 1536-element sky map. Once the required table and the lower resolution map have been initialized , pixel-pair calculations can be performed over a $1536 \times 1537/2$ loop. The two-point correlation simply requires products of pixel temperatures. One can also evaluate difference statistics and produce histograms.

# LUNAR CALIBRATION, POINTING CORRECTION, AND BEAM RECOVERY

The COBE spacecraft orbits in the plane of the planetary terminator with a 60° horn separation angle. In this orientation the Moon comes appreciably into view about 12 days per month during two six-day intervals at the second and fourth quarters of the Moon. At the DMR wavelengths surface features of the Moon do not play a significant role. With the 7° beam widths it is possible to regard the Moon as a point source, the location of which is accurately known, and the brightness temperatures of which are usable as a secondary source of instrument calibration.

A computational model of the lunar microwave emission as function of wavelength, polarization angle, and lunar phase has been developed. The model also produces brightness centroid coordinates to adjust the location as a point source. It is believed to be accurate to within about 5% in terms of absolute brightness temperature. The relative accuracy, as a function of phase, is believed to be accurate to much better than 5%. In flight it will be possible to make fine adjustments to the model using the noise source calibration of the the instruments.

Calibration and pointing correction are accomplished by solving a nonlinear regression problem in terms of instrument offset, gain, and radial and tangential pointing components. The pointing components do not depend upon the intensity values of the model, but they do depend upon the brightness centroid coordinates.

A third software module bins the acquired lunar data and recalculates beam patterns. This enables antenna gain (beam solid angle) to be re-estimated for use in subsequent iterations of the regression algorithm just described.

## GALACTIC EMISSION MODELING AND FITTING

At any one of the DMR frequencies the product of the sparse matrix algorithm discussed above is a map of the sky temperature. These maps include the emission from our galaxy, a component that must be subtracted from the total signal before one can fit cosmological models to the cosmic background radiation maps.

The most important contributions to the galactic emission at the DMR frequencies are synchrotron radiation, emission by ionized hydrogen in interstellar space (HII emission), and emission from galactic dust. Our current synchrotron radiation model is derived from a full-sky survey at 408 MHz by Haslam and colleagues [7], divided into $181 \times 91$ 2° bins. The model for the thermal emission from HII sources is taken from a catalog of 909 sources. The emission from galactic dust is either approximated by a theoretical model or is adapted from actual IRAS data [8]. The theoretical dust model approximates the galaxy as flattened ellipsoid with uniform density of emitting dust. The contribution of each component is compared with the model in a least squares sense in order to find the best coefficients for each one of the model components. Following this, a non-linear fit to a frequency-dependent model of the galactic microwave emission is done, and finally, the galaxy components are subtracted from the input sky map.

## EARTH-SIDELOBE CONTRIBUTIONS

It is also necessary to determine the effect on the signal of the Earth emission diffracted from the edge of the spacecraft shield into the antenna sidelobes. The data are first cleansed of known systematic contributions, such as the galactic background, then accumulated over time into a pixelized grid of weighted average intensities for each DMR horn. The grid's pixel size is consistent with the granularity derived from the sky pixelization. The coordinates of each measurement are defined in terms of the azimuth and elevation of the point on the Earth's limb closest to the DMR horn's line of sight.

The 2-D grid may then be fitted to an empirical or theoretical diffraction model using the measured RMS variations as weights. The fitted grid with error bars represents the Earth systematic contribution to each DMR horn. The estimated signal is 0.1 mK. This magnitude results in a signal to-noise-ratio of approximately 1.0 after one year of data collection. Up to 2 mK or more may contaminate horns at very high Earth-limb angles. Since the spacecraft attitude is correlated with the orbital position, the Earth signal is also significantly correlated with position on the sky. Some large uncertainty in the correction may occur. If the contamination is too large or cannot be effectively subtracted, data may be discarded when the Earth diffraction component exceeds a specified absolute and/or relative magnitude.

## COSMOLOGICAL MODELS

There are a number of instances where, viewing a map, analyzing spherical harmonic coefficients or autocorrelation and two-point correlation functions, is not nearly as efficient or sensitive as fitting to a special cosmological model. It is a general property in signal analysis that correlating to a function that describes a signal is the most sensitive way to find that signal. It is more complicated if the signal has several free parameters that are nonlinear. There are several approaches in this situation – doing the non-linear fit or trying to transform into a coordinate system that results in a linear fit.

For illustration we can consider the case of an anisotropically expanding universe. If the universe is geometrically flat, then anisotropic expansion shows up directly in the spherical harmonic coefficients. In the simplest case of anisotropic expansion – expansion along one axis differing from expansion along the other two axes – all the anisotropy shows up in the coefficient of one quadrupole ($l = 2$) component with the principal axis along the special direction. On the other hand if the universe is open, then the geometry of the universe is not flat, so photons traversing cosmological distances do not travel in straight lines but follow the geodesic of space-time. Novikov [9] pointed out that in such a case of anisotropic expansion there would be an apparent global distortion of the anisotropy from its local character. In this case a simple quadrupole anisotropy would be squashed into a restricted region, the angular extent of which would be characterized by the ratio of the actual density of the universe to the critical density needed to just close (to make flat Euclidean space). If such an anisotropy were to exist it would be most interesting to fit to it and learn the geometry of the universe (and thus the ratio of the actual density of the universe to the critical density) and the parameters of the anisotropy of the universe. One would also like to know how accurately these parameters have been measured and how well the model fits. If the signal-to-noise ratio were good, and there were no systematic errors, then getting this information would be straightforward. An approach to solving this problem is presented by Lukash [10]. He advocates the use of parabolic waves forming the full set of eigenfunctions in the Fourier integral representation of the Lobachevski space – the natural geometry for one type of open universe. The difficulty in this approach is that it has a principal axis built in. For standard spherical harmonics it is not necessary to choose a coordinate system aligned with the anisotropy – one can fit to the data and then solve for the principal axes using moment-of-inertia techniques that are well developed and straightforward to implement. It seems more reasonable to convert to the coordinate system and fit to higher-order moments once one knows the preferred coordinate system.

Bajtlik, Juszkiewicz, Proszynski, and Amersterdamski [11] use the techniques of numerical integration and linearization for small anisotropy. They find a functional form by means of which one can fit to the preferred direction set by the geometry of the universe and the anisotropic expansion, the density ratio, and anisotropy amplitude. This is clearly a nonlinear fit but it provides the preferred direction needed for a linearized expansion or a parabolic wave expansion.

How well can the data be fitted and how well can the model and its parameters be judged correct? This is very difficult to estimate due to the possibility of systematic errors. Suppose we fit our simplest model (4 free parameters plus 3 coefficients for the known dipole anisotropy) to the 6144 pixels making up our standard map. Suppose we find significant values for the parameters - anisotropy amplitude, actual to critical density ratio, and directions and a $\chi^2$ per degree of freedom equal to 1.05 (for the 6143 - 7 = 6136 d.o.f.). Does this mean that: (1) the model is not a good fit? (2) there is a poor fit because the galactic emission is not accounted for correctly, but the model parameters are essentially correct? (3) everything is correct, but we have underestimated the error on each pixel by about 2.5%? How good an upper limit can we put on the values of any parameters or restricted range of values? The exact approach varies depending upon the model we actually use. In this particular case we have to investigate carefully the fit to the portion of the map that contains the main feature of the anisotropy. This can be done in the case where the density ratio is about 0.1 since the primary cosmological effect is concentrated in a region roughly characterized by 0.1 radians (6 degrees) and far from this region we only ask that the effect be negligible, that is, unchanging. Otherwise we have to resort to techniques such as grouping pixels to reduce individual point errors and thus reducing the importance of a small percentage error in estimating the pixel error. We can also do what we call a sign run test, looking to see if the regions of pixels deviating from the model have the expected random distribution of sign of deviation from the model. If the model is not a good representation of the data, the model may well have large regions where it is too low or too high when the $\chi^2$ is minimized.

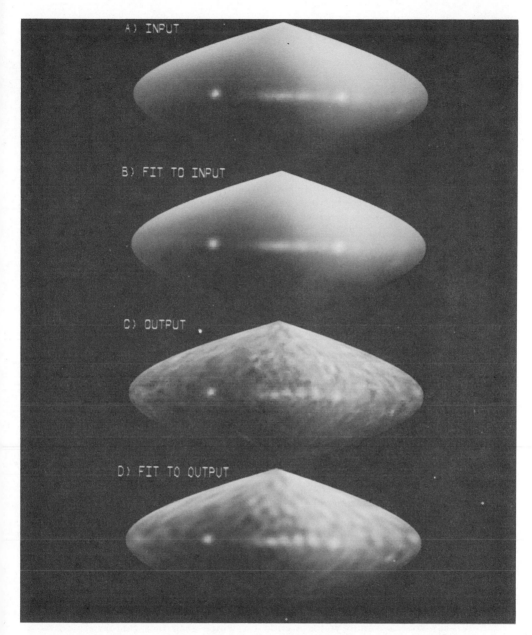

**Figure 4.** a) The input sky map to the DMR pipeline includes: synchrotron, HII, and dust emissions from our galaxy at 31 GHz, and the CMBR (monopole, dipole, and quadrupole). c) Output sky map from the DMR pipeline. b) and d) are sky maps generated from coefficients of fits to maps a) and c) respectively to a spherical harmonic expansion of order 26.

## TESTS AND RESULTS

Using the simulator program and the pipeline, an 'end-to-end' test can be done, and evaluation of the accuracy, efficiency, and robustness of the algorithms is possible. For this purpose we have generated an input sky map (Fig. 4) that includes the galactic dust, synchrotron radiation, HII emission, and the cosmic background monopole, dipole, and a small quadrupole component. The simulator program has been run with a Gaussian noise of 5 mK RMS, while advancing the orbit plane by 60° each time 5 orbits were completed to achieve near full sky observation. The generated data represents 26 hours and 40 minutes of instrument operation. Figure 4 shows the sky map produced by the pipeline. Direct comparison of the input and output dipole coefficients gives an actual error in the value of the dipole anisotropy of 0.8% in the case of fits to the differential temperature, and 0.2% for sky map fits. A second example with no noise adds the effect of the Earth's magnetic field as a vector coupling (three coefficients). The actual error with which the coupling coefficients are recovered is less than 0.7%. The programs performing the data reduction and analysis required 1 hour, 15 minutes of CPU time in a VAX-780.

## ACKNOWLEDGEMENTS

Some of the DMR pipeline software originated from programs written by the authors and their colleagures at LBL, JPL, and UCLA. The final product is being developed and will be produced by STX within the contract NAS5-28561. We would like to thank Dr. Donald A. Krueger, GSFC Chief of Applied Engineering Division, for making possible our participation in this workshop.

## APPENDIX: Quadrilateralized Spherical Cube Data Base and COBE Sky maps

The Spherical Cube is obtained by radially projecting the edges of a cube inscribed within a sphere onto the surface of the sphere. For COBE, the cube is inscribed by the Celestial Sphere.

Each face of the cube is divided into a grid of square elements reflecting the desired level of resolution. Equal area elements on a cube face do not project as equal area elements on the spherical cube surface. Elements near the center of the face have larger projections than those near the edges. To achieve an equal area projection, the rectangular grid must be distorted into a curvilinear one. The derivation of this transformation and its inverse can be found in the references.

A full-sky coverage map yields a cube for which each face can be stored as a completely filled array. A face is divided recursively by a two-dimensional binary grid. At each level of division, the areas are divided into quadrants labeled by a 2-bit binary number. Each level of division is indicated by the addition of two binary bits to the least significant end of a binary number. This binary index is the serial location of a point in the linear data array. The sky map data base, stored and indexed in this fashion, reflects a space-filling binary-tree structure in two dimensions, *i.e.*, a quad-tree.

The (x,y) coordinates of a cell (pixel) on the face grid can be combined into a one-dimensional index to access the array. This is accomplished by allowing the even bits of the binary index to represent one coordinate, while the odd bits represent the other. For example, the index of pixel (1,1) has the binary representation: $01 + 10 = 11 =$ pixel number 3; and pixel $(3,5) = 101 + 100010 = 100111 =$ pixel number 39. The (x,y) coordinates may be recovered from the pixel number by selecting the (odd,even) bits of the binary string and forming two binary numbers.

The DMR resolution of $32 \times 32$ pixels per face can be accommodated by 10 bit indices, ranging over the coordinates (0,0) through (31,31), $i.e.$, pixel numbers 0 through 1023. The faces of the cube are numbered 0 through 5. The pixel numbering convention provides for utilization of pixel numbers 0 through 6143 (stored in two-byte words) by relying on the initial step of subtracting an offset for each face defined by the product: face number $\times$ 1024. All that remains is to adopt a convention for the coordinate system of the cube and the orientation of each face ($i.e.$, relative position of the lower left corner (0,0) pixel) with the others. This pixel numbering scheme provides quick access of a sky measurement for any desired direction within a level of accuracy defined by the equal area partitioning.

REFERENCES

F. K. Chan, E. M. O'Neill, EPRF Technical Report 2-75 (CSC).

E. M. O'Neill, R. E. Laubscher, NEPRF Technical Report 3-76 (CSC).

REFERENCES

1. Smoot, G. F., in "Data Analysis in Astronomy", (1988, this proceedings).

2. Penzias, A. A. and Wilson, R. W., Ap. J., **142**, 419 (1965).

3. Mather, J. C., Opt. Eng. **21** (1982).

4. Brice, C., Luther, H. A., and Wilkes, J.O., "Applied Numerical Methods," Wiley, New York (1969).

5. Golub, G. H., and Van Loan, C. F.,"Matrix Computations," John Hopkins University Press, Baltimore (1983).

6. L. J. Ricardi, IEEE Transactions on Computers, June 1972, p.583.

7. Haslam, C. G. T., $et.$ $al.$ As. Ap. Suppl. Ser. **47**, 1 (1982).

8. Neugebauer, G., $et.$ $al.$, Ap. J. Let., **278**, L1 (1984).

9. Novikov, I. D., Soviet Astronomy **427**, 12 (1968).

10. Lukash, *et. al.*, "Early Evolution of the Universe and its Present Structure," in IAU symposium 104, ed. G. Abell and G. Chincarini, Dordrect, Reidel (1984).

11. Bajtlik, S., *et. al.*, As. J. **300**, 463 (1986).

# SOME STATISTICAL PROBLEMS ENCOUNTERED IN THE ANALYSIS OF ULTRA HIGH ENERGY COSMIC RAY AND ULTRA HIGH ENERGY GAMMA RAY DATA

A A Watson

Department of Physics
University of Leeds
Leeds    LS2 9JT    England

## INTRODUCTION

The subjects of ultra high energy cosmic rays and ultra high energy gamma rays are scientifically connected because, it is widely believed, ultra high energy gamma rays are produced through strong interactions of cosmic ray nuclei. In the context of this workshop the two topics are also connected: because of the paucity of events so far detected at these extreme energies (100 – 10000 TeV for gamma rays and $10^{18}$ – $10^{20}$ eV for cosmic rays) the signals are weak and sometimes embedded in a complex background. In this paper I will outline briefly the detection methods and problems associated with the two types of astronomy and draw attention to recent work which has been done to address the statistical questions which arise. I will try to emphasise some of the problems which are peculiar to this field of study. Many will be well known but it is hoped that what follows will be a useful guide, particularly for newcomers.

## ULTRA HIGH ENERGY COSMIC RAYS

In the context of this article I mean by the phrase 'ultra high (UHE) energy cosmic ray' those cosmic rays with energies above $10^{18}$ eV. The choice of this energy is somewhat arbitrary but above it the transition from cosmic rays coming predominantly from Galactic sources to those which are predominantly from extra-Galactic sources probably takes place. Additionally there has been interest and intense debate for more than 20 years as to whether or not the cosmic ray spectrum at the earth extends beyond about $5 \times 10^{19}$ eV. Above this energy it had been

335

predicted (Greisen 1966: Zatsepin and Kuzmin 1966) that the spectrum would steepen dramatically if the cosmic rays were more than $10^8$ years old.

## Uhe Cosmic Ray Detection

Above about 1 TeV it becomes very difficult to detect individual cosmic rays directly because of their very low flux. Above 1000 TeV ($\equiv$ 1PeV) for example the integral intensity is about $3 \times 10^{-6}$ m$^{-2}$ s$^{-1}$ sr$^{-1}$ or about 90 m$^{-2}$ sr$^{-1}$ per year. At higher energy the primary cosmic ray spectrum steepens and above $10^{18}$ eV the intensity falls to about 90 km$^{-2}$ sr$^{-1}$ per year. Near $10^{20}$ eV, where less than 20 events have been detected, the rate is about 1 km$^{-2}$ sr$^{-1}$ century$^{-1}$.

Measurement of such very low event rates is possible because interaction of the primary cosmic ray in the air generates a cascade shower which, at ground level, is spread out over many hundreds of square metres. At $10^{20}$ eV, for example, particle densities range from $10^5$ m$^{-2}$ close to the axis of the cascade to $\sim 1$ m$^{-2}$ 1.5 km away. Thus a relatively small number of detectors spread over large areas can be used to detect the cascade showers produced by the primary cosmic rays. Arrays with areas greater than 10 km$^2$ have been operated in Australia, the UK, the USA and the USSR for many years: an array of 1000 km$^2$ has been funded in the USSR and may become operational in the mid-1990's. With these giant cosmic ray telescopes the direction of the incoming primary is measured to within 2 or 3 degrees. The size of the shower, from which the primary energy is inferred, is rarely known to be better than 10% and is often much less accurately determined. However, because of the requirement for multiple coincidences between widely spaced detectors, there is virtually no background from which to extract the signal: our knowledge of the cosmic ray spectrum above $10^{19}$ eV is limited by small event numbers and systematic errors.

In an attempt to overcome the problem of low statistics a new technique, first proposed independently by Greisen, Suga and Chudakov, has been developed at the University of Utah. As the air shower travels through the atmosphere the electrons in the cascade excite the nitrogen in the air. The resulting de-excitations are detectable using an impressive system of mirrors and photomultipliers. This device, known as the Fly's Eye, has been operational for some years and is producing data of similar statistical quality to the more conventional particle arrays in the important region above $10^{19}$ eV. The Fly's Eye can monitor a very much larger area than the particle detectors, but operation is restricted to dark, clear nights so that so far the exposures are very comparable. It is, however, particularly useful to have an alternative detection technique for which there will be different types of systematic errors.

## Statistics of Small Numbers

When describing the results of the uhe cosmic ray experiments it is often necessary to draw tentative conclusions from a very small number (or zero!) events. An early discussion of this problem was given by Regener (1951); more recently the question has been considered by Gehrels (1986) who has also discussed related binominal problems. It is clearly important - not least for graphical presentation - to be able to assign upper limits and confidence ranges in a consistent manner. If zero events are observed in a given exposive then the commonly-used 95% confidence level upper limit is 3.00 (Regener 1951, Gehrels 1986); for the 99% level the corresponding figure is 4.61. These assignments of confidence levels are appropriate in the case where the background contamination of the signal is negligible, a condition that is well-satisfied in the uhe cosmic-ray case.

If, in an experiment, the number of events observed is N then it is often desirable and useful to define upper and lower limits such that the probability that the range defined by these limits includes the true expectation value is x%, where x commonly is 84, 95 or 99%. Upper and lower limits, for x = 84% and specimen values of N, are shown in table 1 together with the values which would be assigned if a gaussian distribution, $\pm\sqrt{N}$, is assumed in setting the upper and lower limits. It is important to notice that for N>3 the difference between Gaussian and Poissonian lower limits is unimportantly small. However, for the upper limits, the difference is still 8% at N=10.

Table 1.  Poissonian Upper and Lower Limits for x = 84%
(after Regener 1951)

| N | Lower Limit | | Upper Limit | |
|---|---|---|---|---|
| | Gaussian | Poissonian | Poissonian | Gaussian |
| 1 | 0 | 0.17 | 3.30 | 2.00 |
| 2 | 0.59 | 0.71 | 4.64 | 3.41 |
| 3 | 1.27 | 1.37 | 5.92 | 4.73 |
| 4 | 2.00 | 2.09 | 7.16 | 6.00 |
| 6 | 3.55 | 3.62 | 9.58 | 8.45 |
| 10 | 6.84 | 6.89 | 14.26 | 13.16 |

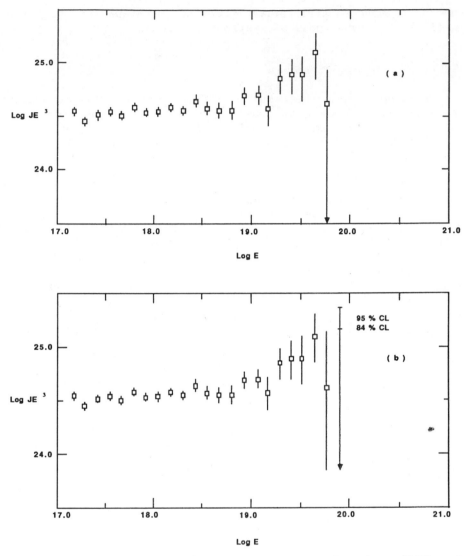

Fig.1a,b.Comparison of Fly's Eye Spectrum using two different
methods of assigning confidence limits to the data at
the highest energies

Particularly at small N, the Poissonian/Gaussian difference
can be important in visual presentation of data. As an example
recent data from the Fly's Eye group (Baltrusaitas et al 1985)
are shown in figure 1 plotted in two different ways. In figure
1a the data points are plotted using errors calculated as $\sqrt{N}$; in

338

figure 1b the same points have been replotted using limits derived according to Regener's method (Watson 1985). With regard to the question of a cut-off above 4 x $10^{19}$ eV the visual impression given by the two presentations is somewhat different and that of figure 1b does not support the contention (Baltrusaitas et al 1985) that there is evidence 'for the development of a spectral "bump" followed by a cut-off at 7 x $10^{19}$ eV'. Of course visual impressions should not be used to settle scientific arguments and methods of evaluating spectral slopes – another area with statistical pitfalls – will now be considered.

## Evaluation of Spectral Slopes

For completeness I will draw attention to what is no longer an error commonly found in cosmic ray work. It is quite wrong to calculate the integral slope of a spectrum using integral data points: the uncertainties on each point are not independent and the error in the slope is grossly underestimated. Evaluation of the differential slope may be carried through using standard linear regression techniques but it should be clear from the discussion of the previous section that this will only be satisfactory procedure if N⩾10 for all intensities. These techniques are unable to accommodate small values of N or upper limits (zero events in a spectral inverval). The recommended approach is to use a method based on the maximum likelihood technique.

An early use of the maximum likelihood (ML) technique was discussed by Jauncey (1967) in the context of the interpretation of the log N – log S plots of radio-astronomy. An adaptation of Jauncey's method has long been used for analysis of Haverah Park uhe cosmic-ray data e.g.Tennant and Watson 1971, Edge et al 1973. The possibilities of the method can be illustrated using the data of figure 1.

In the original paper from the Fly's Eye group, standard linear regression (LR) techniques (assuming a normal error distribution for each point) where used to calculate the spectral slope for all the data and for a restricted energy range from $10^{19}$ to 5 x $10^{19}$ eV. For the whole energy range, slopes found by the LR and ML techniques are negligibly different ($-2.94 \pm 0.02$ (LR) and $-2.95 \pm 0.02$ (ML)) as expected since the high statistic points at low energies dominate in the fit. For the restricted energy range the LR method gives $-2.42 \pm 0.27$, to be compared with $-2.87 \pm 0.36$ for the ML technique. Clearly the uncertainties in slope are too large to draw any useful conclusions about the existence of 'bumps' or spectral cut-offs, as was attempted by Baltrusaitas et al. Moreover, if above $10^{19}$

339

the data, and an upper limit in a bin in which no event has been recorded, are used, the slope found is $-2.56 \pm 0.26$.

A further extension of the maximum likelihood technique in which the range of slopes permitted is obtained on an event-by-event basis is discussed by Linsley at this meeting.

## The Effect of Energy Resolution on Spectral Slope

It is well known that fluctuation in the measured size parameter of a shower, whether these arise from actual fluctuations of shower development or from uncertainties coming in during the process of shower sampling and the ensuing derivation of density values, can give rise to distortion of the spectrum of the measured parameter. The degree and manner of distortion are determined by the functional dependence of the fluctuations and uncertainties on shower size. (A comparable problem was examined by Murzin (1965) in connection with the determination of nuclear interaction parameters in emulsion.) Fluctuations and measurement uncertainties can be formally treated in the same way, because, for a fixed 'input' energy or size, the 'output' is approximately log-normally distributed, the width of this distribution being characterized by $\sigma$, the log-normal standard deviation. If $\sigma$ is independent of primary energy of shower size, then a spectrum which was initially described by a single constant index will be unchanged in slope except close to the threshold of detection, while the intensity will be overestimated at each shower size. The overestimation of intensity is given by $F = \exp (\frac{1}{2} \sigma^2 (\gamma - 1)^2)$, where $\gamma$ is the differential slope index, so that for measurement uncertainties of 20% ($\sigma = 0.08$) the overestimation of intensity is about 2%.

In real experiments on large showers it is important to know how the energy resolution changes with energy. If the energy resolution degrades as the energy increases then false inferences about the shape of the spectrum and the presence or absence of a cut-off may be made at the highest energies. To illustrate this danger, a Monte Carlo calculation has been made (Edge et al 1973) assuming a spectral slope of $-3$ (differential) and an uncertainty described by $\sigma = 0.05 + 0.2 \log (E/E_0)$, ie $\sigma = 13\%$ at $E_0$ and 200% at 100 $E_0$.

Figures 2a,b show the data which would be observed if the true spectrum extended beyond 100 $E_0$ (fig. 2a) and if it cut-off abruptly at 16 $E_0$ (fig. 2b). It is clearly necessary to have accurate knowledge of the errors of energy measurement when interpreting features seen in measured energy spectra.

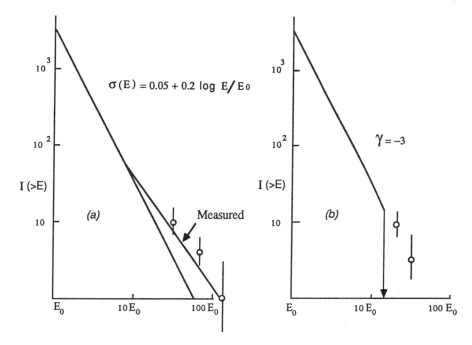

Fig.2a,b.Monte Carlo simulation of measured energy spectra under different assumptions. See text for details.

## Effect of Energy Resolution on Interpretation of Anisotropy Data

An important aim of research on ultra high-energy cosmic ray work is to determine the anisotropy of events as a function of energy. What is commonly measured is the amplitude and phase of the first and second harmonic of the anisotropy in right ascension. A feature of comparisons of results from different experiments has been the poor agreement. A contributory factor to this problem, which I have not seen discussed, is the effect of energy resolution on the measured amplitude (Watson 1981).

Suppose that the assignment of an energy $E_0$ is dispersed over a range of inferred energies with a log-normal spread characterised by $\sigma$, due to fluctuations in shower development and measurement errors. Suppose also that the true dependence of amplitude on energy varies as $kE_0^{-\alpha}$ and that the differential spectral slope is $\gamma$. Then the amplitude in some interval $E_1$ to $E_2$ (in inferred energy) is given by

$$\text{Amplitude} = \frac{\int_0^\infty \left[ \int_{E_1}^{E_2} P_G\ (E,E_0,\sigma)\,dE \right]\ E^{-\gamma}\ kE^{-\alpha}\ dE}{\int_0^\infty \left[ \int_{E_1}^{E_2} P_G\ (E,E_0,\sigma)\,dE \right]\ E^{-\gamma}\ dE}$$

For many experiments $E_2 = 2E_1$. $P_G$ is the Gaussian probability distribution.

It is useful to compare the amplitude measured for typical values of $\alpha$ and $\gamma$ as a function of $\sigma$. Results are shown in figure 3 for $\alpha = 0.5$ and $\gamma = -3$. The effective energy at which the anisotropy is measured is also shown. Thus if $\sigma = 50\%$ (appropriate to many small sea-level arrays tuned to $\sim 10^{16}$ eV), the mean energy, $E_f$, to be associated with an energy interval $E_1$ to $E_2$ is about 0.78 of that which it would be in the absence of fluctuations and the measured anisotropy is 0.86 of that which would have been observed in the idealised case. In practice these factors should be considered when comparing results from different arrays. The effect of the change of phase of the anisotropy with energy in the presence of fluctuations is considerably smaller than the above factors for a realistic variation of 60° per decade.

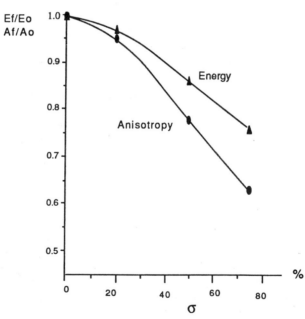

Fig.3.  Comparison of measured energy and anisotropy as a function of energy resolution: $\alpha = 0.5$, $\gamma = 3.0$.

# ULTRA HIGH ENERGY GAMMA RAYS

The term 'ultra high energy gamma rays' refers to gamma rays with energies above ~ 100 TeV. Their study forms a new field which was initiated by the report (Samorski and Stamm 1983a) that Cygnus X-3 was an emitter of gamma rays at 1000 TeV (≡ 1 PeV) and beyond. No theoretical predictions had prepared the astrophysics community for this astonishing result, although several objects (eg the Crab pulsar, Cygnus X-3 and Cen A) were already known to be sources of 1 TeV gamma rays. This new discovery has prompted intense theoretical and experimental activity. Although a much newer field than uhe comic ray astronomy, the associated statistical problems have been given much more detailed attention.

## Detection of UHE Gamma Rays

Like uhe cosmic rays, gamma rays with energies >~ 100 TeV are detected indirectly through the cascades which they produce in the atmosphere. The number of gamma rays is very small by comparison with the charged particle component and as yet it has not proved possible to separate the cascades initiated by the two components in a reliable and reproducible way. It had been anticipated that gamma ray cascades would be deficient in muons by comparison with hadron cascades because of the relatively small value of the cross-section $\sigma(\gamma,\pi)$, but the Kiel workers (Samorski and Stamm 1983b) find an approximately normal muon distribution in events from the Cygnus X-3 direction. There is, however, contradictory evidence from the Akeno experiment (Kifune et al 1986). Furthermore Samorski and Stamm found that the selection of 'old' showers (large value of the age parameter, s, of Nishimura – Kamata cascade theory) enhanced the effect again in contradiction with theoretical expectation (Hillas 1987, Fenyves 1985). The empirical effect of an 'age' cut is supported by recent work from the Ooty group (Tonwar et al 1988).

The uncertainty about the 'gamma ray' nature of the events from Cygnus X-3 has cast some doubt on the reality of the whole field, a doubt which has been further fuelled by claims of time variability in some features of the source emission. An extreme negative position has been adopted by Chardin (1987) and others, but more optimistic views have been taken by Watson (1985) and Protheroe (1987). Many improved experiments are now taking data and very large area telescopes are under construction (see Watson 1987 for details). A feature of all experiments will be the need for careful statistical treatments of the data obtained: a few of the problems are outlined below. The subject has been recently reviewed by Protheroe (1986, 1988).

## Low Rates and High Backgrounds

At 500 Tev the flux of gamma rays from Cygnus X-3 is about $10^{-13}$ $cm^{-2}$ $s^{-1}$ whilst the charged cosmic ray flux at the same energy is $\sim 5.5 \times 10^{-10}$ $cm^{-2}$ $s^{-1}$ $sr^{-1}$. With air shower arrays designed as gamma ray telescopes an angular resolution of $\sim 10^{-3}$ sr $(\sigma(\theta) \sim 1^\circ)$ is now achieved routinely. The flux of gamma rays, therefore, is roughly 20% of the charged cosmic ray flux within a restricted solid angle centred on the source.

Using these figures it is clear that a 4 $\sigma$ detection of an object above the isotropic charged cosmic ray background requires an exposure, At, of $\sim 7 \times 10^{15}$ $cms^2$ or an area of $\sim 10^4$ $m^2$ operating for just over 2 years. However if the aim is to carry out 'short-time' constant astronomy by collecting, say, 10 $\gamma$-rays per pulse ($4.8^h$ in the case of Cygnus X-3) then an area of $\sim 0.5$ $km^2$ or greater is required. Of the devices presently under construction only those of the Chicago (Cronin et al 1987) and Los Alamos groups (Nagle et al 1987) approach this target.

Because of the very low rate of uhe gamma-rays to be expected on even very large detector areas, it is often necessary to assign upper limits to the flux level which can be inferred from a particular sequence of observations. A method of estimating upper limits in this case, and which is claimed to be universally applicable, has been proposed by Protheroe (1984). He has used a Bayesian approach and considered carefully the effect of uncertainty in knowledge of the background. When the background is very well determined, that is when the number of off-source bins used to accumulate background information is $\gg 1$, the problem reduces to solving the following equation (Hearn (1969)) for $S_0$,

$$P(S_0) = \exp(-S_0) \left\{ \sum_{j=0}^{N} (S_0 + B)^j/j! \Big/ \sum_{i=0}^{N} B^i/i! \right\} = 1-C$$

where C is the desired confidence level, and $S_0$ is the corresponding upper limit to the number of source counts consistent with an observed background count B, which is very well-known, and N is the total number of counts in the source-bin. A commonly adopted value for C is 0.95, corresponding to a 95% upper limit. Hearn (1969) has given a very useful plot of $S_0$ vs B for various values of N for the ranges $S_0$, B, N<20. Protheroe has extended these in a very useful way and use of his plots are recommended if very exact evaluation of backgrounds are necessary. However it is often desirable to have a rapid method of evaluating $S_0$ (95%) when such tables are not available and when the background is reasonably well-known, the following formula may suffice

344

$$S_o \ (95\%)/B^{\frac{1}{2}} = 0.5 \ N(\sigma) + 2$$

and can be used with reasonable accuracy (better than 25%) when $B \gtrsim 100$ over the range $- 2 < N(\sigma) < 2$, where $N(\sigma)$ is the 'number of sigma', $(N-B)/\sqrt{B}$, deduced directly from the observations.

## The Problem of the Ephemeris

All attempts so far to establish uhe-gamma-ray sources have relied upon the inherent periodicity of the source, usually the orbital period (~ hours) – rather less often the pulsar period ($\lesssim 5s$). The difficulty in using these periods to enhance the significance of a detection, which is weak without periodic analysis, lies in the poor knowledge of the ephemerides of candidate sources. The period, P, and its first derivative, $\dot{P}$, are usually only obtainable from X-ray astronomy observations but these are rarely contemporary with uhe observations or may never have been measured.

To illustrate the confusion which can arise if the ephemeris of a source is not known or if different ephemerides are used by different authors, it may be useful to recall a couple of examples:–

(i) In 1980 Cygnus X-3 was 're-discovered' at TeV energies through the work of the Dublin/Whipple Observatory group (Danaher et al 1981). For their analysis they adopted an ephemeris proposed by the original discoverers of TeV emission (Stepanyan et al 1977) which was based largely on measurements made in the early 1970's. This emphemeris gave a peak in the $4.8^h$ cycle at a phase of 0.75. Subsequently a more contemporary X-ray ephemeris became available and reanalysis of the same data suggested that the peak was actually at a phase of ~ 0.6. However, when the Soudan group (Marshak et al 1985) began to report anomalous muon signals from Cygnus X-3 they at first used a modern X-ray ephermeris, but compared the position of the peak in their data with that reported by Danaher et al, so adding to the confusion engendered by their unexpected result.

(ii) In the initial Samorski and Stamm report (1983a) of emission from Cygnus X-3 an X-ray ephemeris with P = 0 was used for the phase analysis of a very convincing $4.4\sigma$ DC signal. This emphemeris produced a peak at $\Phi = 0.35$. Subsequently the Haverah Park group (Lloyd-Evans et al 1983) in an attempt to confirm the Kiel result, used the best available X-ray ephermeris for which $\dot{P} \neq 0$ and found a peak at $\Phi = 0.25$. The two results were qualitatively reconcilable but subsequent reanalysis of the Kiel data considerably reduced the significance of the phase distribution peak. However the value of P has since been revised (Mason 1986) and the significance of the Kiel peak substantially restored.

The shape of the 'light curve' associated with the source is also a problem of considerable statistical interest. Considerable work on this topic has been carried out by de Jager et al (1986) using the 'kernel density estimator'.

## Search strategies for serendipitious sources at selected positions

At the present stage of the development of the uhe gamma ray astronomy field workers are concentrating on looking for DC signals and periodic signals from known objects in addition to watching for bursts of activity lasting for a few hours or days. At present methods are being developed to determine the angular resolution of the telescopes: the resolution varies in a complex way and is a function of the zenith angle at detection and of the energy.

An analogous set of problems has been met and solved in the analysis of data recorded by the COS B satellite which recorded gamma rays of 50 to 1000 MeV. This work was greatly aided by the existence of several point-sources which gave clear signals above the ambient background after a relatively short exposure and by the extensive pre-launch calibrations which had been possible with tagged photon beams. The response function of the detector to a point source, as measured in the laboratory, was confirmed by observations on the Vela pulsar (Hermsen 1980). With accurate knowledge of the response function powerful statistical methods were used to search for sources at particular positions (e.g. at positions known to be associated with objects active at other wavelengths) or to look for hot spots on the sky which may or may not be associated with objects active at other wavlengths. These techniques seem capable of adaptation to uhe gamma ray astronomy.

To test for emissioin from specific objects the COS B group (Pollock et al. 1981) developed a maximum likelihood analysis which should be equally applicable to uhe gamma ray data. In a sky-region where the background is uniform the probability of observing a photon in the solid angle $d\Omega$ centred on $(\alpha, \gamma)$ is given by

$$p(\alpha, \delta, A, B)d\Omega = (S(\theta).A + B)d\Omega$$

where $\int p \, d\Omega = 1$, $S(\theta)$ is the response function and A and B the relative source and background contributions. In the COS B analysis all N events within $\theta_{max}$ of the source position were used and the likelihood function obtained as a function of X, the fraction of gamma rays from the source at $\theta = 0$. The natural logarithm of the likelihood is normally used and is given by

$$\ln L(X) = \sum_{i=1}^{N} \ln \left[ \frac{XS(\theta)}{I} + \frac{1 - X}{2\pi(1 - \cos\theta_{max})} \right]$$

where $I = 2\pi\int_0^{\theta_{max}} S(\theta)\sin\theta \, d\theta$ and $X = AI$. It follows that $\ln L(x)$ is a relative measure of how well the data support the hypothesis $X = x$.

The present problem with uhe gamma ray astronomy is that $S(\theta)$ is not well-known because no source of strength equivalent to the Vela pulsar in the COSB data has yet been found and there is, of course, no tagged photon beam available. Two approaches have been adopted. One method is to attempt a detailed Monte Carlo simulation of the array incorporating such parameters as the thickness of the shower disk, the shower front curvature and the timing information in real showers (see, for example, Protheroe and Clay 1984). Alternatively, and most popularly, the angular accuracy is determined from measurements with independent sub-arrays which are used to provide two measures of the shower direction, comparison between which allows an estimate of the precision possible with the full array (see, for example, Eames et al. 1987). A difficulty with the latter method is that it is not clear how to relate the precision achievable with a single sub-array to that which should be associated with the full array. In a recent study, developed for the Bartol/Leeds South Pole array (SPASE), Hillas (1988) has carried out a Monte Carlo analysis of the array performance which agrees excellently with the rms error in space angle found experimentally from the sub-array comparison. He finds that the corresponding rms error in the space angle measured by the full array is between 1·6 and 2·5 times smaller, greater accuracy being obtained with larger showers. This topic merits further study as the optimum size of search bin to maximize the signal is a function (albeit a weak function) of the angular uncertainty.

A further technique developed for COSB analysis and which will surely find application with uhe gamma ray data is the method of cross-correlation analysis. This was used by the COS B collaboration in their detection of emission from the quasar 3C273 (Swanenburg et al. 1978). With this method one searches for a signal of the shape dictated by the response function anywhere in the data. The details of the method are too lengthy to give here but are well set out by Hermsen (1980).

347

## DATA ARCHIVING

It is very likely that most of the uhe cosmic and gamma ray sources which are eventually identified will prove to be variable on a wide variety of time scales. In this context care should be taken to allow proper archiving and documentation of data presently and recently recorded for use in the future. Some effort has been made to document the details of the rather small number of cosmic ray events so far recorded with energies $> 10^{19}$ eV through the efforts of Wada (1980, 1986) at the World Data Center in Tokyo. A much greater effort will be required to collate the very extensive records of uhe gamma ray telescopes as the background of charged cosmic rays is so high and it is necessary to keep events from all parts of the sky for archival purposes.

In creating such archives it seems to me that the necessary use of computers presents a problem. Suppose for example SN1987a had not been 'visible' immediately but was only 'discovered' after several hundred years. Future scientists would surely wish to search the records of proton decay experiments to find the anti-neutrino flux and to establish $t = 0$. Would this in fact be possible? It seems to me that already it would be hard to find ways of reading data recorded say 25 years ago on 5-hole tape and there appears to be no slackening in the pace of computer development. Something analogous to a photographic plate library needs to be developed which can be maintained and readily up-dated as techniques change. Certainly it is unlikely that we will be prescient enough to extract all that is of importance from current data on uhe cosmic and gamma rays.

## REFERENCES

Baltrusaitis R M et al 1985 Phys Rev Lett 54 1875
Chardin G 1988 Physics Reports (in press)
Cronin J W et al 1987 submitted to Nucl Inst and Methods
Danaher S et al 1981 Nature 289 568
de Jager et al 1986 Astron Astrophys 170 187
Eames P J V et al 1987 Proc 20th Int Conf Cosmic Rays
    (Moscow) 2 449
Edge D M et al 1973 J Phys A 6 1612
Fenyves F 1985 Proc Workshop (La Jolla) on Techniques in UHE Gamma
    Ray Astronomy (Editors R J Protheroe and S A Stephens)
    pp 124-12)
Greisen K 1966 Phys Rev Letters 16 748
Gehrels N 1986 Ap J 303 336
Hearn D 1969 Nuclear Instruments and Methods 70 200
Hermsen W 1980 PhD thesis, University of London

Hillas A M 1987 Proc 20th Int Conf Cosmic Rays (Moscow) 2 362
Hillas A M 1988 private communication
Jauncey D L 1967 Nature 216 877
Kifune T et al 1986 Ap J 301 230
Lloyd-Evans J et al 1983 Nature 312 50
Marshak M et al 1985 Phys Rev Letters 54 2079
Mason K O 1986 (unpublished)
Murzin V S 1965 Proc 9th Int Conf Cosmic Rays (London) 2 872
Nagle D et al report at 20th Int Conf Cosmic Rays (Moscow),
    unpublished
Pollock A M T et al 1981 Astron Astrophys 94 116
Protheroe R 1984 Astron Express 1 33
Protheroe R J 1986 in Very High Energy Gamma Ray Astronomy
    Reidel, Dortrecht 1987 (Editor K E Turver) 91-100
Protheroe R J 1987 Proc 20th Int Conf Cosmic Rays (Moscow) 8
    23
Protheroe R J 1988 Proc Astron Soc Australia in press
Protheroe R M and Clay R W 1984 Proc Astron Soc Aust 5 586
Regener V H 1951 Phys Rev 84 161
Samorski M and W Stamm 1983a Ap J 268 L17
Samorski M and W Stamm 1983b Proc 18th Int Conf Cosmic Rays
    (Bangalore) 11 244
Stephanyan A A et al 1977 Proc 15th Int Conf Cosmic Rays 1
    135
Swanenburg B N et al 1975 Nature 275 298
Tennant R M and A A Watson 1971 Haverah Park Internal Note
Tonwar S et al 1988 Ap J, in press
Wada M 1980, 1986 Catalogue of Highest Energy Cosmic Rays
    (World Data Center C2 for Cosmic Rays) Vol 1 and Vol 2
Watson A A 1981 (unpublished)
Watson A A 1985 Proc 19th Int Conf Cosmic Rays (La Jolla)
Zatsepin G T and V A Kuzmin 1966 Sov Phys JETP 4 114

# TRACK PATTERN ANALYSIS FOR STUDY OF POINT-SOURCE AIR SHOWERS

R. Buccheri, C.M. Carollo, J. Linsley, and L. Scarsi

Istituto di Fisica Cosmica e Applicazioni dell'Informatica
Consiglio Nazionale delle Ricerche, Palermo

**ABSTRACT**. We describe and compare two methods of pattern recognition for identifying the arrival direction of a particle shower impacting on a tracking chamber. Results from computer simulations show that both methods suffice to determine the desired direction at a high level of accuracy: thus both provide good alternatives to the traditional method of shower direction determination based on timing of scintillator signals.

## INTRODUCTION

Currently there are several operating or proposed experiments to search for point sources of UHE cosmic radiation. At energies in the UHE range ($E > 100$ TeV), particles reaching the earth from such sources generate air showers capable of penetrating the atmosphere and reaching ground level. The experiments are carried out by means of particle detectors deployed in large arrays. In order to resolve point sources from the nearly uniform background of showers due to ordinary cosmic rays, the arrays used for these searches need to afford the best possible angular resolution. In addition they should provide data on the muon content of the showers they detect, for the purpose of identifying the particles arriving from the point sources. The hypothesis that these particles are gamma rays is favoured by the air shower arrival direction observations, the arrival directions being consistent with rectilinear propagation from known strong sources of non-thermal radiation, and the observed intensities, neutrons or neutrinos requiring much greater source power. It is favored over the hypothesis of massive neutral primaries (more massive, say, than neutrinos) by the air shower arrival time observations, which show correlations between the shower intensity and other emissions by these variable sources. But the gamma ray hypothesis is in strong conflict with evidence that the showers in question contain approximately the same number of muons as one expects, and observes, for air showers produced by hadronic primaries (nuclei). Discussions of this conflict can be found in review articles by Wdowczyk (1986) and Protheroe (1987).

In the usual approach, using scintillators, shower directions accurate within about one degree are determined from relative time delays. To determine the muon content, additional detectors are used, screened by several hundred g cm$^{-2}$ of absorber to remove the dominant electron-photon component of the shower. Adequate muon detectors are likely to cost as much or more than the scintillators. An alternative is to use detectors capable of measuring the shower direction and at the same time distinguishing between shower muons and electrons. It has been suggested (Linsley, 1987) that a detector with the desired properties can be constructed from two tracking chambers separated by a layer of dense material, each of the two chambers being able to produce important information on the local direction of the shower. In the following of the paper we will refer to a tracking chamber consisting of 3 horizontal planes of one centimeter width streamer tubes (Iarocci, 1983). The planes are separated by a distance d = 40 cm and have orthogonal sets of readout strips of width $\alpha$ = 1 cm which give both x and y coordinates of penetrating charged particles.

We report the results of a pattern recognition analysis for the measurement of the shower angle. The work is part of the *PLASTEX* project (*P*alermo-*L*eeds-*A*ir-*S*hower-*T*racking-*EX*periment) to construct and test a prototype of a tracking chamber in the center of a typical scintillator array, the GREX array operated by the University of Leeds at Haverah Park in England. GREX also has near its center a large conventional muon detector operated by the University of Nottingham.


## SIMULATION OF SHOWER IMPACTING ON TRACKING CHAMBER

The case we deal with is that of n particles impacting simultaneously on the detector with an average inclination $\theta$ with respect to the z axis. The impact is simulated by generating pairs of coordinates $(x_{1i}, y_{1i})$ of n points distributed randomly on the first (upper) plane. The $x_{1i}$'s and $y_{1i}$'s are then 'propagated' towards the lower planes by adding a certain amount of spread $\phi$, typical of showers, to straight lines with spatial direction $\theta$. According to detailed air shower simulations by Hillas (1982), the statistical distribution of $\phi$ is given by the following formula:

$$\delta n/\delta r = 2 \, Q \, r \, \exp(-(r-r_0)/\lambda_i) \tag{1}$$

where r is related to $\phi$ and to the shower particle energy E (in MeV) through the equation

$$\phi_i{}^2 = 0.563 \ (21 \ r / E)^2 \ /(1 + 108/E) \tag{2}$$

and $\lambda_i$ has different values $\lambda_1$ and $\lambda_2$ for $r < r_0$ and $r > r_0$ respectively. Particles of low energy are described by $r_0 = 0.59$, $\lambda_1 = 0.478$, $\lambda_2 = 0.380$ and $Q = 0.777$, while high energy particles are fitted instead with $r_0 = 0.37$, $\lambda_1 = 0.413$, $\lambda_2 = 0.380$ and $Q = 1.318$. Guided by Hillas, we take "low energy" to mean $E < 100$ MeV, other energies being regarded as "high".

In our simulations we considered a shower impacting on the tracking chamber at $\theta = 0$ and generated individual values of $\phi_i$ by the use of eq. 2. A random value of E was selected from the shower particle energy spectrum (see below) followed by a random value of r, through an acceptance-rejection method, using eq. 1 with parameter values appropriate to the value of E.

Letting $\pi(E_0, s, >E)$ be the integral energy spectrum of electrons in a shower with primary energy $E_0$ at an athmospheric depth where the shower age is s, a reasonable parametrization is given by the form

$$\pi = K_s \ (E + E_s)^{-s} \tag{3}$$

In the typical case $s = 1$, it was found using the simulation program EGS (Ford and Nelson, 1978) that satisfactory values for the parameters are given by $E_1 = 37$ MeV and $K_1 = 38$ (Linsley and Mikocki, 1986). Thus, solving for E, we obtain the algorithm

$$E = 38 / t - 37 \tag{4}$$

where t is a random number between 0 and 1 representing the value of the integral spectrum at energy E ($> 1$ MeV).

Once having selected the angular deviations $\phi_i$ , points of intersections with planes 2 and 3 are determined, using in addition the distances $d_2 = d$ and $d_3 = 2d$ of the second and third planes from the first, and random values for the azimuthal angle $\varphi$, according to the following formulas (see fig. 1 for the explanation of the symbols):

$$x_{2i} = x_{1i} + d_2 \ tg \ \phi_i \ cos \ \varphi_i \qquad y_{2i} = y_{1i} + d_2 \ tg \ \phi_i \ sin \ \varphi_i$$
$$x_{3i} = x_{1i} + d_3 \ tg \ \phi_i \ cos \ \varphi_i \qquad y_{3i} = y_{1i} + d_3 tg \ \phi_i \ sin \ \varphi_i$$

Finally the projected coordinates $(x_{1i}, y_{1i})$, $(x_{2i}, y_{2i})$, $(x_{3i}, y_{3i})$ are "quantized" by taking integer values, expressing the fact that it is not possible, in the real experimental situation, to discriminate inside a region of one square centimeter.

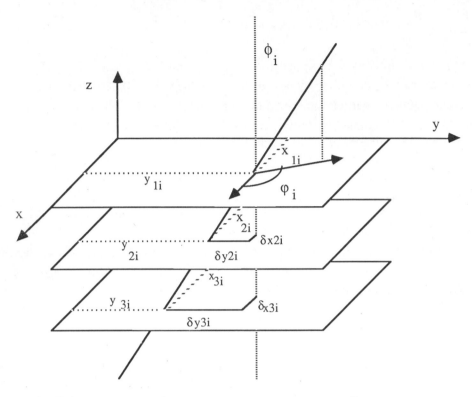

*Fig. 1 - Schematic view of the simulated particle tracks* $(\delta x_{ij} = d_j \, tg \, \phi_i \, cos \, \varphi_i$
*and* $\delta y_{ij} = d_j \, tg \, \phi_i \, sin \, \varphi_i$ ).

## DESCRIPTION OF THE TWO METHODS

The coordinates of the particles impacting the detector will be recorded in just the form described above, as independent xz and yz projections. It is enough to consider only one of them, the xz projection, results for the yz projection being exactly similar. The n simulated particles give rise to h "on-segments" out of the total N, with h less than or equal to n because more than one particle may hit the same streamer tube. The data, for each event, will then consist of a 3 x N array, whose elements $s_{ij}$ (i=1,2,3; j = 1,2,..N) are or 0 or 1 depending on whether the corresponding segment was fired or not.

<u>Method A</u>. For each $s_{1j} = 1$ (i.e., for each fired segment, FS, in plane 1), and each $s_{3k} = 1$ (i.e., each FS in 3), such that $|j-k| < R$, assign a logical value T or F depending on whether $s_{2l} = 1$ or 0, respectively, where $l$ is an integer number between 0 and N given by

$$l = \begin{cases} (j+k)/2 & \text{if } j\text{-}k \text{ is even} \\ (j+k)/2 + 0.5 \quad \text{or} \quad (j+k)/2 - 0.5 & \text{if } j\text{-}k \text{ is odd} \end{cases} \qquad (5)$$

R is an integer chosen to express the fact that air showers are collimated in the vertical direction, typically to within 30 degrees, and that air shower particles are collimated in the shower direction, typically within 20 degrees. To understand (5), note that when j-k is even there is a segment exactly half way between segment j and segment k, but if j-k is odd, the midpoint between j and k falls on the common boundary of two adjacent segments.

If the value is T we say that there is an apparent track through FS with index j of plane 1 and FS with index k of plane 3, and we assign the projected angle to a certain bin with index m, where m = j-k. The possible values of m are -R, -R+1, ..., -1, 0, 1, ..., R-1, R and the total number of bins is 2R + 1.

The "raw outcome" of the procedure, for a given event and for each projection, will be a histogram showing the numbers of apparent tracks with various indices m. The "apparent tracks" will include, in a sense, all simulated tracks passing through planes 1 and 3 (and 2 necessarily), and also some "accidental" tracks. "In a sense" is intended to concede that because of the limited resolution, 1) two or more simulated tracks may correspond to a single apparent track, and 2) the direction of an apparent track, according to eq. 2 will differ slightly from the direction of its corresponding simulated track. Most of the bins of the histogram will contain only accidentals, all of the correctly identified tracks falling in a limited number of adjacent bins centered about a certain bin corresponding to the projected shower direction, which for our simulated events is bin 0. If $m^*$ is the best value of m for describing the location of the peak in the histogram, the corresponding best value of $\theta_x$, the projected shower direction, is given by

$$\theta_x = arctg \, (m^* \, \alpha \, / \, 2 \, d) \qquad (6)$$

The best value of $\theta_x$ will have a certain statistical accuracy which we wish to maximize. The accuracy will tend to increase when the number of correctly identified tracks increases, either through an increase in detector area, or an increase in the density of shower particles such as will occur when the shower is larger or when it strikes nearer to the detector. Increasing the density will increase the number of correctly identified tracks, but will also increase the number of accidental tracks. As the detector approaches saturation, the loss of accuracy due to an increasing number of accidentals is expected to dominate.

Method B. The number of accidentals can be reduced by increasing the number of planes, but this will increase the cost. We became interested in the opposite choice, in the extreme case where the number of planes is only 2. For clarity, we will call the procedure used in this case *Method B*, although there is a close and obvious similarity to the previous case, so we use almost the same nomenclature. Let the 2 planes have indices p and q. Then every pair of fired segments $s_{pi}$ and $s_{qk}$ such that $|i-k|=|m|<R$ gives rise to an apparent track, with R and m having the same meaning as before. The raw outcome will be a histogram of the variable

$$S(m) = \Sigma_i T_i(m) \tag{7}$$

where

$$T(m) = \begin{cases} 1 & \text{if} \quad s_{pi} = s_{qk} = 1 \\ 0 & \text{otherwise} \end{cases} \tag{8}$$

We note that the S-histogram is equivalent to a histogram of the correlation coefficient

$$\rho_m = \Sigma_i w_{pi} w_{qk} / ((\Sigma_i w_{p,i})^2 + (\Sigma_i w_{qk})^2)^{0.5} \tag{9}$$

where $w_{pi}$ and $w_{qk}$ are $s_{pi}$ and $s_{qk}$ referred to their mean values.

As above, if $m^*$ is the best value of m for describing the location of the peak in the histogram, the corresponding best value of $\theta_x$, the projected shower direction, is given by

$$\theta_x = arctg\,(m^* \alpha / d) \tag{10}$$

The width, in bins (m-units), of the peak containing the correctly identified tracks will depend on the value of d, the distance between the 2 planes. If a very small value is chosen for d, the peak will be correspondingly narrow, more easily discernable in spite of the "noise" due to accidentals, whereas if too large a value is chosen, the peak may be lost in this noise. The choice of the pair of planes 1 and 2 or 2 and 3 is therefore more convenient than that of the pair 1 and 3.

## RESULTS FROM COMPUTER SIMULATIONS AND CONCLUSIONS

Figs. 2 and 3 show the histograms derived from the application of respectively methods A and B to a simulated case of 100 shower particles impacting at $\theta = 0$ on a tracking chamber having N = 600 segments. Both histograms refer to the x-coordinates of the impact points. In the case of the method B the pair of planes 1 and 2 has been used. The values used in the two cases for R are respectively 80 and 40, which limit the angle of acceptance of the incoming showers to 45 degrees. In the figures a gaussian fit is also shown; from it we have derived the estimate $m^*$ of the value of m which, treated with eq. 6 (for method A) and eq. 10 (for method B), gives the estimate of $\theta_x$, the projected angle of the impacting

356

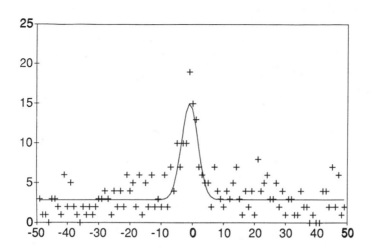

*Fig. 2 - Central part of the distribution of angles $\theta_x$ (in units of m= (2d/α) tg $\theta_x$) resulting from the application of method A to 100 simulated particles impacting on the tracking chamber . Energy and shower angle distributions are given by eq. 1, 2 and 4. A fit with a gaussian shape is also shown.*

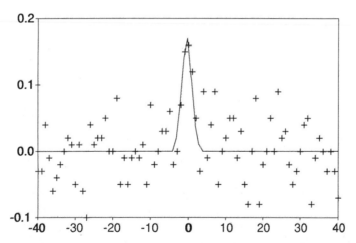

*Fig. 3 - Correlation coefficient $\rho_m$ versus m=(d/α)tg$\theta_x$ resulting from the application of method B to 100 particles in the same conditions as in fig. 2*

shower. The same estimate has been derived in 20 independent simulations of the same case, the results in m-units being shown in Table I. In the last two rows of the table the mean and the rms over the 20 trials are given with, in brackets, the corresponding angles in degrees. It is noted that the correct value of the projected angle is retrieved with an accuracy of the order of 0.4 and 0.8 degrees respectively for the two methods. The combination of the two projections gives, for the estimate of the spatial angle

$$\theta = \text{arctg } (\text{tg}^2\theta_x + \text{tg}^2\theta_y)^{0.5} \tag{11}$$

an accuracy of about 0.5 and 1.0 degrees respectively for methods A and B.

TABLE I

| n | method A | | method B | |
|---|---|---|---|---|
| | xz.proj | yz.proj | xz.proj | yz.proj |
| 1 | -0.22 | -0.04 | 0.13 | -0.54 |
| 2 | 0.37 | -0.07 | 1.66 | 0.34 |
| 3 | 0.03 | -0.65 | -1.84 | -0.22 |
| 4 | -0.61 | 0.64 | 0.17 | -0.13 |
| 5 | 0.19 | 0.68 | 0.62 | 1.07 |
| 6 | -0.46 | -0.14 | -1.28 | -0.19 |
| 7 | 0.06 | 0.63 | -0.66 | -0.02 |
| 8 | 1.00 | -0.46 | 0.88 | 0.13 |
| 9 | -0.81 | 0.26 | 0.11 | 0.58 |
| 10 | -0.83 | 0.22 | -0.32 | 1.07 |
| 11 | 0.60 | 0.26 | 0.76 | -1.30 |
| 12 | -0.72 | 0.84 | -0.17 | 0.16 |
| 13 | 0.31 | -0.85 | 0.91 | 0.13 |
| 14 | -0.22 | -0.04 | 0.13 | -0.54 |
| 15 | -0.28 | -0.18 | 0.02 | 0.09 |
| 16 | -0.01 | -0.70 | -1.15 | 0.89 |
| 17 | -0.38 | 0.51 | -0.48 | 0.59 |
| 18 | -0.89 | 0.48 | -0.34 | 0.13 |
| 19 | 0.62 | 0.87 | 0.05 | 0.91 |
| 20 | -0.85 | 0.06 | -0.17 | 0.47 |
| mean | -0.16 (-.11) | 0.12 (0.09) | -0.05 (-0.07) | 0.18 (0.26) |
| rms | 0.55 (0.39) | 0.52 (0.37) | 0.82 (1.17) | 0.59 (0.84) |

We want to stress that eq. 11 gives the local direction of that part of the shower impacting on our detector and that the accuracies derived above refer to a pessimistic case in which the angular spread of that part is described by the Hillas distribution, valid instead for the whole shower. An array of tracking chambers is therefore expected to have, for the determination of the shower direction, an accuracy consistently better than one degree (typical of conventional scintillators arrays) because it is derived from the combined information from many detectors for which the actual angular spread of the impacting particles is generally smaller than that used in our simulations.

## REFERENCES

- Ford L.R., Nelson W.R., Report SLAC-210,1978 (unpublished)
- Hillas A.M., J. Phys. G: Nucl. Phys., 8, 1461-1473, 1982
- Iarocci E., Nucl. Instr. & Meth., 217, 30, 1983
- Linsley J., Proc. of 20th ICRC (Moscow), 2, 442, 1987
- Linsley J, Mikocki S., 1986, private communication
- Protheroe R., Proc. of 20th ICRC (Moscow), 8, 21, 1987
- Wdowczyk J., "Cosmic Radiation in Contemporary Astrophysics", ed. M. M. Shapiro (Reidel, Dordrecht), p.149, 1986

# DATA ANALYSIS AND ALGORITHMS FOR G.W. ANTENNAS

G.V. Pallottino and G. Pizzella

Dipartimento di Fisica, Università *La Sapienza*,

P.le A. Moro 2, 00185 Roma, Italy

## SUMMARY

The analysis of the sensitivity of a gravitational wave detector shows that the role of the algorithms used for the analysis of the data is comparable to that of the experimental apparatus.

We discuss the response of an antenna to an incoming gravitational wave, deriving a mathematical model of the system, which is used for examining various algorithms of data analysis, including the optimum filter and the matched filter. A brief discussion of coincidence techniques, using two detectors, is also included.

## 1. THE ANTENNA RESPONSE TO A GRAVITATIONAL WAVE

A cylindrical resonant g.w.a. is equivalent to a system of elementary oscillators with angular resonance frequencies

$$\omega_k = (2k + 1)\frac{\pi v}{L} \quad (k = 0, 1, 2....) \tag{1.1}$$

where $v$ is the sound velocity in the bar ($v = 5400$ m/s in Al at a temperature of 4.2 K) and $L$ the bar lenght. As customary we limit ourselves here to the fundamental mode oscillator with resonant angular frequency

$$\omega_o = \frac{\pi v}{L} \tag{1.2}$$

We consider now a polarized g.w. with metric perturbation h(t) impinging on the bar perpendicularly. We indicate with $\eta(t)$ the displacement of the bar end face, equal to the displacement of our equivalent elementary oscillator. We have

$$\ddot{\eta} + 2\beta_1 \dot{\eta} + \omega_o^2 \eta = \frac{2}{\pi^2} L \ddot{h} \tag{1.3}$$

where

$$2\beta_1 = \omega_o/Q \tag{1.4}$$

Q being the merit factor (expressing the losses); the numerical factor on the right of eq. (1.3) is obtained when solving the problem for the continuous bar[1].

Introducing the Fourier transforms H($\omega$) and $\eta(\omega)$ we obtain from (1.3)

$$\eta(\omega) = \frac{2}{\pi^2} L \frac{\omega^2 H(\omega)}{(\omega^2 - \omega_o^2) - 2i\omega\beta_1} \tag{1.5}$$

The quantity

$$T(\omega) = \frac{2L}{\pi^2} \frac{\omega^2}{(\omega^2 - \omega_o^2) - 2i\omega\beta_1} \tag{1.6}$$

is the transfer function of our oscillator.

We get $\eta(t)$ by calculating the inverse Fourier tranform

$$\eta(t) = \frac{1}{2\pi} \int_{-\infty}^{\infty} T(\omega)H(\omega)e^{i\omega t} d\omega \tag{1.7}$$

In the cases that $H(\omega)$ have no poles, the poles of the integrand of (1.7) are

$$\omega_{1,2} \cong i\beta_1 \pm \omega_o \tag{1.8}$$

In most cases $H(\omega_1) = H(\omega_2)$. Then we get

$$\eta(t) = -\frac{2L}{\pi^2} e^{-\beta_1 t} H(\omega_1)\omega_o \sin\omega_o t \qquad (1.9)$$

It is interesting to consider the case of g.w. short bursts that can be expressed with a Dirac function $h(t) = H_o\delta(t)$. In this case $H(\omega) = H_o$ independent of frequency.

Formula (1.9) expresses the two fundamental points of a resonant g.w.a.: a) the antenna detects the Fourier component $H(\omega)$ of the metric perturbation at its resonance frequency, b) the antenna has a "memory" for a time of the order of $\beta_1^{-1}$. The last point allows the use of convenient algorithms for the data analysis. As far as point a), if we want to determine the value of $h(t)$ from $H(\omega_o)$ we must make assumptions on the wave spectrum. The simplest one, very rough, is to consider a short burst with duration $\tau_g$ and obtain $h(t)$ from the measured $H(\omega_o)$ by means of

$$h(t) \simeq H(\omega_o)\Delta\omega \simeq \frac{H(\omega_o)}{\tau_g} \qquad (1.10)$$

considering a bandwidth $\Delta\omega \sim 1/\tau_g$.

Examining eq.(1.9) it is evident that the interesting part of the signal is the amplitude of the response at the antenna resonance frequency. There is no need to record $\eta(t)$ at all t values for determining its sinusoidal behaviour, which would require, according to the Nyquist condition, a sampling time

$$\Delta t \leq \frac{1}{2\omega_o} \ , \qquad (1.11)$$

that is $\Delta t \sim 10^{-4}$ s for the frequencies of interest.

For reducing the amount of recorded data one can eliminate the sinusoidal part at the $\omega_o$ frequency, either by means of an appropriate computer program operating in real time, or by means of lock-ins. Here we shall consider the last case.

## 2. THE BASIC BLOCK DIAGRAM AND THE WIDE BAND ELEC-TRONIC NOISE[2]

The antenna mechanical vibrations are transformed into an electrical signal by means of an electromechanical transducer (i.e.: a piezoelectric ceramic, an inductive or capacitive diaphragm, etc.). We consider

$$V(t) = \alpha \eta(t) \tag{2.1}$$

where $\alpha$ is the transducer constant and $V(t)$ is the output voltage (this is for a voltage-wise signal; the treatment for a current-wise signal is similar).

This voltage is very small and needs to be amplified with a low-noise electronic amplifier. The amplifier noise, although small, is one of the most important limitation to the antenna sensitivity. The noise can be characterized by the voltage and current power noise spectra, $V_n^2$ and $I_n^2$, or by the following combinations of them

$$T_n = V_n I_n / k \tag{2.2}$$

$$R_n = V_n / I_n \tag{2.3}$$

where k is the Boltzmann constant, $T_n$ the amplifier noise temperature ($V_n^2$ and $I_n^2$ in bilateral form) and $R_n$ the amplifier noise resistance.

Indicating with Z the output impedance of the transducer we have the matching parameter between transducer and amplifier

$$\lambda = \frac{R_n}{|Z|} \tag{2.4}$$

whose optimum value will be shown to be of the order of unity.

Another important quantity which will be used later is the ratio of the energy stored in the transducer to the energy stored in the bar of mass M. For a PZT or capacitive transducer, indicating the output capacity with C, we have

$$\beta = \frac{\frac{1}{2}C(\alpha \eta)^2}{\frac{1}{4}M\omega^2 \eta^2} = \frac{2C\alpha^2}{M\omega^2} = \frac{2\alpha^2}{M\omega^3|Z|} \tag{2.5}$$

where the last equality holds in general.

The voltage noise at the amplifier output is of wide-band type

$$S_o = V_n^2 + I_n^2|Z|^2 = V_n^2(1 + 1/\lambda^2) = kT_n|Z|(1/\lambda + \lambda) \tag{2.6}$$

Only if the amplifier input is shorted ($|Z| = 0$) the voltage noise at the output is just $V_n^2$.

The amplifier (with gain G) is followed by the lock-in driven by a synthesizer at $\omega_o$. The lock-in is an electronic device that produces, at its output, the two following quantities

$$x(t) = \frac{1}{t_o} \int_{-\infty}^{t} GV(t')e^{-\frac{t-t'}{t_o}} \text{sign}[\cos\omega_o t']dt' \tag{2.7}$$

$$y(t) = \frac{1}{t_o} \int_{-\infty}^{t} GV(t')e^{-\frac{t-t'}{t_o}} \text{sign}[\sin\omega_o t']dt' \tag{2.8}$$

This is done in two steps. The first step (for $x(t)$) consists in multiplying $V(t)$ by sign $[\cos \omega_o t]$; with the second step the signal is filtered by means of a RC filter with time constant $t_o$, which produces the exponential weighting, i.e. the integration.

For studying mathematically the lock-in operation, in particular for what the noise is concerned, we introduce the autocorrelation function of $V(t)$

$$R_v(\tau) = \lim_{T \to \infty} \frac{1}{2T} \int_{-T}^{T} V(t) \cdot V(t+\tau)dt \equiv E\{V(t) \cdot V(t+\tau)\}$$

(where $E\{\cdot\cdot\}$ stands for Expectation$\{\cdot\cdot\}$). The power spectrum $S_v(\omega)$ is the Fourier transform of $R_v(\tau)$

$$S_v(\omega) = \int_{-\infty}^{\infty} R_v(\tau)e^{-i\omega\tau}d\tau \tag{2.9}$$

Expanding sign($\cos \omega_o t$) in series:

$$\text{sign}(\cos\omega_o t) = \frac{4}{\pi}\left[\cos\omega_o t + \frac{1}{3}\cos3\omega_o t + \cdots\right] \tag{2.10}$$

we calculate the autocorrelation of $V(t)\frac{4}{\pi}\cos\omega_o t$

$$R'(\tau) = \frac{16}{\pi^2}\ \frac{1}{2}\ R_V(\tau)\cos\omega_o\tau \tag{2.11}$$

and the corresponding power spectrum

$$S'(\omega) = \int_{-\infty}^{\infty} R'(\tau)e^{-i\omega\tau}d\tau = \frac{4}{\pi^2}[S_V(\omega+\omega_o)+S_V(\omega-\omega_o)] \tag{2.12}$$

The additive property gives the power spectrum of $V(t)\text{sign}(\cos\omega_o t)$

$$S''(\omega) = \frac{4}{\pi^2}[(S_V(\omega+\omega_o)+S_V(\omega-\omega_o)+$$

$$+\frac{1}{9}(S_V(\omega+3\omega_o)+S_V(\omega-3\omega_o))+\cdots]$$

The contribution due to the higher order harmonics (already at $\pm 3\omega_o$) can be neglected because the amplifier preceding the lock-in has a bandwidth the reduces considerably the content of V(t) at frequencies different from $\omega_o$.

Thus, with a very good approximation, we consider $S'(\omega)$ given by (2.12) as the power spectrum of $V(t) \cdot \text{sign}[\cos\omega_o t]$.

The RC filter of the second step has a transfer function

$$W_2(\omega) = \frac{\beta_2}{\beta_2 + j\omega} \tag{2.13}$$

where $\beta_2 = t_o^{-1}$. According to general theorems, the power spectrum of the signal at the lock-in output is

$$S_x(\omega) = S'(\omega)|W_2(\omega)|^2 = S'(\omega)\frac{\beta_2^2}{\beta_2^2 + \omega^2} \tag{2.14}$$

Its inverse Fourier transform gives the autocorrelation of x(t)

$$R_{xx}(\tau) = \frac{1}{2\pi} \int_{-\infty}^{\infty} S_x(\omega)e^{i\omega\tau}d\omega \tag{2.15}$$

Similar expressions hold also for the quantity y(t). As already mentioned these expressions will be used for evaluating the noise.

As far as the effect of the lock-in on a signal of the type given by (1.9), considering the signal in phase with the synthesizer, $\beta_1 \ll \beta_2$ (which is easily veryfied), $t_o \gg \frac{2\pi}{\omega_o}$ and putting $V_s = \frac{2L}{\pi^2}(e^{-\beta_1 t})H(\omega_1)\omega_o$, we obtain

$$\left\{ \begin{array}{l} y(t) = \frac{1}{t_o}\int_{-\infty}^{t} V_s \sin\omega_o t' e^{-\frac{t-t'}{t_o}} \frac{4}{\pi}\sin\omega_o t' dt' \simeq \frac{2V_s}{\pi} \\ x(t) = \frac{1}{t_o}\int_{-\infty}^{t} V_s \sin\omega_o t' e^{-\frac{t-t'}{t_o}} \frac{4}{\pi}\cos\omega_o t' dt' \simeq 0 \end{array} \right\}$$

Thus, apart for the factor $\frac{2}{\pi}$ that is usually included in the lock-in amplification gain, the lock-in has extracted the amplitude of the signal at $\omega_o$. If the signal is not in phase with the synthesizer both y(t) and x(t) are, in general, different from zero, such that $x^2 + y^2 = (\frac{2V_s}{\pi})^2$.

The lock-in operation can be combined with the resonant bar transfer function for obtaining a much simpler basic scheme, as follows. We apply the result expressed by eq.(2.12) to the bar transfer function (1.6) substituting in (1.6) $\omega$ with $\omega \pm \omega_o$

$$T(\omega \pm \omega_o) = \frac{2L}{\pi^2} \frac{(\omega \pm \omega_o)^2}{\omega(\omega \pm 2\omega_o) - 2i\beta_1(\omega \pm \omega_o)}$$

For the cases of interest we have $\beta_1 \ll \omega_o$; thus the important range of $\omega$ is when $|\omega| \ll \omega_o$, where $T(\omega \pm \omega_o)$ becomes very large. In this region we have

$$T(\omega \pm \omega_o) \simeq \frac{\omega_o^2}{\pm 2\omega\omega_o \mp 2i\beta_1\omega_o} = \frac{\pm i\omega_o}{2\beta_1} \frac{\beta_1}{\beta_1 + i\omega} \tag{2.16}$$

Therefore the bar becomes a low pass filter with time constant $\beta_1^{-1}$, which we represent with the transfer function:

$$W_1(\omega) = \frac{\beta_1}{\beta_1 + i\omega} \qquad (2.16a)$$

The basic scheme, then, is shown in fig. 1. At port 1 the g.w. excitation is applied together, as we shall see in the next section, with the brownian noise. The signal is then integrated by the bar. At port 2 the amplifier electronic noise is applied. The total signal is then integrated by the lock-in RC filter (2.13) providing the two quantities in quadrature x(t) and y(t).

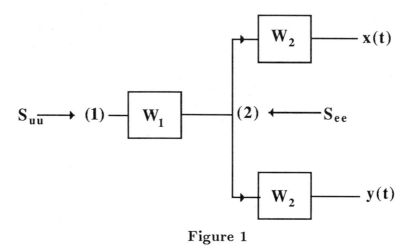

**Figure 1**

*Basic scheme representing the mathematical model of an antenna and the associated instrumentation.*

In what follows, according to the block diagram of fig. 1, we shall consider all signals as referred to the transducer. This means that all the amplifiers (including the lock-in) are assumed to have unity gain.

## 3. THE NARROW BAND NOISE AND THE TOTAL NOISE[2]

There are two types of narrow band noise: one is of brownian origin and is related to the bar thermodynamical temperature T, the other one

368

is due to the electronic amplifier that heats up the mode of the bar at the resonance $\omega_o$ (backaction). In both cases the noise can be represented with a white noise at the bar input, port 1 of fig.1. The effect of the low pass filter that represents the bar consists in narrowing the band of the noise at the bar output, producing a power spectrum of lorentzian type peaked at zero frequency with bandwidth depending on the merit factor Q.

We indicate the white spectrum at port 1 with $S_{uu}$. At the bar output this noise becomes

$$S_{nb}(\omega) = S_{uu}\frac{\beta_1^2}{\beta_1^2 + \omega^2} \tag{3.1}$$

($S_{uu}$ includes the factor $(i\omega_o/2\beta_1)^2$ of eq. (2.16)).

Its autocorrelation is

$$R_{nn}(\tau) = \frac{1}{2\pi}\int_{-\infty}^{\infty} S_{uu}\frac{\beta_1^2}{\beta_1^2 + \omega^2}e^{+i\omega t}d\omega = \frac{S_{uu}\beta_1}{2}e^{-\beta_1|\tau|} \tag{3.2}$$

The mean square value of the voltage noise is

$$V_{nb}^2 = R_{nn}(0) = \frac{S_{uu}\beta_1}{2} \tag{3.3}$$

A direct way to obtain $V_{nb}^2$ for an elementary oscillator with mass m and transducer constant $\alpha$ is by means of the energy equipartition principle at the temperature T

$$\frac{1}{2}m\omega_o^2\overline{\eta^2} = \frac{1}{2}kT$$

from which we get

$$\overline{V^2} = \alpha^2\overline{\eta^2} = \frac{\alpha^2 kT}{m\omega_o^2} = \beta\omega|Z|kT$$

In our case the temperature T has be increased because of the backaction of the electronic amplifier. For a voltage-wise signal the heating is generated by the current noise $I_n^2$ of the voltage amplifier (it would be due to the

369

voltage noise $V_n^2$ of the current amplifier for a current-wise signal). It can be shown that this temperature increase is $4\alpha^2 I_n^2 Q/(m\omega_o^3)$. Thus we define an "equivalent temperature"

$$T_e = T(1 + \frac{\alpha^2 I_n^2 Q}{2m\omega_o^3 kT}) = T(1 + \frac{\beta Q T_n}{2\lambda T})$$ (3.4)

The overall mean square voltage due to narrow band noise is finally

$$V_{nb}^2 = \frac{S_{uu}\beta_1}{2} = \frac{\alpha^2 kT_e}{m\omega^2} = \beta\omega|Z|kT_e$$ (3.5)

In this equation we have used the mass m of the equivalent elementary oscillator; this is also called the reduced mass of the antenna and it is one half of the real mass of the bar

$$m = \frac{M}{2}$$ (3.6)

as it can be immediately recognized from (31), if we write the energy of the bar in the form $\frac{1}{4}M\omega_o^2\eta^2$.

From (3.5) we obtain the white spectrum $S_{uu}$ needed for representing the observed mean square voltage due to the brownian and backaction noises

$$S_{uu} = \frac{2\alpha^2 kT_e}{m\omega_o^2\beta_1}$$ (3.7)

The power spectrum at the lock-in output due to the noise $S_{uu}$ will be

$$S'_{nb}(\omega) = S_{uu}\frac{\beta_1^2}{\beta_1^2 + \omega^2}\frac{\beta_2^2}{\beta_2^2 + \omega^2}$$

The corresponding autocorrelation is found to be

$$R'_{xx}(\tau) = \frac{S_{uu}\beta_1\beta_2}{2}\frac{\beta_2 e^{-\beta_1|\tau|} - \beta_1 e^{-\beta_2|\tau|}}{\beta_2^2 - \beta_1^2}$$ (3.8)

For obtaining the total noise we must add the contribution of the wide band noise at the port 2 due to the amplifier. This noise is given by (2.5). The lock-in, with its first step operation expressed by (2.12), produces a white noise with power spectrum

$$S_{ee} = 2S_o \tag{3.9}$$

At the lock-in output the noise spectrum is

$$S'_{ee}(\omega) = S_{ee} \frac{\beta_2^2}{\beta_2^2 + \omega^2}$$

and the corresponding autocorrelation

$$R'_{ee}(\tau) = S_{ee} \frac{\beta_2}{2} e^{-\beta_2 |\tau|} \tag{3.10}$$

The total noise, in conclusion, has autocorrelation

$$R_{nn}(\tau) = \frac{S_{uu}\beta_1}{2} \left[ \frac{\beta_2}{\beta_2^2 - \beta_1^2} (\beta_2 e^{-\beta_1 |\tau|} - \beta_1 e^{-\beta_2 |\tau|}) + \Gamma e^{-\beta_2 |\tau|} \right] \tag{3.11}$$

where we have introduced the quantity

$$\Gamma = \frac{S_{ee}}{S_{uu}} \tag{3.12}$$

This quantity, as we shall see later, is of fundamental importance for the data analysis algorithms and for characterizing the antenna basic properties (sensitivity and bandwidth).

Other convenient expressions for $\Gamma$ are obtained from (3.5), (2.5) and (2.6)

$$\Gamma = \frac{S_o \beta_1}{V_{nb}^2} = \frac{T_n}{2\beta Q T_e} (\lambda + 1/\lambda) \tag{3.13}$$

## 4. DATA FILTERING

The output data of a g.w.a., as they are provided by the instrumentation, namely the quantities x(t) and y(t) discussed in section 3, do not lend themselves to physical analysis, since the possible signals are deeply embedded in the background noise.

A preliminary filtering is therefore required, which is based on linear techniques, in order to improve as much as possible the signal to noise ratio (SNR). This is done by exploiting the statistical properties of both the signal and the noise, as recognized by Hawking and Gibbons in their pioneering paper[3].

Different filtering technique, of course, are used according to the nature, mainly the duration, of the signals which stems from their physical origin. We can distinguish between short pulses or bursts, whose duration is much smaller than the main time constants of the detection apparatus, and long trains of periodical waves, which maintain coherence over appreciable time intervals.

Both of the above situations will be dealt with in this section, where we shell also consider the intermediate case of transient signals of relatively long duration.

### 4.1 <u>Detection of short bursts</u> [4]

Here we only consider signals whose duration is smaller than both the decay time $\tau_v$ of the mechanical mode used for capturing the incoming gravitational radiation and the dominant characteristic time of the electronics, that is the time constant $t_o$ of the integration filter using in the lock-in amplifier. The latter is also important since a sampling time $\Delta t$ of the same value is usually adopted, with values ranging from some tens of milliseconds to about a second.

Such signals are expected to impart a sudden variation to the energy status of the mechanical mode, which is to be compared with the spontaneous variations due to the various noise effects.

It can be shown that a short burst, independently of its actual shape, can be modeled as a delta function. Being of short duration, its spectral content will cover a frequency range more extended than the spectral osservation window of the detection apparatus, where it will contribute as having the typical white spectrum of a delta function.

The main statistical properties of the background noise, as observed at the output of the antenna instrumentation, have been discussed in section 3. The overall spectrum of the noise can be viewed as the superimposition of two spectra: one representing the white noise of the electronics, which is band limited by the lock-in amplifier, the other one representing the narrow band thermal noise of the mechanical oscillator plus the backaction effect.

The power spectrum of the two components x(t) and y(t), that represents the output of the lock-in amplifier, can be written as:

$$S(\omega) = S_{uu} \frac{\beta_1^2}{\omega^2 + \beta_1^2} \frac{\beta_2^2}{\omega^2 + \beta_2^2} + S_{ee} \frac{\beta_2^2}{\omega^2 + \beta_2^2} \qquad (4.1)$$

where the spectra $S_{uu}$ and $S_{ee}$ have been defined in eqs. (3.7) and (3.9), and the characteristic angular frequencies $\beta_1$ and $\beta_2$, have been defined in eqs. (4) and (2.13). Their ratio

$$\gamma = \beta_1/\beta_2 = \frac{t_o}{\tau_v} \qquad (4.2)$$

in much smaller than unity.

A representation of the noise equivalent to the spectrum (4.1) is provided by its inverse Fourier transform, that is the autocorrelation function, as given by eq. (3.11), which we repeat here for convenience:

$$R(\tau) = \frac{V_{nb}^2}{2} \left[ \frac{e^{-\beta_1|\tau|} - \gamma e^{-\beta_2|\tau|}}{1 - \gamma^2} + \frac{\Gamma}{\gamma} e^{-\beta|\tau|} \right] \qquad (4.3)$$

$V_{nb}^2$ is the narrow band noise variance at the transducer and $\Gamma$ is the spectral ratio dfined by eq. (3.12). We notice that the main contribution is strongly correlated, with characteristic time $\tau_v = 1/\beta_1$.

From (4.3) we obtain the variance of both x(t) and y(t):

$$R(0) = \frac{V_{nb}^2}{2} \left( \frac{1}{1 + \gamma} + \frac{\Gamma}{\gamma} \right) \qquad (4.4)$$

373

the latter term between parenthesis representing the effect of the wide band noise.

The probability density function of both x(t) and y(t) follows the normal law with zero mean and standard deviation $\sqrt{R(0)}$.

The energy status of the antenna mode in the observation band width is described by the variable

$$r^2(t) = x^2(t) + y^2(t) \tag{4.5}$$

whose expected value can be expressed as

$$\bar{r^2} = E\{x^2(t) + y^2(t)\} = 2R(0) \tag{4.6}$$

in the case of absence of any signal as well as of any disturbance of origin different from the noise considered above.

This quantity is distributed according to the Boltzmann function

$$F(r^2) = \frac{1}{\bar{r^2}} e^{-r^2/\bar{r^2}} \tag{4.7}$$

If we denote with $V_s$ the amplitude (referred to the transducer) of the response to a possible gravitational signal, that occurs at t=0, we have (neglecting the noise):

$$r_s^2(t) = \frac{V_s^2 [e^{-\beta_1 t} - e^{-\beta_2 t}]^2}{(1 - \gamma)^2} \tag{4.8}$$

with maximum value

$$r_s^2 = K_d V_s^2 \tag{4.9}$$

where the coefficient $K_d$, usually slightly smaller than unity is given by

$$K_d = \left[ \frac{\gamma^{\gamma/(1-\gamma)} - \gamma^{1/(1-\gamma)}}{1 - \gamma} \right]^2 \tag{4.10}$$

374

Here the signal-to-noise ratio is

$$SNR = K_d V_s^2 / 2R(0). \tag{4.11}$$

This can be considerably improved by detecting the variations of the energy status of the mechanical mod, since most of the noise, being strongly correlated, is not likely to contribute to fast variations of $r^2(t)$ as it does to $r^2(t)$ itself.

An algorithm that is able to detect such variations of energy, indepedently of the energy stored in the mode (as well as of the noise level), is the so called zero-order prediction (ZOP) algorithm:

$$\varrho_z^2(t) = [x(t) - x(t - \Delta t)]^2 + [y(t) - y(t - \Delta t)]^2 \tag{4.12}$$

where $\Delta t$ in the sampling time.

This is sensitive to variations both of the energy level of the mode and of its phase in the phase plane x, y (an incoming pulse may increase or decrease the initial energy, as well as only modify its status in the phase plane).

The noise variance $\sigma_z^2$ of the filtered variables $x_z(t) = x(t) - x(t - \Delta t)$; $y_z(t) = y(t) - y(t - \Delta t)$ can be expressed in terms of the autocorrelation (4.3) as

$$\sigma_z^2 = 2R(0) - 2R(\Delta t) \tag{4.13}$$

which shows the role of the correlation of the noise for reducing the variance.

The above expression can be written, using eq.(4.3), as follows:

$$\sigma_z^2 = \frac{2V_{nb}^2}{e} \left[ \frac{\gamma}{1 - \gamma^2} + \frac{\Gamma}{\gamma}(e - 1) \right] \tag{4.14}$$

We notice, being $\gamma = \beta_1 / \beta_2 = \Delta t / \tau_v$, that the narrow band noise is reduced by the factor $\Delta t / \tau_v$ (which can be rather small) while the wide band noise is increased by the inverse of the same factor.

The variance $\sigma_z^2$ can be minimized with a convenient choice of $\gamma$. With

$$\beta_2^{\text{opt}} = \frac{\beta_1}{\sqrt{\Gamma(e-1)}} \tag{4.15}$$

the two contributions (narrow band and wide band noise, respectively) appearing in the left side of eq. (4.14) are made equal and we have

$$(\sigma_z^2)^{\min} = \frac{4V_{nb}^2}{2}\sqrt{\Gamma(e-1)} \tag{4.16}$$

We notice, recalling the meaning of $\Gamma$, that the limiting noise variance is proportional to the square root of the two noise spectra $S_{ee}$ and $S_{uu}$.

The quantity $\varrho_z^2$, as due to the noise only, follows an exponentially distribution, whose parameter $\overline{\varrho_z^2}$ represents the expected value

$$\overline{\varrho_z^2} = 2\sigma_z^2 \tag{4.17}$$

Also the signal, with this algorithm, undergoes a reduction of amplitude, but of much smaller extent than the noise. We have

$$\varrho_{zs}^2 = K_z V_s^2 \tag{4.18}$$

where the coefficient $K_z$, of the order of $1/e$, is given by

$$K_z = \left[\frac{e^{-\gamma} - e^{-1}}{1 - \gamma}\right]^2 \tag{4.19}$$

The signal to noise ratio, in the optimum condition ($\beta_2 = \beta_2^{\text{opt}}$), is:

$$(SNR)_z = \varrho_{zs}^2 / \overline{\varrho_z^2} = \frac{K_z e V_s^2}{4V_{nb}^2\sqrt{\Gamma(e-1)}} \tag{4.20}$$

The above filtering mehod is both very powerful and very simple. One can, however, think of a more general method, that uses the information provided by a number of data samples larger than two. The framework for the development of an optimum algorithm is provided by the Wiener-Kolmogoroff filtering theory[5], which will be applied, in what follows, to the estimation of the input acting on a g.w. detector, in terms of short impulses.

More specifically, the best linear estimate of the two orthogonal components of the input force is

$$\hat{u}_x = \int_{-\infty}^{+\infty} x(t - \alpha)w(\alpha)d\alpha \tag{4.21}$$

$$\hat{u}_y = \int_{-\infty}^{+\infty} y(t - \alpha)w(\alpha)d\alpha \tag{4.22}$$

Here $w(t)$ in the weighting function, that is the impulse response, of the optimum WK filter, which is determined by minimizing the mean square deviation

$$\sigma_w^2 = E[(u_i(t) - \hat{u}_i(t))^2] \qquad i = x, y \tag{4.23}$$

By applying the orthogonality principle of linear mean square estimation[5] between the deviation and the observation we have for the x component

$$E[(u_x(t) - \hat{u}_x(t)x(t')] = 0 \quad , \quad \forall t' \tag{4.24}$$

that is

$$R_{ux}(\tau) = \int_{-\infty}^{+\infty} R_{xx}(\tau - \alpha)w(\alpha) + \alpha \quad , \quad \forall \tau \tag{4.25}$$

where $\tau = t - t'$, and a similar expression for the y component.

The Fourier transform of the above provides the transfer function of the optimum filter to be applied the data $x(t)$ and $y(t)$

$$W(j\omega) = \frac{S_{ux}(\omega)}{S_{xx}(\omega)} \tag{4.26}$$

Here $S_{xx}$ is the spectrum (4.1) and $S_{ux}$ is the cross spectrum of the signals $u_x(t)$ and $x(t)$, which can be expressed as follows in terms of the transfer functions $W_1$ and $W_2$ of the antenna and of the lock-in (see section 3), respectively:

$$S_{ux} = S_{uu} W_1^* W_2^* \tag{4.27}$$

Using eqs.(4.1) and (4.27) we obtain the following expression

$$W(j\omega) = \frac{W_1^* W_2^*}{|W_1|^2 |W_2|^2 + \Gamma |W_2|^2} = \frac{1}{W_1 W_2} \frac{1}{1 + \Gamma/|W_1|^2} \tag{4.28}$$

which shows that the optimum filter is composed by two parts: the first one is an inverse filter that cancels the dynamics of the apparatus (including any delay). The second provides the smoothing (band limiting) required for minimizing the effect of the wide band electronic noise.

By introducing the characteristic angular frequency of the smoothing section

$$\beta_3 = \frac{\beta_1}{\sqrt{1 + \eta\, \Gamma}} \tag{4.29}$$

we have from eq. (4.28):

$$W(j\omega) = \frac{\gamma}{\Gamma} \frac{(\beta_1 + j\omega)(\beta_2 + j\omega)}{\omega^2 + \beta_3^2} \tag{4.30}$$

which clearly shows the non casual nature of the filter (see eqs. (4.21) and (4.22)) which operates on the past as well as on the future data samples. This is not, of course, a problem, since the analysis is usually performed on data stored in a magnetic tape. A quasi real time filter can be also applied to the output data of a detector, which provides a good approximation of the desired estimate with a delay of the order of a few units of $1/\beta_3$, as required to operate on the "future" data over a time range of this order, beyond which the weighting function $w(t)$ of the filter decays to neglegible values.

The function $w(t)$ is obtained by performing the inverse Fourier transform of eq.(4.30):

$$w(t) = \frac{\gamma}{\Gamma} \left[ (\delta(t) + \frac{(\beta_2 \pm \beta_3)(\beta_1 \pm \beta_3)e^{\pm\beta_3 t})}{2\beta_3} \right] \tag{4.31}$$

378

where the - sign is for $t > 0$ and the + sign for $t < 0$.

The output noise variance of the optimum filter, for each one of the two components, is:

$$\sigma_w^2 = S_{uu} \frac{\beta_1^2}{2\beta_3\Gamma} = V_{nb}^2 \frac{\beta_1}{\beta_3\Gamma} \tag{4.32}$$

Here the quantity considered for the analysis is

$$\varrho_w^2(t) = \hat{u}_x^2(t) + \hat{u}_y^2(t) \tag{4.33}$$

which, in the absence of signal, has expected value

$$\overline{\varrho_w^2} = 2\sigma_w^2 \tag{4.34}$$

and exponential distribution with parameter $\overline{\varrho_w^2}$.

The response to an input delta function signal

$$S(t) = \frac{V_s\beta_1}{2\beta_3\Gamma} e^{-\beta_3|t|} \tag{4.35}$$

is only determined by the smoothing section of the optimum filter (due to the action of the inverse filter): an input delta function, without any delay, is converted into a two-sided exponential. The characteristic time $1/\beta_3$ expresses the available time resolution for optimum SNR. For signals well above the noise, however, one can perform the smoothing with a characteristic time smaller than $1/\beta_3$, thereby improving the time resolution while reducing the SNR.

We remark here that this analysis considers continuous time signals, while the data submitted to filtering are sequences of samples. Therefore $\beta_3$ cannot exceed $1/\Delta t$, that is $\beta_2$.

With the WK filter the signal to noise ratio is, from eqs.(4.32) and (4.35):

$$(SNR)_w = \frac{S^2(0)}{2\sigma_w^2} = \frac{V_s^2}{8V_{nb}^2} \frac{1}{\sqrt{\Gamma(1+\Gamma)}} \tag{4.36}$$

The SNR improvement obtained with this filter can be expressed in terms of reduction of the temperature of the noise. If the noise temperature of the unfiltered data is Te (as given by eq. 3.4) the noise temperature after the optimum filtering is

$$T_{eff\ w} = 4T_e \sqrt{\Gamma(1+\Gamma)} \tag{4.37}$$

The above expression clearly shows the essential role of the electronic noise (expressed by the spectral ratio $\Gamma$) for defining the temperature of the noise, that is the sensitivity of the detector for short bursts.

By substituting eqs. (3.4) and (3.13) into eq. (4.37) we obtain the following expression

$$T_{eff\ w} = 2T_n \sqrt{\left(1 + \frac{1}{\lambda^2}\right)\left(1 + \frac{2\lambda T}{\beta Q T_n}\right)} \tag{4.37a}$$

which is of basic importance as regards the design and the optimization of a detector[2]. From the above, two "matching conditions" can be derived in order to approach the Giffard limit $2T_n$[6], dictated by the noise temperature of the electronic amplifier.

## 4.2 Detection of longer bursts

We come now to a brief discussion of the filtering techniques aimed at detecting bursts of longer duration, that is inputs extended over time intervals larger than one sampling period. These cannot, of course, be modeled as a delta function.

The optimum solution, in this case, is represented by the so called "matched filter"[7] that is a linear filter (which performs on the observed data an operation similar to that of eqs.(4.21) and (4.22), aiming at detecting the occurrence of signals of known shape.

If a signal x(t) with limited temporal support is embedded in white noise n(t), it has been shown that the corresponding matched filter has impulse response equal to the time-reversed signal

$$w(t) = x(-t) \tag{4.38}$$

380

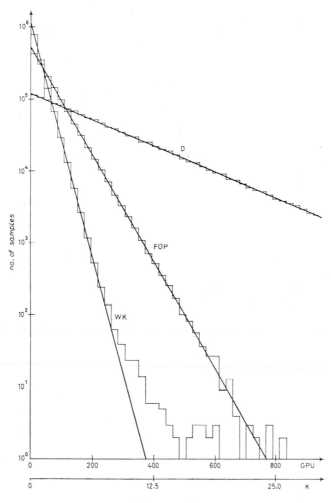

**Figure 2**

*Frequency distributions of the experimental data of a small cryogenic antenna (operating a 7523 Hz). D represents the unfiltered quality $r^2$, FOP the data filtered with an algorithm similar to the ZOP, WK the data filtered with the Wiener-Kolmogoroff filter.*

and transfer function

$$W(j\omega) = X^*(j\omega) \tag{4.39}$$

where $X(j\omega)$ is the Fourier transform of $x(t)$.

The output signal $s(t)$ of this filter is given by the following convolution integral:

$$s(t) = \int_{-\infty}^{+\infty} w(\tau)x(t-\tau) + \tau = \int_{-\infty}^{+\infty} x(-\tau)x(t-\tau)d\tau \tag{4.40}$$

which is maximum for $t=0$

$$s(0) = \int_{-\infty}^{-\infty} x^2(-\tau)d\tau = \int_{-\infty}^{+\infty} x^2(t)d\tau \tag{4.41}$$

where it represents the so called "energy" of the incoming signal. This is to be compared with the noise variance. Denoting by $N_o$ the spectral density of the noise, the variance is

$$\sigma_n^2 = \frac{N_o}{2\pi} \int_{-\infty}^{+\infty} |X(j\omega)|^2 d\omega = N_o \int_{-\infty}^{+\infty} x^2(t)dt \tag{4.42}$$

where we have used the Parseval theorem. The corresponding signal to noise ratio (at $t=0$) is:

$$SNR = \frac{S^2(0)}{\sigma_n^2} = \frac{1}{N_o} \int_{-\infty}^{+\infty} x^2(t)dt \tag{4.43}$$

We note that the matched filter (4.38) is non causal, while a causal realization might be desirable. Furtherly, the signal $x(t)$ can be assumed to vanish outside an interval $(0,T)$. In this case the causal matched filter has impulse response

$$w(\tau) = x(T-t) \tag{4.44}$$

providing the maximum at the delay T.

If the noise spectrum $N(\omega)$ is not white, it can be whitened by passing it through a filter with transfer function

$$|W_w(j\omega)| = \frac{1}{|W_N(j\omega)|} = \sqrt{\frac{N_o}{N(\omega)}} \tag{4.45}$$

This is the inverse filter corresponding to the filter $W_N(j\omega)$ that has shaped a white noise source with flat spectrum $N_o$ providing the actual noise with spectrum $N(\omega)$

$$N(\omega) = |W_N(j\omega)|^2 N_o \tag{4.46}$$

At the output of the whitening filter, however, the shape of signal $x_w(t)$ is distorted: its Fourier transform is

$$X_w(j\omega) = X(j\omega)W_w(j\omega) \tag{4.47}$$

The corresponding matched filter, therefore, has transfer function

$$W_m(j\omega) = X^*(j\omega)W_w^*(j\omega) \tag{4.48}$$

and the overall filter is

$$W(j\omega) = W_w(j\omega)W_m(j\omega) = X^*(j\omega)\frac{N_o}{N(\omega)} \tag{4.49}$$

The maximum signal to noise ratio, at t=0, is given by the celebrated Dwork's formula[8]:

$$SNR = \frac{T_s(0)}{\sigma_n^2} = \frac{1}{2\pi}\int_{-\infty}^{+\infty}\frac{|X(j\omega)|^2}{N(\omega)}d\omega \tag{4.50}$$

We remark that if the signal is a delta function (or the response to a delta function) the matched filter is formally equivalent to the Wiener-Kolmogoroff filter. This can be shown by letting $X(j\omega) = W_1(j\omega)W_2(j\omega)$,

$N(\omega) = S_{uu}|W_1 W_2|^2 + S_{ee}|W_2|^2$ and substituting into eq.(4.49); the result is eq.(4.28).

The W-K filtered data, as given by eqs. (4.21) and (4.22), can be directly applied to the detection of signals of duration longer than a sampling period $\Delta t$. One only has to perform on them a further filtering, matched to the specific shape of the signal considered.

In fact, if the input signal acting on the g.w.a. has a given shape u(t), with Fourier transform $U(j\omega)$, and the observed signal is x(t), with Fourier transform

$$X(j\omega) = U(j\omega)W_1(j\omega)W_2(j\omega) \tag{4.51}$$

the corresponding overall filter has transfer function:

$$W(j\omega) = \frac{U^*(j\omega)W_1^*(j\omega)W_2^*(j\omega)}{S_{uu}|W_1 W_2|^2 + S_{ee}|W_2|^2} \tag{4.52}$$

If a Wiener filter with transfer function (4.28) has already been applied to the observation x(t), what remains to be done is to apply to the data a filter with transfer function $U^*(j\omega)$, that is precisely a filter matched to the specific shape of the input.

The implementation of the above procedure, however, requires the detailed knowledge of the input signal shapes for the two components $u_x(t)$ and $u_y(t)$ of the Rice representation at $\omega_R$ of the actual input u(t), where

$$u(t) = u_x(t)cos\omega_R t + u_y(t)sin\omega_R t, \tag{4.53}$$

as well as of the phase of the reference signal used in the lock-in.

This problem can be approached in two ways. First, one can apply the final matched filtering to the $\varrho_z^2(t)$ quantity, instead of $x_w(t)$ and $y_w(t)$, at the price of a small decrease of the SNR. Here, of course, the template is matched to $u^2(t)$, rather than to u(t). Second, one can use the "direct sampling method", developed by S. Frasca, that consists in processing the data sampled at the output of the amplifiers (without passing through the lock-in). The sampling theorem, in fact, which is usually applied for low pass signals, holds as well for bandpass signals of the same bandwidth.

## 4.3  Detection of periodic signals

We conclude this section by mentioning the detection problem for periodic signals[10]. In this case, both the W-K filter and the matched filter frameworks lead to the same intuitive result: one should filter the data over a vanishing bandwidh centered at the frequency of interest. Here the attainable SNR is maximum, for a given observation bandwidth $\Delta\nu$, if the frequency of the signal coincides with the resonance of the mechanical mode of the antenna, where the contribution of the narrow band noise is dominant (being $\Gamma \ll 1$).

The observation bandwidth depends on the duration of the measuring time $t_m$: $\Delta\nu = 1/t_m$. The SNR, for a sinusoidal signal of amplitude $V_s$ and angular frequency $\omega$, is:

$$SNR(\omega) = \frac{V_s^2/2}{S_{uu}\left\{1 + \Gamma\left[\frac{\omega^2}{\omega_R^2} + Q^2(1 - \frac{\omega^2}{\omega_R^2})\right]\right\} \cdot \Delta\nu} \qquad (4.54)$$

This is maximum at $\omega_R$ and it is larger than $SNR(\omega_R)/2$ in the frequency range

$$B \cong \frac{\omega_R}{2\pi Q\sqrt{\Gamma}} \qquad (4.55)$$

that is much larger than the mechanical bandwidth of the antenna. We notice that the bandwidth (4.55) for sinusoidal signals is just twice the (one-sided) optimum bandwidth of the WK filter for pulse signals, given by eq.(4.29).

This means that the W-K filtered data can be used as well for the detection of periodic signals by performing on them the required spectral analysis. The advantage over using the x(t) and y(t) data is twofold: a) the noise variance of the data is smaller, b) the amplitude of the possible spectral lines thus obtained is not to be corrected according to the Lorentzian shape of the mechanical frequency response curve of the antenna.

## 5. THE CROSS SECTION AND THE ANTENNA SENSITIVITY

In order to detect g.w. it is necessary that the detector absorbes some energy from the wave. As a first step to estimate the cross-section let us consider the energy carried by g.w. It can be shown that the energy carried

385

per unit time across the unit area is given by

$$I(t) = \frac{c^3}{16\pi G}\left[\dot{h}_+(t)^2 + \dot{h}_\times(t)^2\right] \quad \left[\frac{joule}{s \cdot m^2}\right] \tag{5.1}$$

where $h_+$ (t) and $h_\times$ (t) indicate the two polarization states of the g.w. The total energy per unit area is

$$I_o = \int_{-\infty}^{\infty} I(t)dt \quad \left[\frac{joule}{m^2}\right] \tag{5.2}$$

For simplicity we consider one polarization status only, $h_\times(t)$ or $h_+(t)$, and indicate it with h(t). We indicate with H($\omega$) the Fourier transform of h(t),

$$H(\omega) = \int_{-\infty}^{\infty} h(t)e^{-i\omega t}dt \tag{5.3}$$

From (5.1) we get

$$I_o = \frac{c^3}{16\pi G}\int_{-\infty}^{\infty} |\omega H(\omega)|^2 d\nu \tag{5.4}$$

were $\nu = \omega/2\pi$ is the frequency. The quantity

$$f(\omega) = \frac{c^3}{16\pi G}|\omega H(\omega)|^2 \quad \left[\frac{joule}{m^2 Hz}\right] \tag{5.5}$$

is called "spectral energy density", written here in bilateral form (frequencies from $-\infty$ to $+\infty$).

As a special case, we consider a g.w. burst of duration $\tau_g$ that can be described by a sinusoidal wave with angular frequency $\omega_o$ and amplitude $h_o$ for $|t| < \tau_g/2$ and zero value for $|t| > \tau_g/2$. Since we have $H(\omega) \cong h_o\tau_g/2$, from (5.5) we get

$$f(\omega) = \frac{c^3}{64\pi G}\omega_o^2 h_o^2 \tau_g^2 \quad \left[\frac{joule}{m^2 Hz}\right] \tag{5.6}$$

386

Another interesting case is a g.w. burst of the type $h(t) = h_o e^{-\beta_w |t|} \cos\omega_o t$. This wave has duration of the order of $\tau_g = 2/\beta_w$. If $\tau_g \ll 2\pi/\omega_o$ then the Fourier transform is

$$H(\omega) \cong \frac{h_o}{\beta_w} = \frac{h_o \tau_g}{2}$$

and we obtain again the result (5.6).

In order to obtain the total amount of energy per unit area we consider that the frequency bandwith for a duration $\tau_g$ is $1/\tau_g$. Multiplying (5.6) for it we get

$$I_o = f(\omega_o)/\tau_g = \frac{c^3}{64\pi G}\omega_o^2 h_o^2 \tau_g \qquad \left[\frac{joule}{m^2}\right] \tag{5.7}$$

Finally an interesting case is also a g.w. burst of a $\delta$-type, $h(t) = H_o\delta(t)$, $\delta(t)$ being the Dirac function. Since $H(\omega) = H_o$ we obtain from (5.5)

$$f(\omega) = \frac{c^3}{16\pi G}\omega^2 H_o^2 \qquad \left[\frac{joule}{m^2 Hz}\right] \tag{5.8}$$

In order to compute the value of $h_o$ on the Earth due to a g.w. burst of duration $\tau_g$ that occurs at a distance R, indicating with $M_{GW}c^2$ the total g.w. energy, we multiply (5.7) by $4\pi R^2$; we obtain

$$h_o = \sqrt{\frac{16 G M_{GW} c^2}{c^3 R^2 \omega_o^2 \tau_g}} = 1.38 \cdot 10^{-17} \frac{1000 Hz}{\nu} \frac{1000 pc}{R} \sqrt{\frac{M_{GW}}{10^{-3} M_\odot} \frac{10^{-3} s}{\tau_g}} \tag{5.9}$$

If we consider a sinusoidal g.w. of angular frequency $\omega$ with amplitude $h_o$ we obtain the average power per unit area from (5.1)

$$W_o = \frac{c^3}{32\pi G}\omega^2 h_o^2 \qquad \left[\frac{watt}{m^2}\right] \tag{5.10}$$

Indicating with W the total power irradiated by the source, at distance R we obtain

$$h_o = \sqrt{\frac{8G}{c^3} \frac{W}{R^2 \omega^2}} = 2.29 \times 10^{-41} \frac{1000 pc}{R} \frac{1000 Hz}{\nu} \sqrt{W} \tag{5.11}$$

The cross section $\Sigma$ is defined such that, multiplied by the incident spectral energy density $f(\omega_o)$, gives the energy deposited in the bar

$$\varepsilon = \Sigma f(\omega_o) \tag{5.12}$$

The energy $\varepsilon$ is calculated from (1.9)

$$\varepsilon = \frac{1}{4} M \omega_o^2 \left( \frac{2L}{\pi^2} \omega_o H(\omega_o) \right)^2 = \frac{M \omega_o^2 H(\omega_o) v^2}{\pi^2} \tag{5.13}$$

where $v = \omega L / \pi$ is the sound velocity in the bar. Making use of (5.5) for the spectral energy density we obtain the cross section

$$\Sigma = \frac{16}{\pi} \left( \frac{v}{c} \right)^2 \frac{G}{c} M \qquad [m^2 Hz] \tag{5.14}$$

in bilateral form.

We derive now another expression which relates $H(\omega_o)$, the quantity measured with a resonant g.w. antenna, to $T_{eff}$ which expresses the sensitivity of the apparatus. From (5.13) and (4.37) we obtain

$$[H(\omega_o)]_{min} = \frac{L}{v^2} \sqrt{\frac{k T_{eff}}{M}} \qquad [Hz^{-1}] \tag{5.15}$$

In general if we detect an energy innovation $\Delta \varepsilon$, the corresponding value of $H(\omega_o)$ is

$$H(\omega_o) = \frac{L}{v^2} \sqrt{\frac{\Delta \varepsilon}{M}} \tag{5.16}$$

It must be stressed that with a resonant antenna we measure $H(\omega_o)$ and not h(t). If we want to have a feeling about possible values for h(t) we must make assumptions on the h(t) spectrum. For instance for a flat spectrum from 0 to $\nu_g$, which we can think due to a burst of duration $\tau_g \sim 1/\nu_g$, we can put

$$h(t) \simeq H(\omega_o) \nu_g = \frac{H(\omega_o)}{\tau_g} \tag{5.17}$$

# 6. COINCIDENCE TECHNIQUES

When examining the data of a g.w.a. we consider as candidate events the signals whose amplitude exceeds the background distribution.

We do not discuss here the statistical techniques that can be applied to establish the detection of an event, in terms of false alarm probability, false dismissal probability and so on. We just note that any large amplitude signal, observed by an antenna, might be due to local disturbances, of various physical origin, which do not follow the statistical distribution of well behaved noise. It is, therefore, common practice to perform coincidence experiments with two or more g.w.a.'s in order both to improve the detection statistics (by reducing the probability density of the background thereby increasing the SNR) and to reduce drastically the effect of local disturbances.

The standard technique consists in using the two time series provided by two detectors (located, if possible, at great distance) to construct a new time series, where the value of each sample is the smallest of the corresponding samples of the two original time series. The basic idea is that a large g.w. excitation should give rise to large signals in both detectors at the same time. It is, of course, necessary to normalize the values of the two sequences, according to the sensitivity of the detectors, prior to the creation of the new sequence.

It can be easily shown that the probability distribution of the background of the new data is the product of the two distributions of the original data. This means that if both follow the Boltzmann distribution with parameters $T_1$ and $T_2$, the resulting parameter is:

$$T_c = \frac{T_1 T_2}{T_1 + T_2} \tag{6.1}$$

which means a reduction of the noise temperature to one half in the case of $T_1 = T_2$.

This analysis, usually, is performed by first thresholding at suitable levels the original data (thus reducing the amount of data to be processed) and then searching for possible events well above the background.

In order to take into account possible effects of non stationarity of the background, the probability of the candidate events being due to chance is usually evaluated using "local" distributions (obtained by performing the coincidence operation, only near the times of the events, on the two original sequences, one displaced with respect to the other by a small delay).

This procedure, however, improves considerably the SNR only if the responses of the two detector to a given input signal have the same, or nearly the same, amplitude. This is not the case, even for detectors operating at the same frequency and of similar sensitivity, if they have not the same orientation in space.

In order to take into account the effect of the geometrical response pattern of the detectors (this follows the law $sin^4\theta$, being $\theta$ the angle between the axis of the antenna and the direction of the source), as well as other effects, several methods of analysis have been introduced. One of them consists in normalizing the antenna responses with respect to the direction for a given source (as, for instance, the galactic center) before performing the coincidence operation. Each samples of the original time series is divided by the factor $sin^4\theta_i(t)$, where $\theta_1(t)$ and $\theta_2(t)$ are the angles between the axes of the detectors and the direction of the source of the time t.

Other methods, which do not require any assumption about the direction of the source, are based on different ways of combining the original data.

One of them consists in summing the samples of the two original time series[11]. Here the idea is to use at best the signals available, which may differ not only because the detectors have different orientations but also because they operate at different frequencies, where the spectral content of a signal might be considerably different.

The meaning of the resulting sequence, as regards possible signals, is in terms of total energy recorded by the detectors involved. The obvious vulnerability to local disturbances of this method can be, partially, circumvented by accepting as candidate signals only the data whose ratio, in the original sequences, is comprised in a preassigned range.

Another solution consist in creating a new data sequence by performing the product, rather than the sum, of the original time series. This has the additional advantage of being insensitive to possible calibration errors of the detectors. The noise rejection performance of this method is better than the sum method, while the capability of dealing with signals of different amplitude on the two detectors is inferior.

390

# REFERENCES

1. G. Pizzella "Gravitational-Radiation Experiments" <u>Rivista Nuovo Cimento</u> 5:369 (1975).

2. G.V. Pallottino, G. Pizzella "Matching of Transducers to Resonant Gravitational-Wave Antennas" <u>Nuovo Cimento</u> 4C:237 (1981).

3. G.W. Gibbons, S.W. Hawking "Theory of the Detection of Short Burst of Gravitational Radiation" <u>Phys. Rev. D</u> 4:2191 (1971).

4. P. Bonifazi, V. Ferrari, S. Frasca, G.V. Pallottino, G. Pizzella "Data Analysis Algorithms for Gravitational-Wave Experiments" <u>Nuovo Cimento</u> 1C:465 (1978).

5. A. Papoulis "Signal Analysis", Mc Graw Hill, New York (1977).

6. R. Giffard "Ultimate Sensitivity Limit of a Resonant Gravitational Wave Antenna Using a Linear Motion Detector" <u>Phys. Rev. D</u> 14:2478 (1976).

7. A.D. Whalen "Detection of Signals in Noise" Academic Press, New York (1971).

8. B.M. Dwork "Detection of a Pulse Superimposed on Fluctuation Noise" Proc. IRE 38:771 (1950).

9. S. Frasca, G.V. Pallottino, G. Pizzella "Spectral Domain Data Analysis Techniques for a Gravitational Wave Antenna" Signal Processing III, pp. 597-600, North Holland (1986).

10. G.V. Pallottino, G. Pizzella "Sensitivity of a Weber-Type Resonant Antenna to Monochromatic Gravitational Waves", <u>Nuovo Cimento</u> 7C:155 (1984).

11. G. Pizzella "Coincidence Techniques for Gravitational Wave Experiments", Submitted to Il Nuovo Cimento, 1988.

# DETECTING GRAVITATIONAL WAVES BY LASER INTERFEROMETERS

A. Krolak(*)

Department of Applied Mathematics and Astronomy
University College Cardiff,
Cardiff CF1 1XL, Wales, U.K.

## INTRODUCTION

This paper describes detection of gravitational waves
by laser interferometers. These are one of the two
main types of proposed earth-based detectors of
gravitational waves. The other type – bar detector, is
described in papers by Pizzella and Pallotino in this
volume.

At the moment there are four groups around the world
operating prototype laser interferometers at the
California Institute of Technology, the Massachusetss
Institute of Technology, Glasgow University and the Max
Planck Institute for Quantum Optics at Garching near
Munich. These groups and also experimental groups at
Orsay near Paris and Pisa are proposing to build large
size (at least 1 km arm length) detectors.

In Section 1 we shall describe main types of sources of
gravitational waves. In Section 2 we shall explain how
laser interferometric gravitational wave detectors work
and what are the main sources of noise that limit
sensitivity of those detectors. In Section 3 we shall
give calculations of signal-to-noise ratios for some
ofthe sources mentioned in Section 1 that can be
achieved with proposed large scale detectors.

---

(*) On leave of absence from Institute of Mathematics,
Polish Academic of Sciences, Warsaw, Poland.

## 1. EXPECTED GRAVITATIONAL WAVE SIGNALS

Three main types of gravitational wave sources are considered: burst, periodic and stochastic sources. Here we briefly discuss some most important examples of those sources. Full discussion can be found in Thorne (1987) and Hough *et al.* (1987).

a)   Burst sources

i)   Supernovae

These are most often discussed sources of gravitational waves. However because of our incomplete knowledge of the process of gravitational collapse leading to the supernovae explosion we cannot predict the gravitational waveform accurately. The dimensionless amplitude h and time duration $\tau$ of the gravitational wave pulse are given approximately by

$$h \approx 2.7 \times 10^{-22} \left[ \frac{\Delta E_{GW}}{10^{-4} M_{\odot} c^2} \right]^{1/2} \left[ \frac{1kHz}{f} \right]^{1/2} \left[ \frac{10Mpc}{R} \right]$$

(1a)

$$\tau \approx 10^{-3} \left[ \frac{1kHz}{f} \right]$$

(1b)

where $\Delta E_{GW}$ is the total energy radiated in gravitational waves, $M_{\odot}$ is the solar mass, c is the velocity of light, f is the dominant frequency of the pulse, R is the distance to the source. The values of $\Delta E_{GW}$ is very uncertain, it can differ from currently quoted number of around $10^{-4} M_{\odot} c^2$ by orders of magnitude. We expect around 10 such sources per year in Virgo cluster (10Mpc is the distance to the centre of Virgo cluster).

## (ii)    Coalescing binaries

These are binary systems consisting of compact objects like neutron stars or black holes. Such a system radiates gravitational waves (Peters and Mathews 1963) and as a result of radiation reaction the distance between the components of the binary decreases. This results in a sinusoidal signal whose amplitude and frequency increases with time which is called a chirp.

For circular orbits the dominant gravitational wave amplitude h and characteristic time $\tau$ for the increase of frequency ($\tau = f/\dot{f}$) can be obtained from the quadrupole formulas

$$\langle h \rangle = 1.0 \times 10^{-23} \mu_{\odot} m_{\odot}^{2/3} f_{N_{100}}^{2/3} R_{100}^{-1} \qquad (2a)$$

$$\tau = 8.0 \mu_{\odot}^{-1} m_{\odot}^{-2/3} f_{N_{100}}^{-8/3} \text{ sec} \qquad (2b)$$

where $\langle h \rangle$ denotes the average over the orientations of the detector and the source, $m_{\odot}$, $\mu_{\odot}$ are the total and reduced masses in solar masses respectively, $f_{N_{100}}$ is the frequency of the wave in 100 Hz (twice the orbital frequency) and $R_{100}$ is the distance in 100 Mpc. A typical chirp is given in Figure 1.

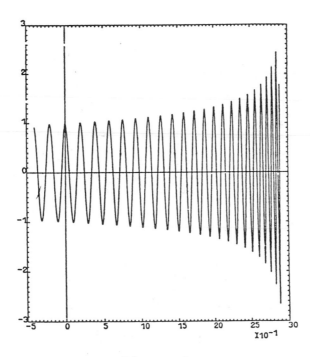

Figure 1.

We express frequencies in ⟨h⟩ and $\tau$ in 100 Hz since only above around that frequency gravitational signals can be detectable by laser interferometers (see next sections). We expect a few such sources per year out to 100 Mpc.

Expressions (2a) and (2b) give the signal to a very good high accuracy. For gravitational wave frequencies of several hundred hertz contributions from tidal effects and eccentricity of the orbit to ⟨h⟩ and $\tau$ are negligible. The first order correction to ⟨h⟩ and $\tau$ come from post-Newtonian effects and these can also be calculated accurately (Krolak and Schutz 1987; Krolak 1987).

REMARK

From formulae (2a) and (2b) taking into account the cosmological expansion one gets the following expression

$$d_{L_{100}} = 8.1 / \left[ f_{GN_{100}}^2 \langle h_{23} \rangle \tau_0 \right] \qquad (3)$$

where $d_L$ is the luminosity distance, $\langle h_{23} \rangle = 10^{23} \langle h \rangle$ and subscript o denotes observed quantities. Since $f_0$ and $\tau_0$ can be determined observationally and ⟨h⟩ can be determined if we have a network of detectors as a result of detection of coalescing binaries we can obtain luminosity distance to the source. This can lead to the Hubble constant determination (Schutz 1986). It is possible that the network of large scale laser interferometers can act as an astronomical observatory providing information complementary to that obtained from optical and radio telescopes (Krolak and Schutz 1987).

b)   Periodic sources

An interesting example of such sources are pulsars. The amplitude of the continuous wave from a pulsar of ellipticity $\varepsilon$ is approximately given by

$$h \simeq 0.8 \times 10^{-26} \left[ \frac{\varepsilon}{10^{-5}} \right] \left[ \frac{f}{100 \text{ Hz}} \right]^2 \left[ \frac{10 \text{ kpc}}{R} \right] \qquad (4)$$

For the well-known pulsars in our galaxy, Crab and Vela

we have the following estimates of parameters

Crab:  $\varepsilon \simeq 3\times 10^{-6}$  $f = 60$ Hz  $R = 2$ kpc  (5a)

Vela:  $\varepsilon \simeq 3\times10^{-5}$  $f = 22$ Hz  $R = 0.5$ kpc  (5b)

The values of ellipticity for pulsars are poorly known. The above values were obtained under the assumption that the observed slow-down rate of the pulsars are entirely due to gravitational radiation reaction.
c)  Stochastic sources
These form a random background at waves produced by all sources.  It is convenient to describe their energy density in the frequency range $\Delta f = f$ about a given frequency by the ratio $\Omega_{GW}$ of their energy density to that required to close the universe.  Then their mean amplitudes $\bar{h}$ are of the order

$$\bar{h} \simeq 0.6 \times 10^{-25} \left[\Omega_{GW}/10^{-10}\right]^{\frac{1}{2}} \left[\frac{100 \text{ Hz}}{f}\right] \qquad (6)$$

For example for cosmic strings one has (Thorne 1987)

$$\Omega_{GW} \sim 10^{-7} \left[\frac{\mu}{10^{-6}}\right]^{\frac{1}{2}} \text{ for all } f \geqslant 10^{-8} \text{ Hz } \left[\frac{10^{-6}}{\mu}\right]$$

$$(7)$$

where $\mu$ is the mass of the string per unit length in geometrical units.

## 2.  LASER INTERFEROMETRIC GRAVITATIONAL WAVE DETECTORS

The basic principle of laser interferometers is shown in Figure 2.  To avoid absolute length measurement the test masses in the interferometer are suspended.  The gravitational wave induces differential displacement in the arms of the interferometer and this is monitored by the photodiode.  This arrangement allows detection of the gravitational wave signal over wide band of frequencies.

There are many sources of noise in such an instrument: imperfections in laser beam, seismic vibrations, thermal motions of the pendula and mirrors.  However above certain frequency one expects that performance of the detector will only be limited by quantum photon-counting noise. One expects this cut-off frequency to be 100 Hz for detectors of 4 km arm-length and 200 Hz for 1 km arm-length.

Present prototypes already achieved photon-counting noise limited performance above 1 kHz over kilohertz bandwidth with sensitivity measured by square root of spectral density of the noise of around $10^{-19}$ $Hz^{-\frac{1}{2}}$. However to detect supernovae in the Virgo cluster or

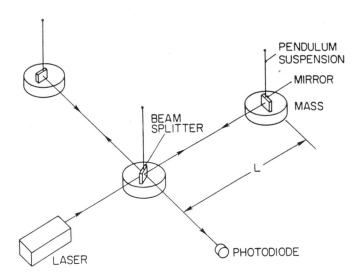

Figure 2. Basic Principle of Laser Interferometer.

coalescing binaries at 100 Mpc one must increase sensitivity by 4 and 5 magnitudes respectively. This will be achieved primarily by an increase of arm-length to one to four km (present prototypes have 10 to 40 metre arm lengths) and by an increase in laser power and quality of the mirrors. There is also an optical technique called light-recycling by which one can increase the effective laser power by more than one magnitude. In the case of sources from which we know the shape of gravitational wave-form accurately (coalescing binaries) we can apply optimal matched filters to increase detectibility considerably.

## 3. EXPECTED SIGNAL-TO-NOISE RATIOS

Thinking about detectibility of gravitational wave sources in terms of signal-to-noise ratios (SNR) rather than amplitudes of the waves started only recently and was pioneered by Kip Thorne (1987).

In the case of supernovae and with the detector in light-recycling configuration whose sensitivity is limited by the photon-counting noise optimised over the dominant frequency at the pulse we have

$$ SNR \simeq 3.3 \; S_o \left[ \frac{\Delta E_{GW}}{10^{-4} M_\odot c} \right]^{1/2} \left[ \frac{1 \; kHz}{f} \right]^{3/2} \left[ \frac{10 \; Mpc}{R} \right] $$

(8a)

$$ S_o = \left[ \frac{0.0818 \; \mu m}{\lambda_L} \right]^{1/2} \left[ \frac{\eta I_o}{100 \; W} \right]^{1/2} \left[ \frac{5 \times 10^{5}}{A^2} \right]^{1/2} \left[ \frac{L}{4 \; km} \right]^{1/2} $$
(8b)

where $\lambda_L$ is the reduced wavelength of the laser light, $\eta$ is the efficiency of the photodiode, $I_o$ is the laser power and $A^2$ are the losses in the mirror. Note the difference between the dependence on frequency of SNR and h (eq.(1a)).

In the case of coalescing binaries we can design the optimal matched filter since we know the wave-form accurately.

Correlating the output of the detector with such an optimal filter we can achieve the signal-to-noise ratio given by the following formula (Helstrom 1968)

$$ (SNR)^2 = 2 \int_o^\infty \frac{|\tilde{h}(f)|^2 \; df}{S_h(f)} $$

(9)

where $\tilde{h}(f)$ is the Fourier transform of the signal and $S_h(f)$ is the spectral density of the noise in $Hz^{-1}$.

In practice we do not know the time of arrival of the chirp and the mass parameter $\mu m^{2/3}$ therefore we shall correlate the data with a two-parameter family of filters. Using fast Fourier transforms and dedicated processors this can be done on line. In the case of a coalescing binary $|\tilde{h}(f)|$ is given to a very good approximation by the following formula

$$|\tilde{h}(f)| \simeq 1.45 \times 10^{-24} \frac{\mu_{\odot}^{\frac{1}{2}} m_{\odot}^{1/3}}{R_{100}} f_{100}^{-7/6} Hz^{-1}$$

(10)

In Cardiff we have investigated three models of noise (Dhurander, Krolak and Lobo, 1988a,b).

a) Detector is in the recycling configuration and above certain frequency $f_s$ the only source of noise is photon-counting noise.

(11a)

$$S_h(f) = \begin{cases} \infty \text{ for } f < f_s \\ \\ 6.8 \times 10^{-51} S_o{}^2 f_k (1+(f/f_k)^2 Hz^{-1} \text{ for } f > f_s \end{cases}$$

$$f_k (knee\ frequency) = \frac{(1-R_c)c}{4\pi L} Hz$$

(11b)

where $R_c$ is the reflectivity of one of the mirrors in the interferometer. The above model was introduced by Thorne (1987).

b) Detector is in the *detuned* recycling configuration. This is a new light recycling technique invented by Brillet and developed by Meers (Vinet *et al.*, 1988,; Meers 1988) which enhances the sensitivity of the detector over a certain bandwidth around a chosen frequency $f_0$. The investigated model is the same as above only for $f > f_s$ spectral density of noise is given by

$$S_h(f) = 6.8 \times 10^{-51} S_o{}^2 f_k [K\beta f_k + 1]\beta$$

$$\times \frac{\left[\beta-(f/f_k)^2\right]^2 + 4(f/f_k)^2}{\beta^2 + (f/f_k)^2} \qquad (12a)$$

$$\beta = 1 + (f_o/f_k)^2 \qquad (12b)$$

where K is a factor determined by the losses in the beam-splitter part of the laser interferometer.

c) Detector is in the standard recycling configuration but also noise due to thermal motions of the pendula and masses is considered. Thermal noise must be taken into account if we consider detection of waves at frequencies below 100 Hz (coalescing binaries and pulsars). There is an experimental effort in Pisa to reduce seismic noise considerably below 100 Hz. Photon-counting and thermal noises are assumed to be statistically indpendent and total noise below a certain cut-off frequency $f_s$ is taken to be

$$S_h(f) = S_h^{PC}(f) + S_h^{TP}(f) + S_h^{TM}(f) \qquad (13)$$

where $S_h^{PC}(f)$ is the spectral density of the photon counting noise given in formula (11a). $S_h^{TP}(f)$ is the spectral density of the thermal motions of the pendulum

$$S_h^{TP}(f) = S_1/f^4 \quad Hz^{-1} \qquad (14a)$$

$$S_1 = 4.2 \times 10^{-41} \left[\frac{T}{300K}\right] \left[\frac{f_p}{1\ Hz}\right] \left[\frac{10^3\ kg}{m}\right] \left[\frac{10^9}{Q}\right] \left[\frac{4\ km}{L}\right]^2 Hz^3$$

$$\qquad (14b)$$

where T is the temperature, $f_p$ is the resonant frequency of the pendulum, m is its mass and Q is its quality factor.

$S_h^{TM}(f)$ is the spectral density of the thermal motion of the masses

$$S_h^{TM}(f) = S_2 \qquad\qquad (15a)$$

$$S_2 = 3.3 \times 10^{-49} \left[\frac{T}{300K}\right] \left[\frac{5 \text{ kHz}}{f_r}\right] \left[\frac{10^3 kg}{m}\right] \left[\frac{10^5}{Q_{int}}\right]$$

$$\times \left[\frac{4 \text{ km}}{L}\right]^2 \text{Hz}^{-1} \qquad\qquad (15b)$$

where $f_r$ is the frequency of the fundamental resonant mode of the mass and $Q_{int}$ is the quality factor of the model.

For coalescing binaries from general formula (9) we obtain the following expression for the signal-to-noise ratio

$$SNR \approx 17.6 \ S_0 \ \frac{\mu_\odot^{\frac{1}{2}} \ m_\odot^{1/3}}{R_{100}} \ \frac{J_z(K, y, \alpha, \ S_1/S_0, S_2/S_0)}{f_{S_{100}}^{7/6}}$$

$$\qquad\qquad (16a)$$

$$y = f_k/f_s \qquad \alpha = f_o/f_s \qquad\qquad (16b)$$

where $f_{S100}$ is the cut-off frequency in 100 Hz and $J_2$ is certain integral depending on the parameters of the detector whose magnitude is of the order of one.

Maximising the value of the integral $J_2$ over the parameters one obtains the following signal-to- noise ratios for a binary consisting of neutron stars of equal masses equal to 1.4 solar mass and coalescing at the distance of 100 Mpc.

The predicted signal-to-noise ratios are very encouraging. In this case optimal filtering enhances the SNR by a factor of around 25 (for $f_s$ = 100 Hz). From the results obtained above one can make definite recommendations to the experimentalists as to how they should prepare the detector. Optimal values for the reflectivity of the mirrors (knee frequency $f_k$) and

| Model | Cut-off frequency | SNR | Optimum values of the parameters |
|---|---|---|---|
| a)<br>$\alpha = 0$<br>$S_1 = S_2 = 0$<br>$S_0 = 1$<br>$K = 0$ | 100 Hz | 13.1 | $f_k = 144$ Hz |
| b)<br>$S_1 = S_2 = 0$<br>$S_0 = 1$<br>$K = 1$ | 100 Hz | 17.9 | $f_k=10$Hz $f_0=144$Hz |
| c)<br>$\alpha = 0$<br>$S_1 = S_2 = 1$<br>$S_0 = 1$<br>$K = 0$ | 100Hz | 10.5 | $f_k = 156$ Hz |
|  | 40 Hz | 27.2 | $f_k = 114$ Hz |

tuning frequency $f_0$ can be obtained. From formula
(16a) it is clear that they should keep the cut-off
frequency as low as possible. Lowering cut-off
frequency below 100 Hz does increase SNR inspite of the
thermal noise. More detailed analysis shows however
that one does not gain much by lowering $f_s$ below 40 Hz.
One can also use the optimum matched filter in the
detection of periodic sources like pulsars (eq.(4)).
In the case of Crab and Vela pulsar using the
parameters given in Section 1b and assuming model (c)
of the noise with $S_0 = S_1 = S_2 = 1$ the correlation
times of optimum filter with the output of the detector
to achieve SNR equal to one are $4.7 \times 10^5$ s and $6.7 \times
10^5$ s respectively.

REFERENCES

Dhurandhar, S.V., Krolak, A. and Lobo, A. (1988)
    Detuned recycling: application to the detection of
    gravitational waves from coalescing binaries.
    Submitted to *Mon.Not.R.astr.Soc.*
Dhurandhar, S.V., Krolak, A. and Lobo, A. (1988)
    Detection of gravitational waves from a coalescing
    binary system: effect of the thermal anoise and
    the efficiency of the detector. submitted to
    *Mon.Not.R.astr.Soc.*

Drever, R. (1982) in: *Gravitational Radiation*, eds.
    Deruelle, N. and Piran, T. Amsterdam: North Holland
Helstrom, C.W. (1968) *Statistical Theory of Signal
    Detection*. Pergamon Press
Hough, J., Meers, B.J., Newton, G.P., Robertson,N.A.,
    Ward, H., Schutz, B.F., Corbett, J.F. and Drever,
    R.W.P. (1986) *Vistas in Astronomy*, <u>30</u>, 109
Krolak, A. (1987) in: *Gravitational Wabve Data
    Analysis*, ed. Schutz, B. D.Reidel Publishing Co.
Krolak, A. and Schutz, B.F. (1987) Coalescing binaries
    – probe of the universe. *Gen.Relativ. and Grav.*,
    <u>19</u>, 1163
Meers, B.J. (1988) On recycling in laser interfero-
    metric gravitational wave detectors.   unpublished
Peters, P.C. and Mathews, J. (1963) *Phys.Rev.*, <u>131</u>, 435
Schutz, B.F. (1986) *Nature*, <u>323</u>, 310
Thorne, K. (1987) in *300 Years of Gravitation*,
    eds. Hawking, S.W. and Israel, W. Cambridge
    University Press

Vinet, J.-Y., Meers, B.J., Man, C.N. and Brillet, A.
    Optimization of long baseline optical inter-
    ferometers for gravitational wave detectors.
    *Phys.Rev.*, to be published.

# DATA ANALYSIS IN NEUTRINO ASTRONOMY

M.Aglietta[a], G.Badino[a], G.Bologna[a], C.Castagnoli[a],
A.Castellina[a], V.L.Dadykin[b], W.Fulgione[a],
P.Galeotti[a], F.F.Kalchukov[b], V.B.Kortchaguin[b],
P.V.Kortchaguin[b], A.S.Malguin[b], V.G.Ryassny[b],
O.G.Ryazskaya[b], O.Saavedra[a], V.P.Talockin[b],
G.Trinchero[a], S.Vernetto[a], G.T.Zatsepin[b], and
V.F.Yakushev[b]

(a)  Istituto di Cosmogeofisica del CNR, Torino, Italy, and
     Istituto di Fisica Generale e Sez.INFN dell'Università di
     Torino, Italy
(b)  Institute of Nuclear Research of the Academy of
     Sciences of USSR, Moscow, USSR

## 1-  Introduction

An Underground Neutrino Observatory (UNO) should have
some very important characteristics in searching for neutrino
bursts. In particular:

1- The best accuracy in measuring the universal time
(UT), in order to perform correlations with other experiments.

2- The capability to detect in real time any burst
candidate, even if the associated optical output is too dim to be
observed or not produced at all during the collapse.

Since the LSD experiment of the Mont Blanc Laboratory
was designed with the main aim to be an Underground
Neutrino Observatory, both these characteristics were fulfilled
since the very beginning: the UT accuracy is better than 2 msec
and  the computer software provides identification of burst
candidates on real time. In addition, the LSD experiment has
been programmed in order to have a good efficiency in data

taking, an high stability over long running times, and a continuous monitoring of the data in real time.

Furthermore, the LSD experiment is located very deep underground (the minimum coverage of rock is 5200 hg/cm² of standard rock) where most of the cosmic ray muon background is suppressed. The apparatus is further shielded with iron slabs and paraphine in order to reduce the low energy radioactive background from the surrounding rock.

Here, we describe some of the on-line procedures of the LSD experiment, used to check if the detector is properly running, and to identify on real time at the occurence any neutrino burst candidate.

## 2- The Mont Blanc LSD Neutrino Observatory

The Mont Blanc neutrino telescope, described in detail elsewhere[1,2], is a liquid scintillation detector (LSD) made of 72 counters (1.5 m³ each) on 3 layers, with a total active mass of 90 tons of liquid scintillator containing $8.4 \ 10^{30}$ free protons. The scintillator is watched from the top of each counter by 3 photomultipliers in a 3-fold coincidence within 150 ns.

The energy calibration, performed by using both cosmic ray muons and a low activity $^{252}$Cf source as a neutron source, shows that a 1 MeV energy loss yields, on the average, 15 photoelectrons in one scintillation counter.

The low background, and the high energy resolution of the scintillation counters allow us to operate the LSD at a very low energy threshold. The electronic system consists of 2 levels of discriminators for each scintillation counter, with two ADCs per counter to measure the energy deposition in the scintillator in 2 overlapping energy ranges. A TDC per counter gives the relative time of the interactions with a resolution of 100 ns. Three memory buffers, 16 words deep, for the 2 ADCs and the TDC of each scintillation counter, allow us to record all pulses without dead time.

This recording system was designed with the purpose to detect both products of $\bar{\nu}_e$ interactions with free protons in the scintillator (i.e. positrons and gammas in a delayed coincidence within 500 μs) through the capture reaction:

$$\bar{\nu}_e + p \Rightarrow n + e^+ \tag{1}$$

followed by:

$$n + p \Rightarrow d + \gamma \tag{2}$$

which gives the main signal in detecting neutrinos from collapsing stars. For positron detection, both the kinetic and the annihilation energies can be recorded in the scintillator. The γ's from the (np,dγ) capture reaction, with $E_\gamma$ = 2.2 MeV, are emitted with an everage delay τ ≈ 200 μs after the main interaction.

## 3- The Data Acquisition System

The LSD experiment started taking data on October 1984 and it is fully running since 1985. From the very begining the experiment had an high running efficiency: the life-time to total time ratio during the last three years (from January 1st, 1986) was ≈ 90% on the average during the entire period, and almost 100% for 250 days in the period encompassing the time when SN 1987a exploded, on February 23, 1987. The short intervals of time, usually less than one day, during which the efficiency is lower then usual, are due to periodic checks of the detector, in particular when we make different tests in order to study the background.

Most of the data are analysed on real time by using on-line procedures, both because our main aim is to search for collapsing star neutrinos (and nobody knows when they would arrive at the Earth), and also because the total trigger rate in LSD is so low (less then 1 event per min) that it is very simple to analyse data on-line.

Whenever a trigger (i.e. a pulse with energy above the 5 to 7 MeV threshold, depending on the position of the counter) occurs in one of the 72 scintillation counters, a 500 μsec wide gate is opened for all the counters. During this time, the LSD experiment records any pulse above the energy threshold 0.8 MeV observed in whatever counter, including the counter fired by the trigger.

On-line checks are also made for the pulse height distribution (both for the high and the low energy threshold), for the pulse height distribution of muons, for the energy spectrum of each counter and for all counters. For monitoring purposes the counting rate is printed every 100 triggers, and at the same time all the histograms from the on-line check procedures are up-dated. Any abnormal behaviour of the experiment is thus signalled by the computer in real time. All these on-line checks allow us to run the experiment in a very quite and safe condition. Finally, an artificial trigger is automatically given every ≈ 7 min in order to test the full electronic system for all the 72 counters.

The absolute time in LSD is recorded by using the signal broadcasted by the italian standard time service (IEN Galileo Ferraris). The clock is periodically tested, and shows that the required accuracy of 2 msec is stable during all these years of operation.


## 4- Search for Bursts on Real Time and Statistical Analysis of Data

In the LSD data acquisition system, on-line software identifies on real time and prints on the computer output any burst candidate. For any burst of N pulses (N > 2) recorded in any interval of time $\Delta t$, between 1 ms and 600 s, this program computes the imitation rate from the background according to the standard distribution:

$$F_{im} = f \sum_{n=N-1}^{\infty} P(n, \Delta t) = f \sum_{n=N-1}^{\infty} \frac{\exp(-f\Delta t)(f\Delta t)^n}{n!}$$

where f is the raw trigger rate (with the present running conditions we have $f = 1.7 \ 10^{-3}$ events/s). If $F_{im} < 0.1$/day, the computer prints the main characteristics of the burst: event number and its time of occurence, fired counters and energy released in each one, and the duration of the burst. In this pre-analysis, all events involving more than 1 counter within the resolution time of the experiment, or with an energy release in excess of 50 MeV, are rejected because due to cosmic ray muons crossing the detector. Off-line analysis is further made in a similar way, but including a more detailed analysis of the single pulses in the burst.

The difference between the off-line and the on-line analysis is less than 0.1%; this very small difference shows that practically even raw data are very clean and no further cut, either in pulse height or/and in statistics, is necessary.

The trigger rate, in most of the 2-years period from January 1986 to January 1988, was 0.84 triggers/min. A fraction of 0.1 triggers/min is due to cosmic ray muons, easily recognizable because they involve at least two counters within the 100 nsec resolution time, or only one surface counter but with an energy release in excess of 50 MeV.

In searching for neutrino bursts, one has to deal with the statistical analysis of trigger pulses in order to have a long term statistics. The experimental time distributions of these pulses, clustered in groups with a multiplicity larger than a given one, and recorded within a given (but variable) interval of time, are plotted in fig. 1 as a function of their duration, and with a bin width of 10 seconds. Fig. 2 shows the same data distributions for pulses with an exact value of the multiplicity. The distributions of fig. 1 and 2 refer to a data taking period of

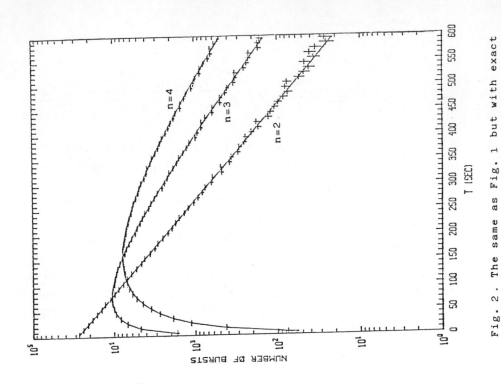

Fig. 2. The same as Fig. 1 but with exact n pulses,2,3 and 4.in their time duration in sec,

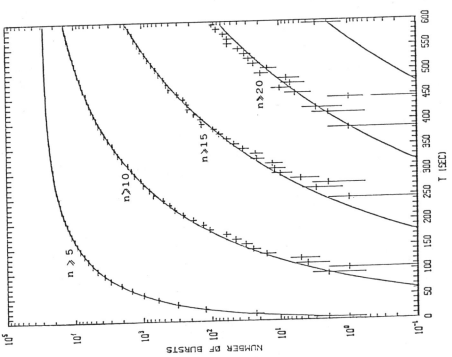

Fig.1. Experimental distribution of burst with n⩾5,10,15 and 20 vs. their duration in sec measured during 217.7 days.

217.7 days that includes the data recorded at the time of SN 1987a. In both figures, the smooth behaviour and the very good agreement with the continuos lines (the corresponding, expected Poisson distributions) show that the trigger rate is stable and that the detector is properly working throughout all this period.

Similar distributions have been obtained for other periods of the detector live-time, since 1985. Compared one to the other, these distributions show slight differences because of slight differences in the trigger rate. This is due to the fact that several tests have been performed in order to study in great detail the background conditions, and consequently the trigger conditions have been changed from time to time. Since the beginning of 1985, no particular burst of pulses was found in disagreement with the statistical expectations, except for the burst recorded on-line on february 23, 1987, and connected[3] with the SN 1987a explosion. Indeed, on February 23.12, 1987 ($2^h52^m36.8^s$ UT) the Mont Blanc computer printed out on real time at the occurence a burst of 5 pulses with a duration of 7 seconds and a background imitation rate of $1.78 \ 10^{-3}$/day. A second burst of 3 pulses with a duration of 0.5 seconds and an imitation rate of $1.74 \ 10^{-2}$/day was printed out at $5^h2^m0.7^s$ UT, the same day.

References

1.- Aglietta M. et al., Nuovo Cim., 9C, 185, 1986
2.- Aglietta M. et al., Proceedings 2nd Rencontres de Physique de la Vallèe d'Aoste, M.Greco Ed., La Thuille, Ed.Frontières, 1988
3.- Aglietta M. et al., Europhys. Lett., 3, 1315, 1987

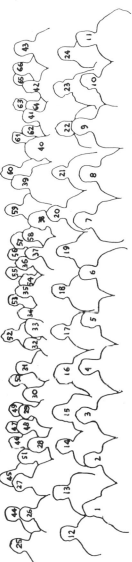

1. M. Carollo, 2. L. Buccheri, 3. J. Linsley, 4. L. Scarsi, 5. S. Serio, 6. V. Castellani, 7. P. Crane, 8. H.M. Adorf, 9. C. Bafia, 10. , L. Morbidelli, 11. M. Fofi, 12. P. Kahabka, 13. P. Ossorio, 14. T. Mineo, 15. O. Di Rosa, 16. M.C. Maccarone, 17. D. Kester, 18. V. Di Gesù, 19. R. Di Gesù, 20. A. Watson, 21. U. Zimmermann, 22. M.me Krolak, 23. G. Bordogna, 24. A. Rampini, 25. N. Perez de la Blanca, 26. A. Strong, 27. F. Murtagh, 28. M. Busetta, 29. A. Di Iorio, 30. E. Palazzi, 31. C. Izzo, 32. O. Schwenker, 33. P. Cabeza Orcel, 34. H. Steinle, 35. W. Voges, 36. R. Hanisch, 37. S. Neff, 38. M. Thonnat, 39. R. Gruber, 40. F. Soroka, 41. R. Albrecht, 42. D. Wells, 43. G. Staw, 44. R. Molina, 45. M. Kurtz, 46. M. Boyarski, 47. A. Etemadi, 48. S. Cortiglioni, 49. S. Torres, 50. R. Diehl, 51. R. Nazirov, 52. S. Sciortino, 53. G. Pizzella, 54. C. Dobson, 55. A. Krolak, 56. V. Pallottino, 57. M. Bougeard, 58. E. Falco, 59. A. Venetoulias, 60. G. Malaguti, 61. A. Accomazzi, 62. A. De Greiff, 63. J. Aymon, 64. G. Hartner, 65. D. Robertson, 66. G. Smoot.

413

# Index